AF587655

CANCER ETIOLOGY, DIAGNOSIS AND TREATMENTS

BRAIN CANCER, TUMOR TARGETING AND CERVICAL CANCER

CANCER ETIOLOGY, DIAGNOSIS AND TREATMENTS

Additional books in this series can be found on Nova's website under the Series tab.

Additional E-books in this series can be found on Nova's website under the E-book tab.

CANCER ETIOLOGY, DIAGNOSIS AND TREATMENTS

BRAIN CANCER, TUMOR TARGETING AND CERVICAL CANCER

ELENA K. SALVATTI
EDITOR

Nova Science Publishers, Inc.
New York

For permission to use material from this book please contact us:
Telephone 631-231-7269; Fax 631-231-8175
Web Site: http://www.novapublishers.com

Additional color graphics may be available in the e-book version of this book.

Library of Congress Cataloging-in-Publication Data

Brain cancer, tumor targeting, and cervical cancer / editor, Elena K.
Salvatti.
p. ; cm.
Includes bibliographical references and index.
ISBN 978-1-61122-738-3 (hardcover)
1. Brain--Cancer--Chemotherapy. 2. Cervix uteri--Cancer--Treatment. I.
Salvatti, Elena K.
[DNLM: 1. Brain Neoplasms--therapy. 2. Drug Delivery Systems--methods.
3. Nanostructures--therapeutic use. 4. Uterine Cervical
Neoplasms--therapy. WL 358]
RC280.B7B6837 2010
616.99'481--dc22

2010042601

Published by Nova Science Publishers, Inc. † New York

Contents

Preface **vii**

Chapter I Targeting Strategies in Photodynamic Therapy for Cancer Treatment **1**
Hamanou Benachour, Céline Frochot, Régis Vanderesse, François Guillemin and Muriel Barberi-Heyob

Chapter II Emerging Targets and Novel Nanotechnology-based Strategies: Changing the Treatment Paradigm in Oncology **39**
Vera Moura, Lígia C. Gomes da Silva, Nuno A. Fonseca, Maria C. Pedroso de Lima, Sérgio Simões and João Nuno Moreira

Chapter III Biological and Clinical Relevance of Neural Stem Cells in the Pathogenesis of Childhood Medulloblastoma **75**
Cornelia M. Bertram, Nicholas G. Gottardo, Ursula R. Kees, Kanti D. Bhoola and Peter B. Dallas

Chapter IV Laparoscopy and "Robotics" in Cervical Cancer Treatment: Current Insights, Novel Approaches and Future Perspectives in the New Era of Gynecological Surgery **127**
Andrea Tinelli, Antonio Malvasi, Maurizio Buscarini, Emilio Ruiz Morales and Stark Michael

Chapter V Malignant Glioma: Challenges, Advances and Perspectives for Treatment **163**
A. Nimer Amr, Sven Kantelhardt and Alf Giese Ella L. Kim

Chapter VI Surgical Treatment of Invasive Cervical Cancer **185**
Sabas Carlos Vieira, Rafael Bandeira Lages, Benedito de Sousa Almeida Filho and Ilanna Naianny Leal Rodrigues

Chapter VII Recent Advances in Glioma Gene Therapy **201**
José Segovia' and Adolfo López-Ornelas

Chapter VIII Congenital Tumors of the Central Nervous System **217**
Melinda Czeh and Angela M. Kaindl

Chapter IX Management of Recurrent Cervical Carcinoma **231**
C. Iavazzo and G. Vorgias

Chapter X Real-Time Tumor Targeting in External Beam Radiotherapy **247**
Marco Riboldi, Matteo Seregni, Andrea Pella, Ahmad Esmaili Torshabi, Roberto Orecchia and Guido Baroni

Chapter XI Review of Tumours of the Paranasal Sinuses Based on 14 Cases **261**
Alaitz Santamaría Carro, Sergio Pinar-Sueiro, Adolfo Vergez Sánchez and Roberto Víctor Fernández Hermida

Index **277**

Preface

This book gathers current research from across the globe in the study of brain cancer, cervical cancer and tumor targeting. Some topics discussed in this compilation include emerging targets in cancer therapy and novel nanotechnology-based therapeutic strategies that have been changing the paradigm in cancer treatment; laparascopy and robotics in cervical cancer treatment; the history and current status of glioma therapy and glioma gene therapy; congenital tumors of the central nervous system; a review of tumors of the paranasal sinuses and real-time tumor targeting in external beam radiotherapy.

Chapter I - Despite recent advances in surgery, chemotherapy, and radiation treatment, survival of patients with advanced malignancy remains suboptimal. Photodynamic therapy is a relatively new cytotoxic treatment, predominantly used in anti-cancer approaches, that depends on the retention of photosensitizers in tumor and their activation after light exposure in the presence of molecular oxygen. Photosensitizers are photo-sensible compounds that upon photoactivation, effect strongly localized oxidative damage within the target cells. The ability to confine activation of the photosensitizer by restricting illumination to the tumor allows for a certain degree of selectivity. Nevertheless, the targeted delivery of photosensitizers to defined cells is a major challenge in photodynamic therapy of cancer, and one area of importance is photosensitizer targeting. In this sense, an arsenal of targeting strategies has been developed recently. Alterations or increased levels in receptor expression of specific cellular type occur in the diseased tissues. Therefore, photosensitizers can be covalently attached to molecules such as specific peptides, leading to a receptor-mediated targeting strategy. These active-targeting approaches may be particularly useful for vascular-targeted photodynamic therapy. The present chapter will focus on recent and significant advances and developments in targeting strategies in photodynamic therapy with the emphasis on target specificity.

Chapter II - The entwined mechanisms of survival that allow tumors to grow and spread in the human body have been intensively studied, in the attempt to recognize specific patterns and create effective clinical approaches. New insights into the molecular basis of tumor development, progression and metastases gave rise to new targets in cancer therapy and opened new opportunities for science to create and identify novel targeting strategies.

Tumors start to grow and survive along with the supportive role of the microenvironment, sustained by processes such as angiogenesis (formation of new blood vessels from preexisting ones). Cancer cells interact with different players and promote the organization of an immature vessel network to access nutrients and oxygen, but also to colonize distant areas from the tumor primary site. The thorough knowledge of this process has opened a portal for understanding cancer associated mechanisms and design treatment strategies, accounting for the interaction of therapeutic agents directly with the unique features of tumors and their supporting cells. Moreover, the discovery of the regulatory machinery behind tumor progression provided a rationale for the development of therapies concerning cancer stem cells, dormant cells, circulating tumor cells, metastases suppressor genes and non-coding RNAs. The understanding of tumor cell interaction with the stroma, inflammatory infiltrates, cell-cell interaction and even mechanical forces also gave rise to the integration of new key regulators and possible targets for interference with tumor progression and metastases.

Alongside with the discovery of efficient and selective molecular targets for the development of novel ligands with high specificity towards cancer cells and/or their microenvironment are the means for reaching such targets. Many non-invasive methods for delivering therapeutic agents have been designed and improved to minimize inherent risk association. Nowadays, delivery systems are engineered to comprise a multitude of features, such as safety, efficacy, target accumulation, cargo delivery, remote action and, more importantly, therapeutic and diagnostic features. As a consequence of the urgency in creating novel treatment solutions within cancer therapy, the number of targeted nanoparticles in clinical trials has increased over the last years and is expected to present major benefits in cancer therapeutic approaches, as it leads to a more personalized medical care.

In this chapter, the authors discuss emerging targets in cancer therapy and novel nanotechnology-based therapeutic strategies that have been changing the paradigm in cancer treatment.

Chapter III - Paediatric brain tumours have overtaken leukaemias as the leading cause of cancer-related death in children despite being only half as common. Medulloblastoma (MB), the most common type of malignant paediatric brain tumour, is a heterogeneous group with at least five histological and four molecular subtypes. Despite major progress in the understanding of the genetic and molecular pathogenesis of MB, relapse rate and side effects remain major concerns. The existence of multiple MB subtypes suggests that the current clinical categorisation of two risk stratification groups is inadequate and highlights the need for more tailored therapeutic strategies. The best understood subtype comprises sonic hedgehog (SHH)-dependent MB. Several components of the SHH pathway are viable drug targets that may translate to effective subtype-specific treatment options in the near future. Similarly, subtype-specific targets for clinical intervention are being investigated in other molecular variants of MB. Some MB may rely on a small population of CD133+ tumour cells termed brain tumour stem cells (BTSCs) that phenotypically and functionally resemble normal CD133+ neural stem cells (NSCs) in the developing foetal brain. The resemblance suggests that BTSCs could arise from transformed NSCs residing in the cerebellum. BTSCs are more resistant to therapy and are not effectively targeted by current treatments. New methodologies allow the generation, maintenance and manipulation of human NSCs and BTSCs, providing valuable tools and reagents for studying the mechanisms of neoplastic deregulation. The development of NSC- and BTSC-based models may be of benefit for elucidating the underlying molecular mechanisms of MB pathogenesis and facilitate transition of new knowledge into clinical setting in the future.

Chapter IV - Cervical cancer is the second most common cancer in women worldwide and is a leading cause of cancer-related death in women in underdeveloped countries. More than 99% of cervical uterine cancer cases show HPV presence. Minimal invasive surgery is increasingly its popularity in surgical oncology. Robotic surgery with its numerous advantages over conventional laparoscopy has assumed an ever-expanding role in minimally invasive surgery. This chapter focuses on the recent adoption, experience, and applications of laparoscopy and robot-assisted laparoscopy in cervical cancer surgery, focusing on radical endoscopic treatments, such as hysterectomy, bilateral salpingo-oophorectomy, pelvic and para-aortic lymphadenectomy.

A computer aided and manual search for clinical and systematic reviews, randomized controlled trials, prospective observational studies, retrospective studies and case reports published between 1980 and April of 2010, assessing robotic surgery was carried out defined by search strings including: laparoscopic surgery; robotics or robot or robot-assisted or da Vinci; radical hysterectomy; cervical cancer, minimally invasive surgery.

Literature has shown the feasibility of a radical resection by laparoscopic surgery and documented an equivalent number of pelvic nodes harvested by laparoscopy and open surgery. With technical advances and emerging devices, as well as accumulating experiences in laparoscopic surgery, radical laparoscopic hysterectomy was achieved as a feasible and safe surgical procedure, with long-term data on the morbidity and survival. Robotic assisted gynecologic surgery is, recently, worldwide increased, as the number of dedicated scientific articles. Few retrospective and prospective studies have demonstrated the feasibility of robotic assisted surgery in radical hysterectomy. In general, robot-assisted gynecologic surgery is often associated with longer operating room time but generally similar clinical outcomes, decreased blood loss, and shorter hospital stay. Robotic-assisted procedures are not, however, without their limitations: the equipment is still very large, bulky, and expensive, the staff must be trained, specifically on draping and docking the apparatus to maintain efficient operative times. Functional limitations include lack of haptic feedback, limited vaginal access, limited instrumentation, and larger port incisions. Exchanging instruments becomes more cumbersome and requires a surgical assistant to change the instruments. Additionally, the current robotic instruments do not include endoscopic staplers or vessel sealing devices. Finally, laparoscopic radical hysterectomy is a feasible and safe procedure that is associated with fewer intraoperative and postoperative complications than abdominal radical hysterectomy. The role of robotic-assisted surgery is continuing to expand, but well-designed, prospective studies with well-defined clinical, long-term outcomes, including complications, cost, pain, return to normal activity, and quality of life, are needed to fully assess the value of this new technology in radical hysterectomy.

Chapter V - From the earliest records of surgical resection of a cerebral lesion in 1885 till the present time, neoplasms of the CNS have posed an interdisciplinary therapeutic challenge. While benign brain tumours have been treated with relatively satisfactory results since the early decades of the last century, malignant neoplasms continue to elude cure despite advances in diagnostics and steadily evolving neurosurgical techniques. In fact, mean survival rates for the largest group of malignant primary brain tumours, glioblastoma multiforme (GBM), have only slowly improved in the last four decades. The major underlying reason for the limited clinical success of therapeutic surgical options is that GBMs, while rarely capable of metastasizing outside of the central nervous system (CNS), possess the propensity for infiltrative spread within the CNS, a phenomenon known as “local glioma invasion”. Another prominent feature of gliomas is their heterogeneity with respect to the cell type composition, biological phenotypes and responses to cytotoxic treatments. Over the last two decades, much progress has been made towards understanding of the GBM biology, including the realisation that malignant gliomas possess an hierarchical structure, whereby various types of cells might exist within the same tumour to exert different functions. Recently, the potential research breakthrough in understanding the biological basis for the less than satisfying response of malignant gliomas to current therapies has been made through the identification of a specialized subpopulation of cells generally termed brain tumour initiating cells (BTICs) that might be largely responsible for post-treatment recurrence of malignant glioma. In this chapter, the authorsdiscuss the history and current status of glioma therapy as well as some of the newly opened research avenues with special reference to BTICs.

Chapter VI - The standard treatment of stages 0-Ia1-Ia2-Ib1 cervical carcinoma is surgery, unless the patient has no clinical conditions for this procedure. There is no standard management of stages Ib-IIa. Both radical surgery and radical radiotherapy have proven to be equally effective, but differ in associated morbidity and complications. Concurrent chemo-radiation with cisplatin-based chemotherapy is important for patients with advanced disease, lymph node invasion, as well as parametria or surgical margins involved by the tumor, because this method improve disease-free and overall survival rates. The aim of this chapter was review the current approach to surgical management of cervical cancer. The standard surgical treatment for invasive cervical carcinoma stage Ib-IIa is radical hysterectomy class III with pelvic lymphadenectomy. Radical hysterectomy was firstly described by Wetheim more than one hundred years ago and was then modified and re-popularized by Meigs in 1950s. This operation yields 5-year survival rates of 75%-90% in most cases. The authors also reviewed studies that discuss the use of radical hysterectomy class II as a technique with same class III survival rates. The possibility and the advantages of ovarian preservation, laparoscopic radical hysterectomy, trachelectomy, and robotic approach were also reviewed. Finally, the authors discussed the use of sentinel lymph node in cervical carcinoma, a promising technique for management of this cancer and an area in which the authors have some important researches.

Chapter VII - Almost half of the primary tumors of the Central Nervous System (CNS) are gliomas, and the most malignant form of these tumors, glioblastoma multiforme (GBM), aggressively infiltrates into the brain parenchyma. Current treatments for GBM include surgical resection, chemotherapy and radiotherapy, but even after combined therapies, the prognosis remains very poor, with a median survival of roughly one year after diagnosis. Although the knowledge regarding the genetic basis, and the molecular signaling pathways involved in the origin and progression of the tumors has increased significantly in the last few years, allowing the generation of new chemotherapeutic agents that are used together with sophisticated surgical and radiation techniques, new therapeutic approaches are necessary for GBM. Gene therapy appears to be a relevant strategy for the treatment of gliomas, because it permits the delivery of therapeutic genes to tumors that rarely metastasize outside of the CNS. The authors have proposed the use of Growth arrest specific1 (Gas1) as an adjuvant for the treatment of gliomas, because of its capacity acting as a tumor suppressor. In this chapter, the authors will discuss the main characteristics of gliomas, the rationale for the application of gene therapy for their treatment, and the effects of the expression of Gas1 reducing the growth of these tumors.

Chapter VIII - Congenital tumors of the central nervous system are rare but have a poor prognosis due to their size and location as well as the contraindication of radiotherapy at this age. Various classifications have been described based on the age of the patient at clinical manifestation. The short latency between tumor onset and clinical signs suggests a genetic predisposition, as has been described for some hereditary tumor syndromes that manifest in infancy (astrocytoma in tuberous sclerosis, choroid plexus tumor in Li Fraumeni syndrome, medulloblastoma in Gorlin syndrome, atypical rhabdoid/teratoid tumor in rhabdoid tumor predisposition syndrome). Here, the authors describe the clinical signs, pre- and postnatal diagnostic paradigms, histopathological and genetic characterization of the most frequent congenital tumors of the central nervous system: teratomas, astrocytomas, choroid plexus papillomas, primitive neuroectodermal tumors, medulloblastomas, atypical rhabdoid/terotoid tumors, craniopharyngiomas. Treatment options and pitfalls will be discussed.

Chapter XI - Background:

Cervical carcinoma recurrence is defined as a de novo tumor development, at least three months after the radical excision of the disease or after radiotherapy treatment. One third of cervical carcinomas are going to recur after initial treatment.

Method:

The authors present a review on the field of recurrent cervical carcinoma regarding the diagnosis, treatment options and prognosis of such an entity.

Results:

The risk factors for recurrence include: stage, tumor size, histopathologic characteristics. 80% of recurrences occur in the first two years after treatment for the initial disease. 2/3 of the local recurrences after surgical treatment occur in the pelvis and are central. Classification of patients is based on the Shingelton and Hatch criteria:

a. Adherence to the pelvic side wall
b. Tumor < or > of 3 cm
c. Time interval between the primary treatment and the relapse.

The indication for pelvic exenteration is non-metastatic persistent or recurrent cervical malignancy after prior pelvic irradiation. The disease must be confined to the central pelvis in order to be completely respectable. The so-called 'triad of trouble' (including ureteral obstruction, neural sheath involvement and venous or lymphatic compromise of the iliac vessels) is used for the determination of respectability (possible when only one of them is present). Radiotherapy is used in pelvic recurrences after surgery and include a combination of external radiotherapy and brachytherapy plus chemotherapy with platinum-based regimens. The possibility of treatment of patients with recurrences outside the pelvis is rather small and for this reason such patients are candidates for palliative treatment.

Conclusion:

Treatment options in patients with persistent or locally recurrent cervical cancer are limited. Less than 20% of such patients are going to survive. Patient selection is crucial. Such patients should be treated in an organized oncologic centre under the continued supervision of surgeon, radiotherapist, internist, and psychiatrist.

Chapter X - Tumor targeting in external beam radiotherapy is a crucial issue, as the accuracy that can be achieved is directly related to local tumor control and side effects. This applies both to conventional radiation therapy, that uses photons and electrons, and to particle therapy, where protons and ions provide higher geometrical selectivity and enhanced biological effects. Current strategies prescribe the massive use of image guidance technologies to maximize the targeting accuracy (Image Guided RadioTherapy, IGRT).

When IGRT is applied to moving tumors, such as in the lung or liver, image guidance becomes challenging, as motion leads to increased uncertainty. Furthermore, the need of continuously monitoring tumor motion to achieve adequate tumor targeting accuracy limits the applicability of IGRT technologies due to imaging dose constraints. Therefore, minimally invasive technologies for continuous monitoring, such as optical tracking and surface detection, have been proposed for real-time tumor targeting.

In this chapter the authors focus on real-time tumor targeting of moving tumors, investigating the applicability of using external surrogates combined with periodic imaging to model tumor motion. The authors first present the theoretical background of computing targeting errors with moving targets. Then the authors review the state of the art technologies and quantify the actual targeting accuracy based on published data and retrospective clinical studies. Finally the authors compare different strategies to model tumor motion in a patient database, discussing the use of different correlation models, the need of model updating, and the best trade off between targeting accuracy and imaging dose.

Chapter XI - Purpose

To review tumours of the paranasal sinuses and to analyse the characteristics of the surgical approach using lateral orbitotomy.

Material and Methods

The authors present a series of 14 patients with various types of tumours of the paranasal sinuses, treated in the authors centre. The authors present a descriptive, retrospective study including the clinical picture of the patients, diagnostic methods used in consultations and imaging scans, medical-surgical management and long-term prognosis.

Results

All the subjects included in the study underwent paranasal tumour resection through a lateral approach. In two cases, this was complemented by superomedial and transpalpebral orbitotomy, and one of the patients required a bilateral approach. Three patients had tumour relapses; two of them being successfully operated on, with subsequent remission of the tumours, while the other patient did not undergo surgery due to the degree of systemic involvement and prognosis.

Conclusion

The lateral approach allows good surgical access to tumours of the paranasal sinuses. Acceptable aesthetic and functional outcomes are obtained with the current techniques of reconstruction of the facial skeleton after radical resection.

In: Brain Cancer, Tumor Targeting and Cervical Cancer
Editor: Elena K. Salvatti
ISBN: 978-1-61122-738-3

Chapter I

Targeting Strategies in Photodynamic Therapy for Cancer Treatment

***Hamanou Benachour*[a], *Céline Frochot*[b,d], *Régis Vanderesse*[c,d], *François Guillemin*[a,d] and *Muriel Barberi-Heyob*[a,d,*]**

[a]Centre de Recherche en Automatique de Nancy, Nancy-University, CNRS, Centre Alexis Vautrin, Vandœuvre-lès-Nancy, France

[b]Laboratoire Réactions et Génie des Procédés, Nancy-University, CNRS, Nancy, France

[c]Laboratoire de Chimie Physique Macromoléculaire, Nancy-University, CNRS, Nancy, France

[d]GDR CNRS 3049 "Médicaments Photoactivables - Photochimiothérapie", France

Abstract

Despite recent advances in surgery, chemotherapy, and radiation treatment, survival of patients with advanced malignancy remains suboptimal. Photodynamic therapy is a relatively new cytotoxic treatment, predominantly used in anti-cancer approaches, that depends on the retention of photosensitizers in tumor and their activation after light exposure in the presence of molecular oxygen. Photosensitizers are photo-sensible compounds that upon photoactivation, effect strongly localized oxidative damage within the target cells. The ability to confine activation of the photosensitizer by restricting illumination to the tumor allows for a certain degree of selectivity. Nevertheless, the targeted delivery of photosensitizers to defined cells is a major challenge in photodynamic therapy of cancer, and one area of importance is photosensitizer targeting. In this sense, an arsenal of targeting strategies has been developed recently. Alterations or increased levels in receptor expression of specific cellular type occur in the diseased tissues. Therefore, photosensitizers can be covalently attached to molecules such as

* Address correspondence. Centre de Recherche en Automatique de Nancy, Centre Alexis Vautrin, Avenue de Bourgogne, Brabois, F-54511 Vandœuvre-lès-Nancy, France, Telephone : +33 3 83 59 83 76.

specific peptides, leading to a receptor-mediated targeting strategy. These active-targeting approaches may be particularly useful for vascular-targeted photodynamic therapy. The present chapter will focus on recent and significant advances and developments in targeting strategies in photodynamic therapy with the emphasis on target specificity.

Targeted Photodynamic Therapy: Preliminary Information

Photodynamic therapy (PDT) is now well established as a clinical treatment modality for various diseases, including cancer. It involves a photosensitizer, light and molecular oxygen, whose combined action results in the formation of reactive oxygen species and singlet oxygen, which is thought to be the main mediator of cellular death induced by PDT. PDT is being developed as a treatment for cancer of the oesophagus, bronchi, brain and bladder, as well as for other non-oncological applications, such as the treatment of age-related macular degeneration. PDT is also used as a successful non-invasive therapeutic modality for treating cutaneous neoplasm. A limitation of PDT is that it cannot cure advanced disseminated disease because irradiation of the whole body with appropriate light doses is not possible. Nevertheless, for advanced disease, PDT can enhance quality of life and lengthen survival. For early or localized disease, PDT can be a selective and curative therapy with many potential advantages over available alternatives. The successf of any PDT concept will crucially depend on the nature of the photosensitizer, classically defined as a chemical entity, which, upon absorption of light, induces a chemical or physical alteration of another chemical entity. Early preparations of photosensitizers for PDT were based on a complex mixture of porphyrins called haematoporphyrin derivative (HpD). Nowadays, Photofrin®, a purified fraction of HpD derivative, is the most commonly used photosensitizer and the only drug approved by the Food and Drug Administration for the treatment of superficial bladder cancer in Canada and early lung and advanced oesophageal cancers in the Netherlands and Japan. Despite its effectiveness porfimer sodium has several drawbacks. The drug induces protracted skin photosensitivity, and the initial selectivity for the tumor tissue remains low. With the exception of porfimer sodium, the only other PDT compound currently approved for systemic use in cancer treatment is temoporfin. Temoporfin (*meta*-tetra(hydroxyphenyl)chlorin, *m*-THPC, Foscan®) is effective for the palliative treatment of head and neck cancers and it was approved in Europe for this indication in 2001. Nevertheless, like porfimer sodium, temoporfin is also associated with a pronounced skin photosensitivity and exhibits a slow elimination from the blood compartment (between 2 and 3 days) and finally a weak selectivity between tumor and healthy tissue. In contrast, Tookad® (Pd-bacteriopheophorbide, also known as WST09) and its derivative WST11 are believed to be purely vascularly mediated. Indeed, the clearance of Tookad® from the circulation is very fast and its plasma half-life is less than a few hours. The benzoporphyrin derivative verteporfin (Visudyne®) was also recently approved, but it has only been developed for the treatment of age-related macular degeneration, and not indicated for cancer. Concerning photosensitizers for topical application to treat skin lesions, protoporphyrin IX (PpIX) can be produced in nucleated cells after topical application of either aminolevulinic acid (ALA) or methyl aminolevulinate (mALA) to the site of a skin cancer or precancerous lesion. Several thousand patients have already been treated with PDT for a variety of advanced neoplasms, and have shown an improvement in their quality of life and a lengthened survival. However, a limitation of the conventional PDT for cancer treatment remains usually the lack of selectivity of photosensitizers; the ideal drug delivery system should enable the selective accumulation of the photosensitizer within the diseased tissue and the delivery of therapeutic concentrations of photosensitizer to the target site with little or no uptake by non-target cells. Unfortunately, the majority of the photosensitizers are taken up non-selectively by all cell types and consequently they can accumulate into tumor tissue but also in normal cells. The accumulation of a photosensitizer in neoplastic tissue relative to normal tissue depends on the photosensitizer molecule, the normal tissue being considered and the animal tumor model being chosen. The reason for the preferential accumulation in tumor compared with certain normal tissues not belonging to the reticulo-endothelial system may be a result of the greater proliferative rates of neoplastic cells, poorer lymphatic drainage, leaky vasculature, or some specific interaction between the photosensitizer and marker molecules on neoplastic cells. Observations suggested a possible specific interaction of the photosensitizers with tumor vasculature however; there is also evidence that at longer times after injection of the photosensitizer, the treatment effect becomes distinctly less vascular. One suggested specific interaction has been the low-density lipoprotein receptor-photosensitizer interaction.

This chapter will describe recent and significant advances and developments in strategies for achieving selective enhanced photosensitizer concentration in neoplastic target tissue with the emphasis on target specificity. Description will be focused on the recent and significant published articles describing each targeting strategy and references of related reviews will be indicated.

1. Introduction

Despite recent advances in surgery, chemotherapy, and radiation treatment, survival and quality of life of patients with advanced malignancy remain suboptimal. PDT is gradually becoming a promising novel therapeutic approach for the treatment of many tumors. It is now regarded as an effective, non-invasive and relatively non-toxic therapeutics for both cancer and many other diseases.

The concept underlying PDT is the use of light-absorbing non-toxic molecules, called photosensitizers which, when delivered to target cells, and activated by light of an appropriate wavelength and dosage, induce local production of singlet oxygen ($^1O_2$) and other reactive oxygen species (ROS). So, the presence of adequate amount of molecular oxygen in the target tissue is also required. After photoactivation, the photosensitizer transfers its energy to nearby molecular oxygen to generate cytotoxic ROS that cause cellular and molecular oxidative damage, leading to target cells death by apoptosis or necrosis, with minimal damage to the surrounding tissue [Levy *et al.* 1996; Chen *et al.* 2009].

Clinical Applications of PDT

Cancer was the first target for PDT and over the last three decades PDT has been proved to be effective in early lung cancer, Barrett's oesophagus, bladder cancer, head and neck cancers, and to be the ideal treatment for skin cancers that are easy to irradiate [for a recent review, see Chatterjee *et al.* 2008]. PDT-mediated neovascular targeting is used to treat tumors by destroying blood vessels that feed them [Abels. 2004] and is now in common use for the treatment of neovascular age-related macular degeneration [Wormald *et al.* 2007]. There are increasing broad clinical applications of PDT for the treatment of various tumors in different localisation in the body, thanks in part to the development of light delivery systems which allow specific illumination of diseased tissue. The effect of PDT on any tumor depends on a number of factors. These include the light energy absorbed by the target tumor tissue, the local concentration of the photosensitizer, and the inherent sensitivity of the tissue to the photodynamic effect. While easy accessible tumor lesion is illuminated from the surface, the development of fibre optic-based interstitial, intravesical and endoscopic light delivery systems have greatly facilitated the delivery of light in difficult to reach sections, thus making possible to treat basal and parenchymal lesions of virtually any part of the body [Sibani *et al.* 2008].

PDT Advantages

PDT has become an increasingly recognized alternative to cancer treatment in clinic. Three decades of PDT have revealed several advantages of this approach over more conventional therapies in the treatment of cancers. Due to its dual-selectivity, PDT presents limited side-effects [Brown *et al.* 2004] because phototoxicity is limited to photosensitized cells in the area illuminated and because photosensitizers preferentially accumulate in cancer tissues [Brown *et al.* 2004; Collaud *et al.* 2004; Kessel, 1992]. In addition, as ROS have very short lifespan and travel at very short distances, due to their extreme reactivity, PDT-induced oxidative damage is largely limited to the site of ROS generation [Redmond *et al.* 2006], thus sparing the extracellular tissue matrix that allows post-PDT regeneration of healthy tissue [Grant *et al.* 1997]. Moreover, PDT treatment is non-invasive and can be repeated without cumulative toxicity. It can be also applied in parallel to other treatment modalities. Furthermore, in contrast to most other cancer therapies, PDT can induce immune processes, even against less immunogenic tumors, and thus contribute to long-term tumor control [Chatterjee *et al.* 2008].

Biodistribution and Natural Selectivity of the Photosensitizers

As indicated before, PDT has inherent dual selectivity due to control of light delivery and to some extent selective photosensitizer accumulation in tumors. Biodistribution and cellular incorporation of photosensitizers depend partly on the interactions between these molecules and transport molecules in plasma, particularly albumin and lipoproteins. This means that most photosensitizers when injected into the blood circulation behave as macromolecules either because they are more or less firmly bound to large protein molecules or because they have formed intermolecular aggregates. Thus, understanding of the interactions of photosensitizer with plasma proteins is fundamental to determine their pharmacological behaviour, and from there, the time span between the photosensitizer's injection and irradiation of the tumor site (drug-light interval, DLI) during PDT [Sasnouski *et al.* 2006; Bonneau *et al.* 2002; Brault, 1990; Bonneau *et al.* 2004]. As it is a pre-requisite for PDT, selective photosensitizer accumulation and retention in tumors has been the subject of much discussion and speculation, leading to a number of theories, which aimed to explain this tumor-specific biodistribution. It is thought that the increased vascular permeability to macromolecules typical of tumor neovasculature is responsible for the preferential extravasation of the photosensitizer in tumors. In addition, it is thought that tumors have poorly developed lymphatic drainage, and those macromolecules, which extravasate from the hyperpermeable tumor neovasculature, are retained in the extravascular space. This means that therapeutic agents that gain interstitial access to the tumor present higher retention times than normal tissues. The combination of leaky vasculature and poor lymphatic drainage, which characterise tumor vasculature, result in what is known as the Enhanced Permeation and Retention (EPR) effect [Byrne *et al.* 2008]. Another theory suggested that the lower pH commonly found in tumor tissues microenvironment, routinely around 0.5 pH unit lower than in normal tissues [Mordon *et al.* 1995], has the effect of trapping some of the anionic photosensitizers, which are ionized at normal physiological pH. All these characteristics typical of tumor tissues are chiefly responsible of passive targeted delivery of clinical approved photosensitizer for PDT application.

The effectiveness of PDT is widely determined by the efficiency of the local generation of $^1O_2$ [Chen *et al.* 2006], the degree of efficiency and selectivity to which therapeutic concentrations of the photosensitizer is delivered to the target site with minimal uptake by non-target cells [Konan-Kouakou *et al.* 2005] and subsequent localized light irradiation.

Recent advances in lasers result in the development of fibre optic-based light delivery systems, which have improved the control of light delivery into diseased tissues. This has provided significant progress in the improvement of the irradiation specificity, especially in tumor lesions affecting internal organs and cavities of the body. It is very important to note that such applications for PDT require systemic delivery of the photoactivatable drugs. However, biodistribution and bioavailability of the photosensitizer can be compromised by their insufficient selectivity and low water solubility [Chen *et al.* 2009]. Indeed, most of the elaborated photosensitizers are highly hydrophobic and thus poorly water soluble and prone to aggregation in aqueous solutions [Ricchelli, 1995; Boyle *et al.* 1996], which compromise efficiency ROS generation and thus the therapeutic effect. Furthermore, currently clinically approved photosensitizers often display poor bioavailability and unfavourable biodistribution, resulting in suboptimal tumor specificity and, consequently, undesirable side effects such as prolonged cutaneous photosensitivity and damage to surrounding healthy tissues [Moriwaki *et al.* 2001]. Low bioavailability of the delivered photosensitizer make it difficult to achieve significant drug concentrations in diseased sites without distributing drugs throughout healthy tissues. This means that larger amount of photosensitizer than necessary have to be injected and that healthy tissues get to be exposed to the potential adverse effects of the cytotoxic drugs [Sun *et al.* 2009].

By taking note of these photosensitizer-related limitations, the selective targeted delivery of photosensitizers to diseased cells is one of the major problems in PDT of cancer that still a challenge to take up, and one area of importance is photosensitizers targeting. In this sense, to overcome the drawbacks encountered in passive photosensitizer targeting, an arsenal of targeting strategies have been recently developed, including conjugates and supramolecular carrier platforms like dendrimers, polymeric micelles, liposomes and nanoparticles. Further, numerous interesting works have clearly demonstrated that more specific drug targeting, cellular uptake and bioavailability can be achieved by binding various ligands, known as targeting moieties, to the surface of the carrier platforms, such as peptides, growth factors, transferin, antibodies or antibody fragments, oligonucleotide aptamers and small compounds such as folate that can recognize tumor cell markers [Rifkin *et al.* 2006]. The rational of all these strategies is taken from biologic and molecular characteristics of tumor tissues. For instance, alterations or increased levels in receptor expression of specific cellular type occur in diseased tissues. Therefore, photosensitizers can be covalently attached to targeting moieties such as peptides, leading to a receptor-mediated targeting strategy. These active-targeting approaches may be particularly useful for vascular-targeted PDT. Indeed, PDT effects are mediated not only through direct killing of tumor cells but also through indirect effects, involving both initiation of an immune response against tumor cell antigens and destruction of the tumor vasculature [Solban *et al.* 2006].

The present chapter will focus on recent advances and developments in targeting strategies in PDT with the emphasis on target specificity. We will discuss the tumor neovasculature-associated targeting and the uncontrolled cell proliferation-associated markers targeting.

2. PDT – Tumor Neovascularisation-Associated Targeting

The targeting of tumor vasculature has become a large area of focus for the development of new cancer therapeutics [Tozer, 2005]. It is clearly known that when a tumor grows, its need for nutrients and oxygen increases, and thus the number and size of blood vessels increase proportionately (angiogenesis). This implies that destroying the tumor neovasculature can be an effective approach for the control of the tumor growth.

Angiogenesis is characterized by the invasion, migration and proliferation of smooth muscle and endothelial cells, leading to the formation of new microvessels. This process is mostly facilitated by a variety of proangiogenic factors such as growth factors and their respective receptors, which are overexpressed by infiltrating tumor and host cells. Moreover, angiogenesis is promoted by the secretion of proteolytic enzymes, namely metalloproteinases, which facilitate the propagation of tumor through the basement membrane and into the extracellular matrix of the local connective tissue [Ahmad *et al.* 2010]. In addition to its important role in tumor growth, angiogenesis appears to be one of the most crucial steps in tumor translation to the metastatic form, and thus promotes tumor propagation in other tissues [Folkman, 1996]. Thereby, by attacking the growth of the neovasculature, the size and the metastatic capabilities of tumors can be regulated [Folkman, 1996].

Tumor neovasculature targeting appears as an approach of significant research interest for the development of active photosensitizer delivery systems able to enhance selectivity and efficiency of vascular PDT for cancer treatment. The main molecular targets explored in the vascular-targeting PDT for cancer includes the vascular endothelial growth factor receptors (VEGFRs such as neuropilin-1 (NRP-1)), receptor tissue factor (TF), $\alpha_v\beta_3$ integrins, and matrix metalloproteinase (MMPs) activities.

2.1. Vascular Endothelial Growth Factor Receptors (VEGF Receptor-2(VEGFR-2) and NRP-1)

Vascular endothelial growth factor (VEGF) is one of the most potent direct-acting angiogenic proteins known [Ferrara, 2004]. This growth factor is upregulated in the majority of neoplastic cells, mainly as a response to hypoxia and many oncogens, and its overproduction is correlated with high microvascular density and poor prognosis [Ishigami *et al.* 1998]. The overexpression of VEGF results in the upregulation of VEGFR-1 (also known as fms-like tyrosine kinase (flt-1)) and VEGFR-2 (foetal liver kinase-1, flk-1 or KDR (kinase domain region)) [Kremer *et al.* 1997; Plate *et al.* 1992; Bando *et al.* 2005]. Of the two receptors, the VEGF receptor of interest for most active targeting strategies is VEGFR-2, as it is generally believed to be the main receptor that mediates VEGF biological activities and thus, play a major role in tumor-associated angiogenesis. Moreover, it is highly expressed on endothelial cells in tumor neovasculature [Brown *et al.* 1995; Couffinhal *et al.* 1997]. Therefore, VEGFR-2 is seen as promising molecular target for anti-angiogenic drug delivery, and that specific targeting of VEGFR-2 could provide an interesting approach for selective and efficient photosensitizer delivery to tumor neovasculature.

Although targeting VEGFR-2 has been widely investigated for the selective delivery of therapeutic drugs for conventional therapies such as radiotherapy, a limited number of studies have been performed for the selective delivery of photoactivatable drugs. Targeted verteporfin (Visudyne®) have been the subject of the first research works that have explored VEGFR-2 as a molecular target for vascular PDT. By using verteporfin conjugated to a peptidic motif (ATWLPPR) known to bind VEGFR-2, Renno *et al.* have showed that PDT using targeted verteporfin was more effective in causing choroidal neovascularization closure than untargeted verteporfin. Importantly, as expected targeted verteporfin also resulted in more selective treatment than untargeted drug [Renno *et al.* 2004]. These studies suggest that targeting VEGFR-2 can be an effective approach to confine sufficient doses of photosensitizers within the tumor neovasculature and thus potentiate the vascular photodynamic effect with minimal side effects. In another hand, further studies have not confirmed the findings that ATWLPPR peptide recognizes KDR [Tirand *et al.* 2006]. Our group demonstrated in agreement with others [Perret *et al.* 2004] that this peptide targeted NRP-1 and not KDR. Although VEGFR-2 still a potential target to explore, the latter observations have then attracted a great interest on the potential of NRP-1 as promising target for targeted vascular PDT.

Neuropilins (NRP) have also been identified as receptors for the $VEGF_{165}$ isoform [Soker *et al.* 1998]. The transmembrane protein NRP-1 has been described as a positive modulator of VEGFR-2 and is thereby a crucial inducer of angiogenesis [Hong *et al.* 2007; Nasarre *et al.* 2010]. It was reported that, when co-expressed with VEGFR-2, NRP-1 enhanced the binding of $VEGF_{165}$ to VEGFR-2 and $VEGF_{165}$-mediated chemotaxis [Soker *et al.* 1998]. Such effects may result from the formation of a ternary complex between $VEGF_{165}$, VEGFR-2 and NRP-1 [Whitaker *et al.* 2001; Soker *et al.* 2002]. Neuropilins are expressed specifically in tumor angiogenic vessels and some tumor cells, and promote tumor angiogenesis and progression [Miao *et al.* 2000]. Hence, targeting NRP-1 can lead to the selective vascular localization of photosensitizers, and thus enhance the vascular photodynamic effects. In accordance with this, the conjugation of a photosensitizer (5-(4-carboxyphenyl)-10,15,20-triphenyl-chlorin (TPC)) to the heptapeptide (ATWLPPR), specific for NRP-1 has been described by our group. This targeted photosensitizer proved to be very efficient in endothelial cells, compared to its nonconjugated counterpart. Our group evidenced *in vitro* by competition experiments in human umbilical vein endothelial cells (HUVEC) [Tirand et *al.* 2006] and *in vivo* by biodistribution studies in nude mice xenografted with U87 human malignant glioma cells (Figure 1) [Tirand *et al.* 2007; Thomas *et al.* 2008] that a part of the accumulation of the conjugated photosensitizer (noticed TPC-Ahx-ATWLPPR) was related to NRP-1-dependent mechanisms but also to non-specific mechanisms. Taking advantage of RNA silencing techniques known as RNA interference, we have selectively silenced NRP-1 expression in MDA-MB-231 breast cancer cells to provide the evidence for the involvement of NRP-1 expression in the conjugate cellular uptake. We also observed *in vivo* the vascular effect by measuring the tumor blood flow during PDT using both conjugated and nonconjugated PS. The conjugate-mediated vascular targeted PDT produced a selective vascular effect, leading to vascular shutdown and tumor growth delay [Thomas *et al.* 2009; Bechet *et al.* 2010].

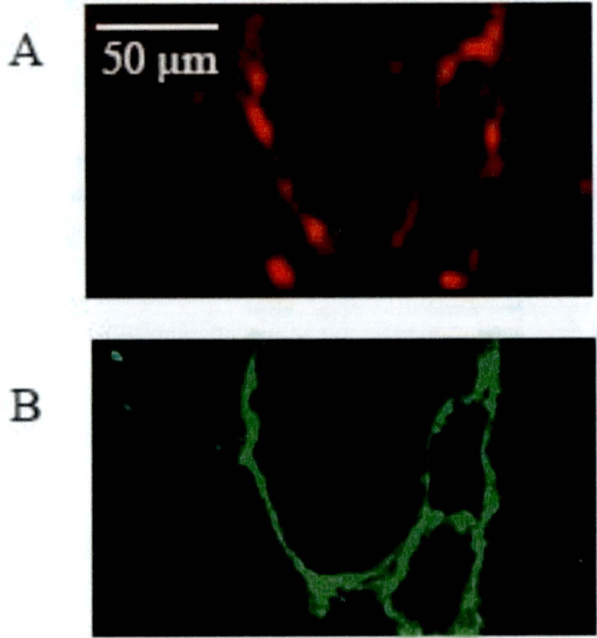

Figure 1. Localization of the photosensitizers 4 hours after the intravenous injection. (A) Color composite image of TPC-Ahx-ATWLPPR fluorescence [red]. (B) Color composite image of CD31-staining [green] in the same region as (A). Analysis of the tumor sections 4 h after TPC-Ahx-ATWLPPR administration showed that the photosensitizer was mainly colocalized inside the vascular endothelium. [Thomas *et al.* 2008].

From the biological mechanism point of view, the conjugate-mediated vascular effect implies the induction of tissue factor (TF) expression that may lead to the thrombogenic effect within the vessel lumen (Figure 2) [Bechet *et al.* 2010].

Thus, this original targeting strategy, using a peptide competing with $VEGF_{165}$ binding on NRP-1, led to a selective accumulation in endothelial cells lining tumor vessels. Nevertheless, using this approach, affinity for NRP-1 remains low and could be improved. Reasons for this may include, *(i)* intramolecular interactions between chlorin and peptide, and steric hindrance due to the TPC moiety, *(ii)* aggregation of the photosensitizer molecules, *(iii)* low stability of the peptide moiety that, may be due to sensitivity to circulating peptidases action. The conception of peptide-targeted delivery system, capable to overcome these drawbacks, would lead to an improvement of affinity. In this sense, the development of multifunctional nanoparticles as photosensitizer carriers is under investigation.

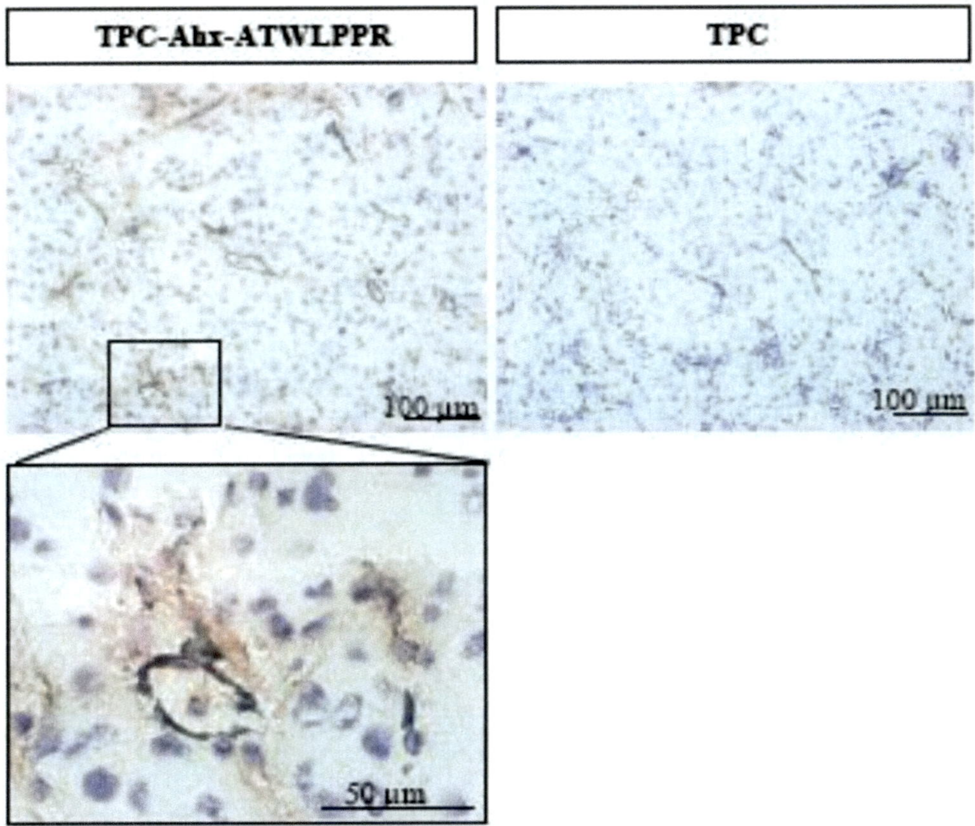

Figure 2. Immediately after PDT, TPC-Ahx-ATWLPPR-induced photodynamic activity induced TF expression (brown staining) compared to the nonconjugated photosensitizer (TPC). Tissue factor staining appeared to be non-uniform and was not limited to vessel *lumen* areas but also present in tumor tissues. Bottom, enlarged view of the corresponding specimen. [Bechet *et al.* 2010].

2.2. Receptor Tissue Factor (TF)

Several studies have generated promising results in inhibiting tumor growth in animal models by targeting receptor TF in tumor cells and vascular endothelial cells [Shoji *et al.* 2008]. Tissue factor is a transmembrane receptor that forms a high-affinity and specific complex with its endogenous ligand, coagulation factor VII (fVII), as the initial step of the blood coagulation pathway [Nemerson, 1988]. Accumulating evidence suggest that receptor TF is aberrantly overexpressed on endothelial cells of the tumor vasculature and various tumor cells [Contrino *et al.* 1996; Hu *et al.* 1999; Chen *et al.* 2001] but not on endothelial cells of normal blood vessels [Drake *et al.* 1989; Semerano *et al.* 1997]. Moreover, it has been suggested that VEGF protein secreted by tumor cells induces endothelial cells in tumor vasculature to express receptor TF [Clauss *et al.* 1990; Zucker *et al.* 1998]. All the aforementioned characteristics suggest that receptor TF can serve as specific therapeutic target for targeted drug therapy directed at tumors and tumor angiogenesis. Accordingly, several methods have been developed for selectively delivering drugs to receptor TF-expressing tumor vasculature and tumors using fVII as a drug carrier. An immunotherapy strategy for destroying the tumor vasculature by targeting receptor TF on tumor vascular endothelial cells with an immunoconjugate (icon) molecule has been described [Hu *et al.* 1999]. More recently, this group has described an approach targeting receptor TF on tumor cells and angiogenic vascular endothelial cells by fVII-targeted verteporfin PDT for breast cancer *in vitro* an *in vivo* in mice. Authors showed that *(i)* fVII protein could be conjugated with verteporfin without affecting its binding activity; *(ii)* fVII-targeted PDT could selectively kill receptor TF-expressing breast cancer cells and VEGF-stimulated angiogenic HUVECs but no side effect was observed on non-receptor TF expressing unstimulated cells; *(iii)* fVII targeting enhanced the effect of verteporfin PDT by three to four-folds; *(iv)* fVII-targeted PDT induced significantly stronger levels of apoptosis and necrosis than non-targeted PDT; and *(v)* fVII-targeted PDT had a significantly stronger effect on inhibiting breast tumor growth in mice than non-targeted PDT [Hu *et al.* 2010]. Since receptor TF is expressed on many types of cancer cells and selectively on angiogenic vascular endothelial cells, these findings suggest that fVII-targeted PDT could have broad therapeutic applications.

2.3. Integrin $\alpha_v\beta_3$

The $\alpha_v\beta_3$-integrin, a heterodimeric transmembrane glycoprotein receptor, is highly expressed on neovascular endothelial cells and is important in the calcium-dependent signalling pathway, leading to endothelial cell migration. It is also overexpressed in many tumor cells, such as osteosarcomas, neuroblastomas and lung carcinomas, but also in actively proliferating endothelial cells and around tumor tissues. Moreover, $\alpha_v\beta_3$-integrin is an endothelial cell receptor for extracellular matrix proteins harboring the RGD peptidic sequence. Therefore, $\alpha_v\beta_3$-integrin is considered as an attractive molecular target for antivascular therapies. In accordance with this, several investigations have described the potential of such receptors in vascular-targeted PDT. Chaleix *et al.* reported the solid-phase synthesis of four porphyrins bearing the $\alpha_v\beta_3$-integrin ligand RGD (H-Arg-Gly-Asp-OH) tripeptide [Chaleix *et al.* 2004]. Three of the conjugates prepared displayed photodynamic activity on the K562 leukaemia cell line to a degree comparable to that of Photofrin II®. The authors later described the synthesis of a cyclic peptide containing the RGD motif and adopting conformations showing an increased affinity for integrins [Chaleix *et al.* 2004]. In the ring closing metathesis method used for synthesis, the enzymatically-sensitive disulfide bond of the cyclic CRGDC thiopeptide previously described is replaced by a carbon-carbon double bond in order to increase the plasmatic lifetime of the peptide. Carboxy-glucosylporphyrins coupled to this peptide *via* a spacer arm showed the same efficiency for $^1O_2$ production as hematoporphyrin but to the best of our knowledge, no *in vivo* study on these compounds has been reported yet.

To increase the selectivity of phthalocyanines, Allen *et al.* have evaluated the use of viral proteins to deliver the photosensitizer to the selected tissue [Allen *et al.* 1999]. Adenoviruses cause illnesses such as gastroenteritis, conjunctivitis and respiratory diseases. They have received a great deal of attention as gene therapy vectors, owing to the facility and the safety to manipulate and to stock them. It has also been shown that adenoviruses can infect several cells and tissue types and gain access into a cell *via* receptor-mediated endocytosis. Adenoviruses efficiently break the endosomes upon infection in the cell and therefore, it was anticipated that adenovirus-photosensitizer dyads would target the cell nucleus more quickly than free photosensitizer. This targeting of the nuclear cell resulted in a 2.5 fold increase of the photodynamic activity of photosensitizer in comparison with a cytoplasmic localization [Allen *et al.* 1999]. Adenovirus penton base proteins contain the RGD peptide sequence motif. Adenoviruses type 2 structural proteins; the hexon, penton bases and fiber antigen were isolated and purified. The adenovirus tetrasulfonated aluminum phthalocyanine derivatives were tested both *in vitro* and *in vivo*. The penton base conjugate was the most efficient *in vitro,* as measured in two positive cell lines (A549, Hep2) expressing integrins [Allen *et al.* 1999]. According to this study, it appears that adenoviral proteins can be used to target tumor cells. Despite a non-optimal photodynamic activity, *in vivo* results were encouraging. Nevertheless, tumor targeting using adenoviral protein vehicles may promote inflammation and anti-protein cellular immunity, which could limit their usefulness.

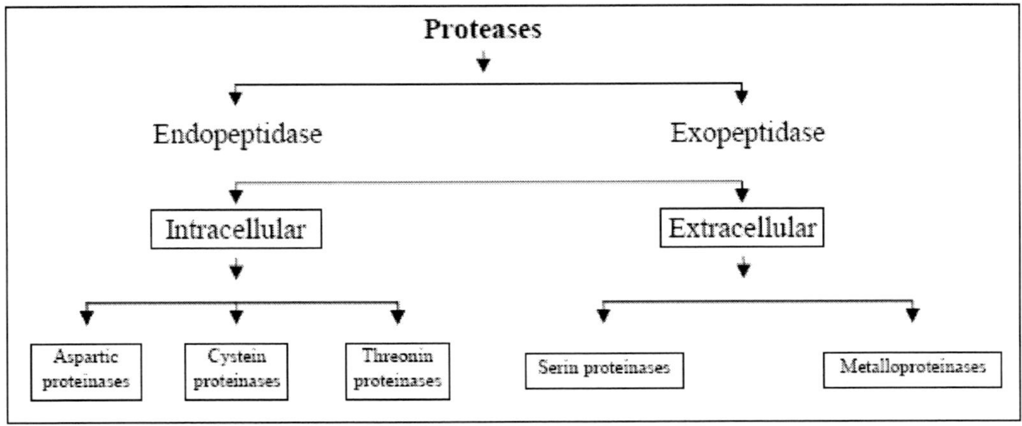

Figure 3. Summary of different proteases.

Our group conjugated a photosensitizer (5-(4-carboxyphenyl)-10,15,20-triphenylchlorin or porphyrin) to the RGD motif as a common sequence [Frochot *et al.* 2007]. We reported an efficient solid-phase synthesis of a new family of peptidic photosensitizers with linear or cyclic [RGDfK] RGD motif. Chlorins containing linear and constrained RGD motif were incorporated up to 98- and 80-fold more, respectively, than the unconjugated photosensitizer over a 24 h exposure in HUVEC overexpressing $\alpha_v\beta_3$ integrin. Peptidic moiety also led to a non-specific increased cellular uptake by murine mammary carcinoma cells (EMT-6), lacking

RGD binding receptors. The higher photodynamic efficiency was related to conjugates greater cellular uptake [Frochot *et al.* 2007]. Survival measurements demonstrated that HUVEC were greatly sensitive to conjugates-mediated photodynamic therapy. This study also showed that peptidic moiety was obviously an interesting means to introduce such a balance between hydrophilicity and hydrophobicity, previous observations suggesting the requirement of amphiphilicity for efficient photodynamic activity [Sobolev *et al.* 2000]. More recently, Boisbrun *et al.* have described the synthesis, characterization, fluorescence, and singlet oxygen quantum yields of tetraphenylporphyrin and tetraphenylchlorin coupled to RGD type peptides. They reported that some of theses compounds are very promising for potential PDT applications [Boisbrun *et al.* 2008].

2.4. Matrix Metalloproteinases (MMPs)

Similar to $\alpha_v\beta_3$ integrin, the matrix metalloproteinases (MMPs) are well known targets that interact with the extracellular matrix (ECM). MMPs are a family of structurally related zinc-dependent endopeptidases capable of degrading the ECM [Vihinen *et al.* 2005]. Proteolytic degradation of the ECM is crucial for cancer development, invasion and metastasis, all of them associated with increased expression and activities of several different proteases [Van Kempen *et al.* 2006]. Numerous studies have documented increased expression of proteases (Figure 3) such as MMPs in many human malignant tissue types, often correlating with poor prognosis [Egeblad *et al.* 2002].

Thus, these tumor-associated proteases could act as activators of the protease-mediated PDT agent [Choi *et al.* 2006]. Indeed, protease-sensitive macromolecular prodrugs have attracted interest for bio-responsive drug delivery to disease sites with up-regulated proteolytic activities, and many strategies emerged from protease targeting to convert inactive PDT prodrug into active drug when activated *in vivo* [Law *et al.* 2009]. Extracellular matrix proteolysis is also involved in the angiogenesis necessary for the continued growth of solid tumors. Prior reports demonstrated that the expression of multiple MMPs by tumor cells and/or host cells is a common feature of both experimental and clinical models of head and neck squamous cell carcinoma [O-charoenrat *et al.* 1999]. Indeed, MMPs are tumor promoting enzymes playing an important role in angiogenesis [Sharwani *et al.* 2006; Chetty *et al.* 2010] and thus, are potential interesting target in PDT for cancer treatment. MMPs are well known for their ability to cleave components of the ECM, but they also cleave other proteinases, latent growth factors and growth factor binding proteins, chemotactic molecules, cell surface receptors and cell-cell adhesion molecules [Fiore *et al.* 2002]. They are classified into, at least, five main groups according to their substrate specificity, primary structure and cellular localization (collagenases, gelatinases, stromelysins, the matrilysins and membrane-type MMPs). MMPs are also regulated by tissue inhibitors of metalloproteinases (TIMPs). It is likely that distinct MMPs play a role at different stages of tumour development; a growing number of studies provide strong evidence that MMP-2 (gelatinase-A) and MMP-9 play important roles in the process of tumour growth and angiogenesis, making them interesting targets to trigger tumour environment. In addition, gelatinases are responsible for the final degradation of collagens after initial cleavage by collagenases. Consequently, protease-sensitive macromolecular prodrugs have attracted interest for bio-responsive drug delivery to sites with up-regulated proteolytic activities. Among them, proteases are of a great interest because of their unique ability to cleave selectively amide bonds in peptides. Prodrugs activated by proteases are not recent but applied to PDT, proteases become attractive biological triggers in drug development to control $^1O_2$ production; many strategies emerged from protease targeting to convert pharmacological inert prodrug into active pharmacophore when activated *in vivo* [Law *et al.* 2009]. Before being an interesting field in PDT, protease activity has been used for bio-imaging [Funovics *et al.* 2003; Lovell *et al.* 2008; Pham *et al.* 2004]. Among these, matrix metalloproteinase-7 (MMP7) has been a particularly important target, because of its high expression in pancreatic, colon, breast, and non small-cell lung cancer [Shiomi *et al.* 2003, Leinonen *et al.* 2006]. Considering this, Pham *et al.* designed a peptide-based near-infrared (NIR) fluorescence probe consisting of a NIR fluorescence emitter linked *via* a MMP7 substrate peptide linker to a NIR fluorescence absorber for sensing tumor-associated MMP7 activity [Pharm *et al.* 2004]. The underlying principle is that in a fluorescence resonance energy transfer (FRET) between the donor and the acceptor, the absence of proteases results in the quenched fluorescence for the donor; however, in the presence of proteases, the substrate peptide linker is cleaved, releasing the FRET interaction between the donor-acceptor fluorophore, which result in a four-fold increase in the fluorescence signal for an initially quenched molecular dye [Verma *et al.* 2007].

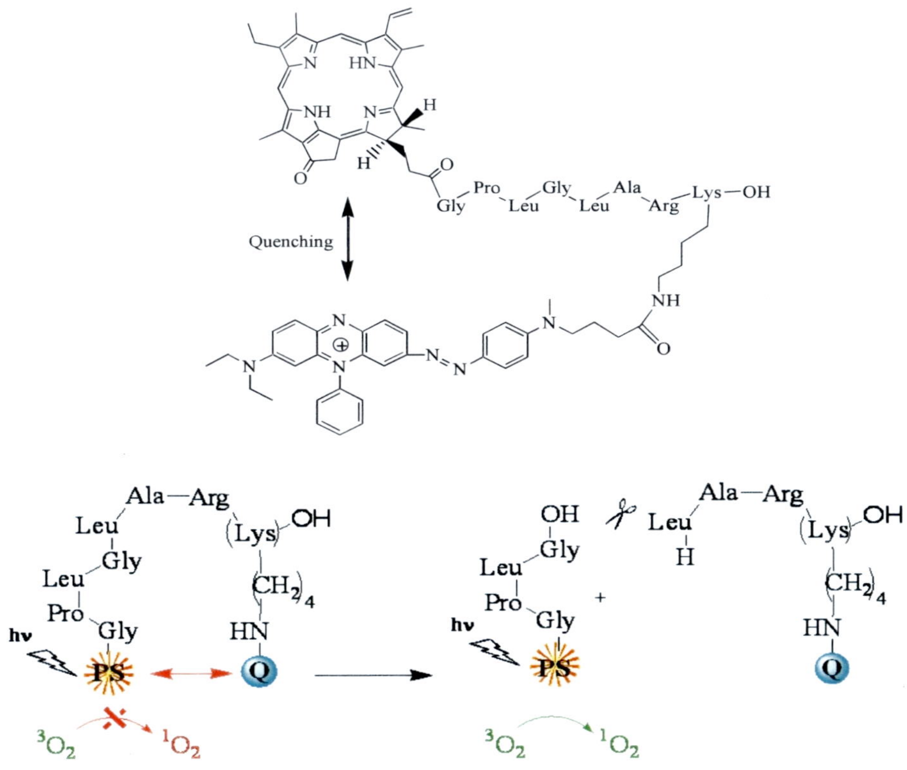

Figure 4. MMP7-triggered photosensitizing molecular beacon concept. Q: quencher.

As most photosensitizers used in PDT are porphyrins derivatives that absorb light energy and then transfer to oxygen or emitted fluorescence, it has been reported that $^1O_2$ generation is closely correlated with fluorescent intensity of the photosensitizing agent [Law *et al.* 2009]. So, when a photosensitizer is in a fluorescent quenched state, its ability to generate $^1O_2$ is also reduced. Molecular beacons are FRET-based target-activatable probes, offering control of fluorescence emission in response to specific targets such as proteases, as a useful tool for *in vivo* cancer imaging. By combining the two principles of FRET and PDT, Zheng *et al.* suggested a concept of photodynamic molecular beacon for controlling the photosensitizer's ability to generate $^1O_2$, and thus, for controlling its PDT activity. They described the synthesis and characterization of a MMP7-triggered photodynamic molecular beacon, using (i) pyropheophorbide as the photosensitizer (PS) ; (ii) black hole quencher 3 (Q) as a dual fluorescence and $^1O_2$ quencher; and (iii) a short peptide sequence, GPLGLARK, as the MMP7-cleavable linker. As depicted in Figure 4, the photosensitizer and the quencher are attached to the opposite end of the MMP7-specific cleavable peptide linker to keep them in close proximity, enabling FRET and $^1O_2$ quenching to form inactive prodrug. Thus, the photosensitizer's photoactivity is silenced until the linker interacts with the target tumor-associated MMP7.

After validating the MMP7-triggered production of $^1O_2$ in solution, the authors demonstrated the MMP7-mediated photodynamic cytotoxicity in cancer cells. *In vivo* studies also revealed the MMP7-activated PDT efficacy [Zheng *et al.* 2007]. This study showed that specificity and selective PDT-induced cell death can be achieved efficiently by not only targeting a photosensitizer to a target diseased tissue, but also by creating an active form of a photosensitizer from an inactive (quenched) prodrug by exploiting specific cellular functions (such as upregulated enzymes) at the site of action.

3. PDT - Neoplastic Cells Targeting - Uncontrolled Cell Proliferation Markers

This second part deals with photosensitizers covalently bound to different biomolecules (sugars, steroid hormones, amino acids, proteins…), *via* a direct coupling or using linkers/spacers. These biomolecules, as targeting moieties are used to direct the photosensitizing agents against neoplastic cells-associated specific antigens. The coupling of a vector can either *i)* modulate the amphiphilicity and enhance the solubility of these compounds in biological media and prevent self-aggregation (passive targeting), or *ii)* promote cellular recognition (active targeting), with the aim to increase the biological efficiency. Several of these targeting strategies offer the advantage of transporting the photosensitizer across the cellular plasma membrane, resulting in intracellular accumulation of the photosensitizers, which may allow for targeting photosensitive intracellular sites, and thus improve photodynamic efficiency.

3.1. Epidermal Growth Factor (EGF)-receptors

The ligands that bind the EGF receptor (EGFR [HER1, Erb-B1]) include EGF, transforming growth factor-α (TGFα), heparin-binding EGF-like growth factor (HB-EGF), amphiregulin (AR), betacellulin (BTC), epiregulin (EPR), and epigen. Each of the mature peptide growth factors is characterized by a consensus sequence consisting of six spatially conserved cysteine residues. This consensus sequence is known as the EGF motif and is crucial for binding members of the HER receptor tyrosine kinase family. In addition to binding EGFR, HB-EGF, BTC, and EPR are also reported to bind HER4. Since a high EGF receptors expression frequently accompanies development of several tumor types such as squamous carcinomas, its natural ligand EGF was an attractive candidate for the conception of drug targeting strategies. One binding to its receptor, EGF is internalized in the cell through receptor-mediated endocytosis, enabling the intracellular accumulation of photosensitizers [Gijsens *et al.* 2000]. Based on this concept, EGF receptors provide an interesting approach for selective photosensitizer delivery to tumor tissues.

Lutsenko *et al.* have conjugated EGF with aluminum and cobalt disulfonated phthalocyanines, the EGF:photosensitizer *ratio* being 1:1. The study compared first their photocytotoxic activities with those of nonconjugated phthalocyanines for the human breast carcinoma cell line MCF-7 [Lutsenko *et al.* 1999]. Two different types of activation mechanisms have been used: photoactivation and activation by ascorbic acid. Indeed, the phototoxicity of phthalocyanines is not limited only to the generation of $^1O_2$ (a type II energy transfer reaction) but, photoinduced electron abstraction from appropriate biomaterial (a type I electron transfer reaction) could also give one-electron oxidized primary radicals, which can provide the precursors of oxidative damage in phthalocyanines photosensitization. It is well known that complexation of phthalocyanines with open shell or paramagnetic metal ion gives dyes with shortened triplet lifetime, which renders it photoinactive [Chan *et al.* 1987]. Anyway, phthalocyanines catalyse the formation of ROS such as superoxide radical ($O2^{\cdot-}$), hydrogen peroxide (H_2O_2) and hydroxyl radical ($OH^{\cdot}$) from molecular oxygen and reducing molecules. EGF-phthalocyanines exhibited a much higher photoactivity than their nonconjugated analogous. The antitumoral activity was also investigated *in vivo* on the growth of solid tumors (murine melanoma cell line B16) implanted in mice. The administration of the conjugate strongly increased mean life spans and mean survival time of animals bearing tumors whereas free phthalocyanines had practically no effect on these parameters [Lutsenko *et al.* 1999]. According to the aurhors, the maximal therapeutic effect needed to be optimized in relation with different schedules for the drug administration.

Targeting the EGFR with antibody(Ab)-delivered photoactive molecules may serve as a two-armed feedback-controlled approach: the photoactivatable compound together with illumination may target and destroy the EGFR, and the fluorescent dye may allow the progress of the treatment to be monitored. Soukos *et al.* studied an anti-EGFR monoclonal (m)Ab (C225) coupled to either the near-infrared fluorescent dye *N,N'*-di-carboxypentyl-indodicarbocyanine-5,5'-disulfonic acid (indocyanine Cy5.5) for detection or a photochemically active dye (chlorin$_{e6}$, Ce6) for therapy: (1) to determine whether the intra-venous injection of a conjugate between C225 and the fluorescent dye indocyanine Cy5.5 could deliver sufficient amounts of the fluorophore into clinically unrecognizable early premalignant lesions and, (*2*) to test the photodynamic effect of a conjugate between C225 and Ce6 [Soukos *et al.* 2001]. Authors demonstrated that immunophotodiagnosis using C225-Cy5.5 conjugate might have potential both as a diagnostic modality for oral precancer and as a surrogate marker. As suggested, the use of a NIR emitting fluorophore as a label may be an alternative to use radioisotopes, with consequent increased patient safety. To extend the analogy with radioisotopes, the principle of using an anti-EGFR mAb to deliver photoactive molecules applies equally well to therapeutic photosensitizers as to fluorophores, and furthermore, immunophotodiagnosis can be used to assess response. HER2, also called human EGF receptor-2, is highly upregulated target on tumor cell surfaces such as breast cancer. Bhatti *et al.* described complex of multiple photosensitizers (verteporfin or pyropheophorbide-a) with anti-HER2 single chain Fv (scFv). They demonstrated that the photoimmunoconjugates were more potent than either free photosensitizer, effectively killing tumor cells *in vitro* and *in vivo*. Treatment of human breast cancer xenografts with the photoimmunoconjugate comprising an anti-HER2 scFv with about 8-10 molecules of the pyropheophorbide-a led to a significant tumor regression [Bhatti *et al.* 2008]. It is important to note that the clearance of the immunoconjugate was still more rapid that free pyropheophorbide-a, suggesting that the time from administration of scFv-pyropheophorbide-a to exposure to laser light could be therapeutically attractive with very little danger of skin photosensitivity.

3.2. Cholesterol and Low Density Lipoprotein (LDL) Receptors

The steroid hormones are all derived from cholesterol. They are an interesting class of biomolecules to target photosensitizers to cancer cells. In particular, as previously described, cholesterol is a vital component of eukaryotic cell membranes and can be taken up quickly by cancer cells [Candide *et al.* 1986]. It thus appears that the covalent coupling of cholesterol to a molecule could favour its association with LDL and increase its photodynamic efficiency. Furthermore axial substitution of suitable central metals (Si, Ge, Al) by cholesterol, which is a lipophilic and bulky ligand, could decrease phthalocyanine self-aggregation (inhibition of the ππ-stacking by addition of steric hindrance on the sides of the macrocycle) and increase the crossing of the lipophilic membrane. Photodynamic activities of the new cholesterol-phthalocyanine derivatives were investigated on two pigmented melanoma cell lines: M3Dau and SK-MEL-2 [Barge *et al.* 2004]. M3Dau cells were of interest as they led to melanoma tumors in nude mice, allowing further *in vivo* validations. SK-MEL-2 cells were also used in order to elucidate whether the photocytotoxicity of the phthalocyanines could vary depending on the pigmented melanoma cell line used. Segalla *et al.* demonstrated that cholesterol-phthalocyanine derivatives, injected systemically into mice bearing an intramuscularly implanted MS-2 fibrosarcoma, was quantitatively transferred to serum lipoproteins and localized into the tumor tissue. The selectivity of tumor targeting was expressed by the ratio of photosensitizer concentration in the MS-2 fibrosarcoma to the muscle. It was similar to that observed for other liposome-delivered phthalocyanines [Cuomo *et al.* 1991]. Another grafting of cholesterol was achieved by Maree *et al.* [Maree and Nyokong. 2001]. They performed the photophysical study and showed that unfortunately it was highly aggregated at low concentrations. No biological evaluation was performed on this compound yet.

The LDL particle is the principal carrier of cholesterol in human plasma and delivers exogenous cholesterol to cells by endocytosis *via* the LDL receptors. As indicated before, cholesterol is absolutely essential for the cell's growth and viability. Consequently, rapidly proliferating cells have a great demand for cholesterol for membrane synthesis; hence, many cancer cells express an abnormally high amount of LDL receptors [Vitols, 1991]. Therefore, LDLs have been suggested to be able to play an important role in transport and release of photosensitizers in tumor cells [Firestone, 1994; Konan *et al.* 2002]. A *plethora* of studies have showed that photosensitizers such as phthalocyanine, Visudyne® and chlorin e6 mixed with LDL before administration led to an increase in photodynamic efficiency [for a review, see Konan *et al.* 2002].

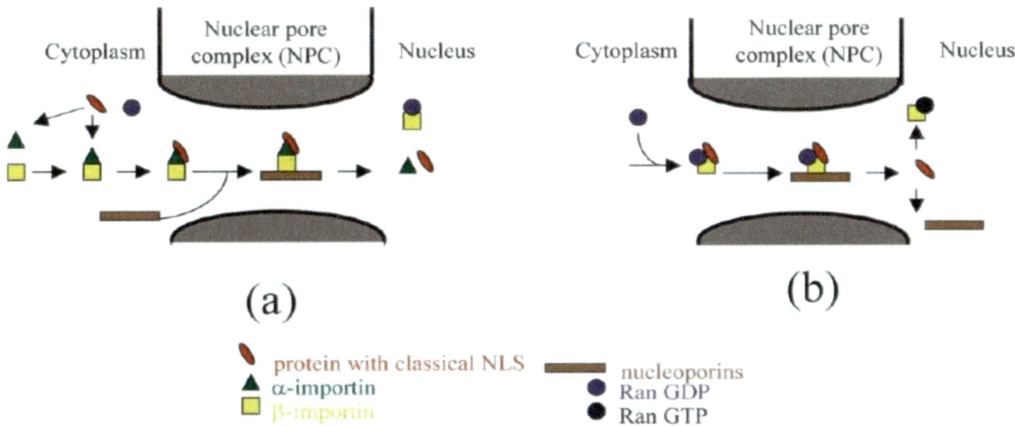

Figure 5. Mechanisms of protein nuclear transport. (a) First step of proteins transport with classical NLS, involving the α/β-importin heterodimer (*e.g.*, SV40 large T antigen); (b) second step-energy-dependent translocation of the importin complex with the NLS-carrying protein toward the nucleoplasm, requiring a Ran protein mostly in the GDP form in the cytoplasm and in the GTP form in the nucleus [Schneider *et al.* 2006].

Zheng and co-workers prepared a delivery system comprising a pyropheophorbide cholesterol oleate photosensitizer incorporated into LDL with a modest photosensitizers payload. Interestingly, laser scanning confocal microscopy studies demonstrated that such a LDL-based photosensitizer was internalized exclusively by LDL receptors overexpressing human hepatoblastoma G2 (HepG2) tumor cells [Zheng *et al.* 2002]. To reduce the dose for more efficient cancer detection and treatment, Li *et al.* designed a novel strategy to improve LDL's photosensitizers payload using a silicon-phthalocyanine as new NIR dyes. The choice of such compound was based mainly on the fact that the central silicon atom of silicon-phthalocyanine allows axial coordination of two bulky ligands on each side of the phthalocyanine ring to prevent stack aggregation usually encountered in solution for the planar molecular structure; since such aggregation is thought to be the major limiting factor for achieving high probe/LDL payload. Confocal microscopy studies demonstrated that the internalization of the conjugate by human HepG2 cells was still mediated by LDL receptors pathway. Using clonogenic assay, the *in vitro* PDT response of HepG2 cells revealed a significant enhanced efficacy of LDL receptors-targeted PDT [Li *et al.* 2005]. More recently, Song *et al.* described the preparation and characterization of the naphtathalocyanine reconstituted LDL nanoparticles as well as the *in vitro* and *in vivo* validation of the LDL receptors-targeting of the conjugate to cancer cells and in mice bearing human HepG2 tumors. A confocal microscopy study revealed that the photosensitizer reconstituted LDL nanoparticles retain their LDL receptors-mediated uptake by cancer cells. Its preferential uptake by tumor *vs.* normal tissue was also confirmed *in vivo* by non-invasive optical imaging technique [Song *et al.* 2007].

3.3. Oestrogen Receptor

Always in the aim to improve the uptake of the phthalocyanine derivatives by receptor-rich endocrine tumors, van Lier *et al.* have reported the synthesis of phthalocyanine-oestradiol conjugates [Khan *et al.* 2003; Ali *et al.* 1997], and their photocytotoxic activities. Phthalocyanine-oestradiol conjugates were prepared with both aliphatic and aromatic alkynyl spacers in order to modulate the overall amphiphilicity of the molecule. Photocytotoxic activity and relative binding affinity for oestrogen receptors were measured on murine EMT-6. The highest receptor binding affinities were observed with lipophilic conjugates coupled *via* a relatively long spacer while the sulfonated analogues showed little binding affinities. Surprisingly, the most hydrophilic trisulfonated phthalocyanine-oestradiol presented the highest photocytotoxicity, comparable to those reported for the nonconjugated phthalocyanine. The nature of the spacer did not seem to influence the biological activity. Grafting of estrone, the biosynthetic precursor of oestradiol receptors overexpressed in breast cancer cells, was achieved by Maree *et al.* [Maree *et al.* 2002]. From a photophysical point of view, aggregation was prevented due to the presence of axial ligands in all complexes. From the biological point of view, the octaoestrone phthalocyanine showed the best promise for both i) detection through fluorescence and ii) treatment of tumor cells since this complex has the longest triplet lifetime.

3.4. Nucleus and DNA

Initial attempts to increase photosensitizers delivery focused on improving targeted to non-targeted tissue *ratios*. However, it was demonstrated that elevated tumor to normal tissue *ratios* did not ensure improved efficiency *in vivo* [Rosenkranz *et al.* 2000]. Photoactivated photosensitizer generates ROS, which are able to damage DNA, membranes, and other cell structure and macromolecules. It has been postulated that $^1O_2$ is the most important of these reactive species with a life span of 200 ns and a short migrating circumference about 45 nm. The site of $^1O_2$ production is therefore the site of photosensitization. The cell nucleus is the most sensitive site, in contrast to cell membranes and other cytoplasmic organelles [Sobolev *et al.* 2000]. Therefore, creation of a molecule of $^1O_2$ in close proximity to the cancer cell DNA would dramatically increase the odds of cell death whether by induction of apoptosis or by necrosis. None of the currently available photosensitizers is known to localize in the nucleus. Efforts have therefore been made to enhance delivery of the photosensitizer to the cell nucleus. Several peptide sequences have nuclear localizing properties. The entire transport into the nucleus, either passive or active, proceeds through the nuclear pore complex (NPC, Figure 5). In accordance with its name, NPC is a molecular sieve admitting free passive diffusion of proteins under 40-45 kDa, while larger proteins require a special addressing signal (amino-acid sequence); thus the proteins to be imported must have a nuclear localization signal (NLS) [Jans *et al.* 1997] or an analogous sequence. The proteins imported into the nucleus do not undergo proteolytic processing because, in contrast to the signals for targeting to the endoplasmic reticulum, mitochondria, peroxisome, etc., the NLS can be used repeatedly through a number of cell divisions, which involve dissolution of the nuclear envelope. The process of NLS-mediated transport of proteins into the nucleus can be described in two basic steps (Figure 5).

NLS mediated transport of proteins into the nucleus is temperature- and energy-dependent and is suppressed by antibodies against the cytoplasm-soluble NPC components called nucleoporins. The role of NLS resembles that of a ligand receptor interaction: NLS is recognized, i.e., specifically bound by importins [Jans *et al.* 2000]. Several peptides sequences have nuclear localizing properties. Different NLS sequences have been tested to favour intranuclear delivery of photosensitizers into the target cells [Schneider *et al.* 2006]. Gariépy's group has designed branched peptides that act as multitasking intracellular vehicles [Sheldon *et al.* 1995; Singh *et al.* 1998; Singh *et al.* 1999a]. These vehicles incorporate eight identical peptide arms coding for two functional domains, namely a pentalysine domain acting as a cytoplasmic transport sequence and the simian virus SV40 large T antigen NLS which guide their nuclear uptake. These squid-like intracellular vehicles are referred to as loligomers. By solid-phase synthesis, the coupling of chlorin e6 (Ce6) to a nucleus-directed linear peptide (Ce6-peptide) or a branched peptide (Ce6-loligomer) composed of eight identical arms, displaying the sequence of the Ce6-peptide was achieved. The authors have shown that cellular distributions of the two conjugates into radiation-induced fibrosarcoma cells (RIF-1) clearly differed. Nevertheless, an enhanced (1.5-fold) nuclear localization was only significant for Ce6-loligomer, compared to Ce6-peptide. Photodynamic activity of both conjugates against Chinese Hamster Ovary (CHO) cells was markedly increased compared to Ce6 (respectively 400 and 40 times higher for Ce6-loligomer and Ce6-peptide) but the phototoxicity was dependent on the cell line targeted (much better results were obtained with CHO than with RIF-1 cells, no details given). The mechanism of Ce6-loligomer-mediated toxicity was shown to involve the generation of ROS, confirming that the activation of Ce6 is not hindered upon its attachment to the loligomer. Although it was clearly demonstrated that Ce6-loligomer reached the cytosol and accumulated in the nucleus, it remained unclear if the enhanced phototoxicity observed was predominantly due to nuclear localization, or a simple consequence of the improvement of Ce6 uptake in relation with the peptide construct. Thus, results obtained by Gariépy *et al.* clearly demonstrated that the incorporation of Ce6 into peptide-based intracellular vehicles enhanced its phototoxicity and this approach highlights the utility of designing peptides as vehicles for regulating the intracellular distribution of photosensitizers such as Ce6 in order to maximize their efficacy in PDT. According to initial observations, Ce6 has a significantly higher photosensitizing activity when present in conjugates containing specific ligands such as insulin or concanavalin A and, is thus able to be internalized by receptor-expressing cells [Sobolev *et al.* 1992; Akhlynina *et al.* 1993; Akhlynina *et al.* 1995]. With the goal to enhance the photodynamic activity of Ce6 by the directed delivery of photosensitizers, Sobolev *et al.* have also shown that photosensitizers can be successfully redirected within cells by using cross-linked modular polypeptide transporters possessing (Figure 6): (i) an internalizable ligand providing a cell-specific delivery; (ii) a module enabling escape from endosomes; (iii) a NLS conferring interaction with importins, the cellular proteins mediating active translocation into the nucleus; (iv) and, a module allowing attachment of the photosensitizer. Lysosomes are the final destination for many macromolecules like photosensitizers taken up by endocytosis from the cell surface. To target the nucleus, ligands need to either circumvent the lysosomal uptake or favour an escape from endosomes. Thus, variegated approaches have been undertaken to expedite intra-cellular delivery prior to lysosomal trafficking. Endolysosomal escape may be exemplified by endosomolytic peptides. To enhance the photosensitization effected by these conjugates,

Sobolev *et al.* linked variants of the SV40 large tumor antigen NLS to target Ce6 to the nucleus [Akhlynina *et al.* 1997].

They set out to compare the photodynamic activity of conjugates without or with the T antigen NLS together with the T antigen N-terminal flanking sequence which have been shown to enhance T antigen nuclear transport. The authors demonstrated that the casein-kinase II phosphorylation site within this flanking region increased the rate of T antigen nuclear import, while phosphorylation by the cyclin dependent kinase *cdc2* at another site reduced the maximal level of transport by about 70%. BSA or β-galactosidase/ β-galactosidase fusion proteins were used as carriers to which insulin, Ce6 and the NLS were included either as peptides covalently coupled to BSA or within the coding sequence of P10 (NLS-containing β-galactosidase fusion protein). The most efficient photosensitizing agent was found to be the P10-Ce6-insulin adduct. Subsequent work of this group demonstrated that co-incubation of cells with human adenovirus serotype 5 synergized with NLS containing P10-Ce6-insulin conjugate. The innovation of this concept was to include with the conjugate, attenuated adenoviruses, which possess endosomolytic function [Akhlynina *et al.* 1999]. This very original targeting strategy demonstrated that the use of adenoviruses in conjunction with nuclear-targeting photosensitizers could have interesting implications for achieving efficient cell-type-specific PDT.

A large number of nucleobase porphyrins, which are of interest for nucleus targeting, have also been described in the literature [Sessler *et al.* 2006; Cormia *et al.* 1995]. Koval *et al.* have first reported in 2001 the synthesis of oligonucleotide-phthalocyanines conjugates for specific DNA modifications *in vitro* and *in vivo* for the development of sequence specific gene targeting reagents [Koval *et al.* 2001; Hammer *et al.* 2002; Li *et al.* 2001]. Due to the presence of a phthalocyanine core and adenine substituent, these compounds present strong intermolecular interactions, resulting in a poor solubility in common organic solvents and unusual spectral features (observation of fluorescence emission despite strong aggregation). No biological evaluation was performed.

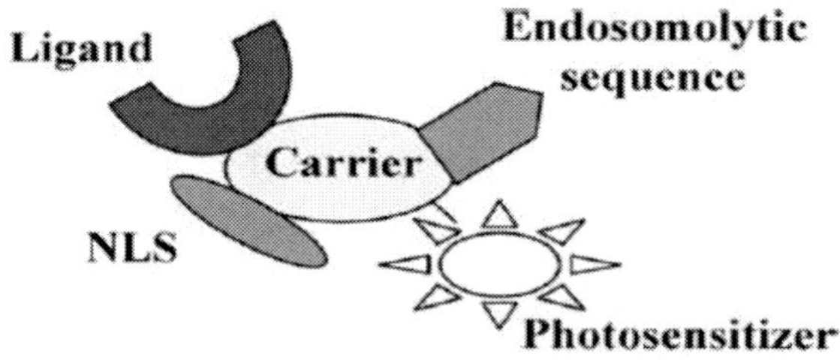

Figure 6. Scheme of distinct modules conferring cell-specific targeting for photosensitizers. The various sequence modules for cell surface binding and uptake (a ligand), an endosomolytic sequence, and a NLS as well as the photosensitizer, are linked to the carrier molecule[Schneider *et al.* 2006].

3.5. Folic Acid Receptor

The vitamin folic acid is a ligand able to target covalently attached bioactive agents quite specifically to folate receptor (FR)-positive cancer cells [Garin-Chesa *et al.* 1993; Lu *et al.* 2002]. The FR, which is a well-known cancer cell-associated protein, can actively internalize bound folates *via* endocytosis [Leamon *et al.* 1991; Kamen *et al.* 1986]. Folate receptors exist in three major forms namely FR-α, FR-β and FR-γ. The FR-α form is overexpressed by many types of tumor cells, including ovarian, endometrial, colorectal, breast, lung, renal, neuroendocrine carcinomas and brain metastases [Garin-Chesa *et al.* 1993; Parker *et al.* 2005]. Upon receptor interaction, the folic acid-receptor complex is taken up by cells and moves into many organelles involved in endocytotic trafficking, providing for cytosolic deposition [Turek *et al.* 1993]. Proper synthesis procedures have been pointed out to link folic acid to molecules in order to develop targeting delivery systems. It was demonstrated that the glutamate γ-carboxyl group modification does not induce significant loss of folic acid affinity for the receptor [Leamon *et al.* 1991]. Moreover, due to its high stability, compatibility with both organic and aqueous solvent, low-cost, non-immunogenic character, ability to conjugate with a wide variety of molecules, and low molecular weight, folic acid has attracted wide attention as a targeting agent for tumor detection. Numerous examples that take advantage of folic acid uptake to promote targeting and internalization include inorganic nanoparticles [Dixit *et al.* 2006; Bharali *et al.* 2005], polymer nanoparticles [Wang *et al.* 2005; Soppimath *et al.* 2007], polymeric micelles [Bae *et al.* 2005], lipoprotein nanoparticles [Zheng *et al.* 2005], tumor imaging agents [Okarvi *et al.* 2006], and dendrimers [Myc *et al.* 2007]. Our group synthesized two new conjugates composed of folic acid coupled to 4-carboxyphenylporphyrin *via* two short linkers that were different in nature but similar in size. Both conjugated photosensitizers showed improved intracellular uptake in KB cells acting as a positive control due to their overexpression of FR-α. Using a short PEG, 2,2'-(ethylenedioxy)-bis-ethylamine, as spacer arm, our group demonstrated that the uptake of this conjugate was on average 7-fold higher than tetraphenylporphyrin used as a reference and that the cells were significantly more sensitive to folic acid-conjugated porphyrin-mediated PDT [Schneider *et al.* 2005]. More recently, we demonstrated that *in vivo* selectivity of meta tetra(hydroxyphenyl)chlorin (*m*-THPC)-like photosensitizer conjugated to folic acid exhibited enhanced accumulation in KB tumors *in vivo* compared to *m*-THPC, 4 hours after intravenous injection (Figure 7) [Gravier *et al.* 2008]. Tumor-to-normal tissue ratio exhibited a very interesting selectivity for the conjugate (5:1) in KB tumors. An aspect that is important for folic-acid-mediated drug delivery concerns the rate of FR recycling between the cell surface and its intracellular compartments. Accumulation of folate conjugates in KB cells will depend upon not only the number and accessibility of FR on the cell surfaces but also the time required for unoccupied receptors to recycle back to the cell surface for additional conjugated QDs uptake. Using radioactive conjugates, Paulos *et al.* found empty FR+ to unload their cargo and return to the cell surface in about 8 to 12 hours [Paulos *et al.* 2004]. Using folic acid-linked drugs in a FR-targeting strategy could be considered as an efficient method to improve selectivity of anti-cancer treatment, specifically to FR+ positive cancer cells.

Stefflova *et al.* prepared a construct (pyro-peptide-folate, PPF) consisting of three compounds: (1) pyropheophorbide, (2) peptide as linker, (3) folate as targeting moiety [Stefflova *et al.* 2007]. They observed a higher accumulation of PPF in KB cells compared to HT1080 used as negative control, leading to a more effective photodynamic effect. The effect of folate was also confirmed *in vivo* with an accumulation of PPF in KB tumors (KB *versus* HT1080 tumors 2.5:1). It was interesting to note that the short peptide used as spacer considerably improved the delivery efficiency with a reduction (50-fold) in PPF uptake in liver and spleen.

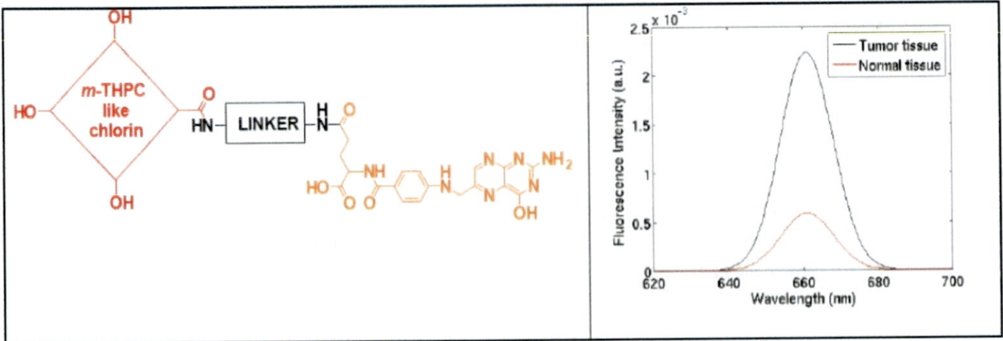

Figure 7. *In vivo* selectivity of *meta*-tetra(hydroxyphenyl)chlorin (*m*-THPC)-like photosensitizer conjugated to folic acid. Using optical fiber fluorimetry, *m*-THPC-like photosensitizer conjugated to folic acid exhibited enhanced accumulation in KB tumors compared to normal tissue, 4 hours after intra-venous injection of the conjugate (2.2.10-6 mol/kg) [Gravier *et al.* 2008].

3.6. Cholecystokinin A Receptor (CCK_A)

A peptide hormone cholecystokinin CCK8 (H-Asp-Tyr-Met-Gly-Trp-Met-Asp-Phe-NH_2) is known to interact with CCK_A, a receptor that is overexpressed in some colonic, gastric and brain cancers [de Luca *et al.* 2001; Reubi, 2003]. De Luca *et al.* described the synthesis of 5,10,15-tris(2,6-dichlorophenyl)-20-(4 carboxyphenyl)porphyrin (abbreviated PK) covalently bound to CCK8, using a lysine residue as a spacer. A molecular dynamics analysis indicated that the conformational characteristics of the CCK8/CCK_A receptor-model complex and of the PK-CCK8/ CCK_A receptor-model complex were similar. This evidence supports the hypothesis that the introduction of the porphyrin-Lys moiety does not influence the mode of ligand binding to the CCK_A receptor model. A NMR conformation study of the PK-CCK8 conjugate was also carried out in dimethylsulfoxide. No Nuclear Overhauser Effect correlation between the porphyrin ring protons and the peptide chain of the molecule were observed, indicating that the porphyrin macrocycle does not lie close to the peptide moiety of PK-CCK8, thus confirming the results of the preliminary computational study. The authors concluded that the presence of the porphyrin substituent at the N-terminus of CCK8 does not interfere with the conformation of the peptide residues supposed to be involved in receptor binding. This study suggested that peptide-photodynamic drug conjugates approach might be useful for targeting the tumor cells expressing CCK_A receptor.

3.7. Gonadotropin-releasing Hormone (GnRH) Receptors

Gonadotropin-releasing hormone (GnRH) is considered as a key integrator between the nervous and the endocrine systems and plays a pivotal role in the regulation of the reproductive system. Several studies revealed that GnRH and its receptors also play extra pituitary roles in numerous normal and malignant cells or tissues [for review, see Aguilar-

Rojas *et al.* 2009]. Thus, GnRH analogues, both agonists and antagonists, have attracted great interest because of their potential application in the treatment of diseases such as prostate and breast cancers [Emons *et al.* 1994; Schally, 1999; Gründker *et al.* 2002]. Their mechanism of action is believed to be related to gonadal steroid deprivation. This phenomenon results from GnRH analogue action that leads to down-regulation of the GnRH receptors (GnRH-R) and to desensitization of the pituitary gonadotropes [Rahimipour *et al.* 2003]. GnRH-R expression was identified on different tumors (breast, ovarian, endometrial, prostate, renal, brain, pancreatic, melanomas, non-Hodgkin's lymphomas, etc.) [Eidne *et al.* 1987; Yano *et al.* 1994; Fister *et al.* 2007], and the limited number of GnRH-R in normal tissues provided a basis for the development of diagnostic and therapeutic approaches of cancer. Since the conjugation of bulky moieties, such as tetramethylrhodamine, to the ϵ-amino group of [D-Lys6]GnRH does not significantly affect the bioactivity of GnRH analogues or their internalization by gonadotropes [Hazum *et al.* 1983], several agonists and antagonists of GnRH were attached to various cytotoxic compounds, and the resulting conjugates were evaluated for their anticancer activity. Rahimipour *et al.* described the coupling of protoporphyrin IX to peptides acting as GnRH agonists with the aim to selectively target GnRH-R that are overexpressed in many prostate and breast cancers [Rahimipour *et al.* 2003]. The peptide-photosensitizer conjugates obtained exhibited binding affinity for GnRH receptors, as assessed *in vitro* by displacement assays using 125I[D-Lys6]GnRH as the radioligand. The affinity of the photosensitizer-peptide conjugates was found to be lower than that of the corresponding peptide. Photodynamic activity was however enhanced (about 1.5-fold) in GnRH expressing-pituitary gonadotrope aT3-1 cells, compared to unconjugated protoporphyrin IX. The photodynamic activity of the conjugates was attenuated by co-incubation with the parent peptide, indicating that phototoxicity was receptor-mediated [Rahimipour *et al.* 2003].

3.8. Transferrin Receptors

One attractive protein-receptor system for the selective delivery of anti-tumor drugs utilizes the high affinity interaction between the iron-transporter transferrin (Tf) and its cognate cell-surface receptor (TfR). Tf is a serum non-heme iron-binding glycoprotein that plays a crucial role in transport iron from the sites of absorption or storage to proliferating cells [Singh, 1999b; Derycke *et al.* 2004]. Upon binding to TfR upon the cell surface, the Tf is endocytosed into acidic endosome compartment where the iron dissociates. Thus, cells that require iron express TfR on their surface, which mediates the cellular uptake of iron from the circulating Tf. As iron is an essential element for cell proliferation and metabolism, and because the increased proliferative activity of malignant cells requires a higher iron supply, it is not surprising that TfR is overexpressed on a variety of malignant cells compared to normal cells [Ponka *et al.* 1998; Derycke *et al.* 2002]. Indeed, the magnitude of TfR expression and turnover is proportional to the proliferative activity of the tumor tissue, because a higher proliferation rate requires more iron [Singh *et al.* 1999b]. Therefore, the Tf-TfR system has been used in several formats to target photosensitizer molecules to different types of malignant cells [Hamblin and Newman, 1994; Rittenhouse-Diakun *et al.* 1995; Cavanaugh, 2002; Gijsens *et al.* 2002; Li and Qian, 2002; Derycke *et al.* 2004].

Gijsens *et al.* prepared a Tf conjugated polyethylene glycol (PEG)-liposome that contain the photosensitizer aluminum phthalocyanine tetrasulfonate compound. The anti-proliferative activity of the targeted liposomes was evaluated and compared to the native photosensitizer and the non-targeted liposome on Hela tumor cells. After irradiation, the conjugate was 10 times more photocytotoxic than free photosensitizer whereas the conjugated liposome displayed no photocytotoxicity at all [Gijsens *et al.* 2002]. The high photocytotoxicity of targeted liposome was shown to be the result of a high intracellular concentration in HeLa cells, which could be lowered dramatically by incubating the conjugate with a competing transferrin concentration [Gijsens *et al.* 2002]. Later, this group confirmed the selective uptake of the targeted liposome in human AY-27 transitional-cell carcinoma cells and in an orthotopic rat bladder tumor model [Derycke *et al.* 2004]. Biodistribution studies revealed that among rats bearing AY-27 cell-derived bladder tumors, intravesical instillation with the targeted liposome resulted in a significant selective accumulation in tumor tissue compared to surrounding healthy tissues (normal urothelium and submucosa/muscle) [Derycke *et al.* 2004]. More recently, Laptev *et al.* demonstrated that bioconjugates composed of Tf and haematoporphyrin photosensitizer, significantly improve the specificity and efficiency of PDT for erythroleukemic cells [Laptev *et al.* 2006].

3.9. Lectins Saccharide Receptors

In order to enhance the delivery of photosensitizers to the target tumor site, a variety of carbohydrate-conjugated photosensitizers have also been synthesized and evaluated for anti-cancer PDT efficacy. Since oligosaccharides play important roles in cellular communication through saccharide-receptor interactions [Bertozzi *et al.* 2001], which are usually specific and multivalent, their conjugation with a photosensitizer can confer high selectivity and specificity as shown by different groups [Chen *et al.* 2004; Zhang *et al.* 2003; Di Stasio *et al.* 2005]. Among the carbohydrate-binding proteins, galectins (a subfamily of lectin) share a highly conserved domain with a high affinity for β-galactoside [Liu and Rabinovich, 2005]. It has been reported that compared to their expression in healthy cells, galectin-1 and galectin-3 expression may increase or decrease on the surface of tumor cells depending upon the cancer types [Van den Brule *et al.* 2004], which may play an important role in selective targeting for PDT *via* carbohydrate-photosensitizer conjugates. Pandey *et al.* in 2007 have described the synthesis of a series of carbohydrate-photosensitizer conjugates with different chemical characteristics such as types of carbohydrate moieties, position of conjugation, types of linkers and their impact on intracellular localization and *in vitro/in vivo* efficacy using the RIF tumor cells and CH3 mice bearing RIF tumors. The comparative *in vitro/in vivo* studies of the series of positional isomers of the lactose-purpurinimide conjugates showed a substantial difference in their photosensitizing efficacy. A noticeable difference between lactose conjugates in cell uptake (RIF tumor cells) was also observed, depending upon position of conjugation. The galectin-1 and -3 binding values for the carbohydrate conjugates obtained by ELISA test showed higher affinity than the corresponding nonconjugate purpurinimides [Pandey *et al.* 2007]. These findings provided interesting basis for the development of carbohydrate conjugates for specific photosensitizer delivery in tumor cells overexpressing galectins. In this sense, D'Aurias *et al.* in 2009 have reported that human galectin-1 can be characterized as a porphyrin-binding protein based on its tight interactions with Zn/Mn and

Au-porphyrins, indicating that galectin-1 may have potential as a delivery molecule to target tumor cells [D'Aurias *et al.* 2009]. Interestingly, galectin-1 has been recently identified as a proangiogenic factor, as tumor can stimulate tumor angiogenesis by secretion of galectin-1 [Thijssen *et al.* 2010].

It was also reported that asialoglycoprotein (ASGP) receptors were abundantly expressed in the surface of hepatoma cells and mammalian hepatocytes. Several studies suggested that targeting could be accomplished through introduction of galactose residues, which can bind specifically to the ASGP receptors on the cells [Fallon and Schwartz, 1985; Eisenberg *et al.* 1991]. Wu *et al.* in 2010 described the synthesis of porphyrin and galactosyl conjugated micelles based on amphiphilic copolymer incorporating galactosyl and porphyrin, and they evaluated their targeting and PDT efficiency in HEp2 and HepG2 cells. Since HepG2 cells express abundantly the ASGP receptors, the authors concluded that porphyrin and galactosyl conjugated polymer micelles displayed higher targeting and photodynamic efficacy in HepG2 cells than in receptor-negative Hep2 cells, corroborating the view of ASGP receptors-mediated delivery [Wu *et al.* 2010]. Brevet *et al.* in 2009 described the synthesis of mesoporous silica nanoparticles combining covalent anchoring of the porphyrin to the mesoporous silica matrix and targeting of cancer cells with mannose attached on the surface of the nanoparticle. They demonstrated that these nano-objet presented a much higher *in vitro* photo-efficiency in MDA-MB-231 through mannose-dependent endocytosis than non-functionalized nanoparticles [Brevet *et al.* 2009]. The authors also speculated that the covalent coupling of the photosensitizer inside the nanoparticles could enable to avoid premature release of the photoactivable compound from the carrier and thus the related risk of side effects.

Ferreira *et al.* in 2009 reported the design of synthetic DNA oligonucleotide aptamers selected to bind specifically to O-glycan-peptide signatures of mucin glycoproteins which are specifically and abundantly expressed on the surface of a broad range of epithelial cancer cells (breast, colon, lung, ovarian and pancreatic). These surface antigens are only internalized by epithelial cancer cells. They showed that when conjugated to the PDT agent Ce6 *via* their 5' end and delivered to a number of epithelial cancer cells, these DNA aptamers displayed a significant enhancement (>500-fold increase) in toxicity upon light activation, compared to the free drug and have not cytotoxic effect on cell types lacking such O-glycan-peptide signatures. Authors concluded that targeting phototoxic DNA aptamers to under glycosylated determinants of mucin 1 offers great potential in directing therapeutic agents to and into a broad range of epithelial cancer cells [Ferreira *et al.* 2009]. Others recent works described different glycodendrimeric structures and their potential for targeting photosensitizers through saccharide-lectin interactions [Ballut *et al.* 2009; Makky *et al.* 2010b; Makky *et al.* 2010a]. This group reported that glycodendrimeric porphyrins interacted more significantly with specific lectin, concavaline A, than the drug compound devoid of any sugar. They also revealed that glycodendrimeric porphyrins can be incorporated into liposome membranes and that such liposomes bearing the drug derivatives could constitute an efficient carrier for drug targeting in PDT [Ballut *et al.* 2009; Makky *et al.* 2010b].

3.10. Tumor Specific Antigens - Targeting Antibodies

It is now widely accepted that neoplastic transformation generates new and specific antigenic biomarkers not present on normal cells. A large amount of studies has proved the effectiveness of anti-cancer PDT using mAb-conjugated photoactivatable molecules as drug delivery systems into tumor tissues [for review see Solban *et al.* 2006, and Verma *et al.* 2007]. Using mAb as antigen-specific carriers for selective delivery of photosensitizer to the tumor is also known as photoimmunotherapy. Such approach combines phototoxicity of the photosensitizers with the selectivity of mAb directed against tumor-associated antigens.

HER2 is a highly upregulated target receptor on tumor cell surfaces such as breast cancer. To enhance photosensitizer immunoconjugate uptake by tumor cells, Savellano MD *et al.* in 2005 suggested multiepitope HER2 targeting strategy. Anti-HER2 photosensitizer immunoconjugates were produced *via* the conjugation of two mAb, HER50 and HER66, to pyrophorbide-a. Uptake and phototoxicity experiments using human cancer cells revealed selective uptake and potential photodynamic effect of the immunoconjugates on the HER2-overexpressing target cells. Moreover, the multiepitope targeted photoimmunotherapy was significantly more effective than single-epitope targeted photoimmunotherapy with a single anti-HER2 photosensitizer immunoconjugate [Savellano *et al.* 2005], suggesting that multitargeted photoimmunotherapy should also be useful against cancers that overexpress others antigenic receptors.

Nevertheless, it has been observed, particularly *in vivo* that the large size of the mAb could limit the ability of the immunoconjugate to penetrate solid, deep-seated and poorly vascularized tumors [Traid *et al.* 2003]. To overcome this drawback, accumulating recent data suggested that an alternative can be the use of a single chain (sc) Fv mAb fragments (scFv), which are both more efficient at penetrating tumor tissues because of their smaller size and more effectively cleared from the circulation because of the lack of the Fc domain [Staneloudi *et al.* 2007]. Staneloudi and co-workers described the development and characterization of an immunoconjugate comprising two isothiocyanato-porphyrins attached to colorectal tumor-specific scFv. They demonstrated that these conjugates had a selective photocytotoxic effect as showed by *in vitro* assays against colorectal cell lines [Staneloudi *et al.* 2007]. As indicated in the section 3.1 of the present chapter, more recently Bhatti *et al.* described complex of multiple photosensitizer molecules (verteporfin or pyropheophorbide-a) with anti-HER2 scFv and demonstrated that the photoimmunoconjugates are more potent than either free photosensitizers, effectively killing tumor cells *in vitro* and *in vivo* [Bhatti *et al.* 2008]. Indeed, many evidences suggest that the photoimmunoconjugates produced by coupling photosensitizers to recombinant mAb fragments scFv can not only target specifically and destroy tumor cells, but it is possible to load much more photosensitizing agents on an scFv than to the larger whole mAb [Milgron, 2008], and thus increase the amount of delivered drug molecules in the tumor tissues.

Conclusion

Selective delivery of therapeutic amounts of photosensitizers in diseased tissues is recognized as an absolute requirement for efficient and safe PDT for the treatment of cancers.

This is also the challenge to extend the application of PDT for the treatment of a broad ranges of tumor types, as such modality present many advantages over the conventional therapies. For this goal, development in targeted PDT continues to take advantage of the advances in the characterization of molecular mechanisms of tumor development. A large number of specific molecular targets have been identified and exploited in tumor targeting. Many photosensitizing agents have been elaborated and evaluated through *in vitro* and *in vivo* studies; however, only very few have reached clinical evaluation phases. Each of the handful of photosensitizers has specific characteristics, but none includes all the properties of an ideal photosensitizer. Although third-generation photosensitizers have been widely described for selective targeting, very few have been evaluated for clinical applications as the *in vivo* selectivity was not sufficiently high. Nanoparticles, as multifunctional platforms could represent emerging photosensitizer carriers that show great promise for PDT. In bio-nanotechnology, their development can overcome most of the shortcomings of classic photosensitizers.

References

Abels, C.Targeting of the vascular system of solid tumours by photodynamic therapy (PDT). *Photochem Photobiol Sci.* 2004, 3,765-71.

Aguilar-Rojas, A; Huerta-Reyes, M. Human gonadotropin-releasing hormone receptor-activated cellular functions and signaling pathways in extra-pituitary tissues and cancer cells. *Oncol Rep.* 2009, 22, 981-90.

Ahmad, MZ; Akhter, S; Jain, GK; Rahman, M; Pathan, SA; Ahmad, FJ; Khar, RK. Metallic nanoparticles: technology overview & drug delivery applications in oncology. *Expert Opin Drug Deliv.* 2010, 7, 927-42.

Akhlynina, TV; Jans, DA; Statsyuk, NV; Balashova, IY; Toth, G; Pavo, I; Rosenkranz, AA; Naroditsky, BS; Sobolev, AS. Adenoviruses synergize with nuclear localization signals to enhance nuclear delivery and photodynamic action of internalizable conjugates containing chlorin e6. *Int J Cancer.* 1999, 81, 734-40.

Akhlynina, TV; Jans, DA; Rosenkranz, AA; Statsyuk, NV; Balashova, IY; Toth, G; Pavo, I; Rubin, AB; Sobolev, AS. Nuclear targeting of chlorin e6 enhances its photosensitizing activity. *J Biol Chem.* 1997, 272, 20328-31.

Akhlynina, TV; Rosenkranz, AA; Jans, DA; Sobolev, AS. Insulin-mediated intracellular targeting enhances the photodynamic activity of chlorin e6. *Cancer Res.* 1995, 55, 1014-9.

Akhlynina, TV; Rosenkranz, AA; Jans, DA; Gulak, PV; Serebryakova, NV; Sobolev, AS. The use of internalizable derivatives of chlorin E6 for increasing its photosensitizing activity. *Photochem Photobiol.* 1993, 58, 45-8.

Ali, H; Van Lier, JE. Synthesis of monofunctionalised phthalocyanines using palladium catalysed cross-coupling reactions. *Tetrahedron Letters.* 1997, 38, 1157-60.

Allen, CM; Sharman, WM; La Madeleine, C; Weber, JM; Langlois, R; Ouellet, R; van Lier, JE. Photodynamic therapy: tumor targeting with adenoviral proteins. *Photochem Photobiol.* 1999, 70, 512-23.

Bae, Y; Jang, WD; Nishiyama, N; Fukushima, S; Kataoka, K. Multifunctional polymeric micelles with folate-mediated cancer cell targeting and pH-triggered drug releasing properties for active intracellular drug delivery. *Mol Biosyst.* 2005, 1, 242-50.

Ballut, S; Makky, A; Loock, B; Michel, JP; Maillard, P; Rosilio, V. New strategy for targeting of photosensitizers. Synthesis of glycodendrimeric phenylporphyrins, incorporation into a liposome membrane and interaction with a specific lectin. *Chem Commun (Camb).* 2009, 8, 224-6.

Bando, H; Weich, HA; Brokelmann, M; Horiguchi, S; Funata, N; Ogawa, T; Toi, M. Association between intratumoral free and total VEGF, soluble VEGFR-1, VEGFR-2 and prognosis in breast cancer. *Br J Cancer.* 2005, 92, 553-61.

Barge, J; Decréau, R; Julliard, M; Hubaud, JC; Sabatier, AS; Grob, JJ; Verrando, P. Killing efficacy of a new silicon phthalocyanine in human melanoma cells treated with photodynamic therapy by early activation of mitochondrion-mediated apoptosis. *Exp Dermatol.* 2004, 13, 33-44.

Bechet, D; Tirand, L; Faivre, B; Plénat, F; Bonnet, C; Bastogne, T; Frochot, C; Guillemin, F; Barberi-Heyob, M. Neuropilin-1 targeting photosensitization-induced early stages of thrombosis via tissue factor release. *Pharm Res.* 2010, 27, 468-79

Bertozzi, CR; Kiessling, LL. Chemical glycobiology. *Science.* 2001, 291, 2357-64.

Bharali, DJ; Lucey, DW; Jayakumar, H; Pudavar, HE; Prasad, PN. Folate-receptor-mediated delivery of InP quantum dots for bioimaging using confocal and two-photon microscopy. *J Am Chem Soc.* 2005, 127, 11364-71.

Bhatti, M; Yahioglu, G; Milgrom, LR; Garcia-Maya, M; Chester, KA; Deonarain, MP. Targeted photodynamic therapy with multiply-loaded recombinant antibody fragments. *Int J Cancer.* 2008, 122, 1155-63.

Boisbrun, M ; Vanderesse, R ; Engrand, P ; Olié, A ; Hupont, S ; Regnouf-de-Vains, JB ; Frochot, C. Design and photophysical properties of new RGD targeted tetraphenylchlorins and porphyrins. *Tetrahedron,* 2008,64, 3494-3504.

Bonneau, S; Morlière, P; Brault, D. Dynamics of interactions of photosensitizers with lipoproteins and membrane-models: correlation with cellular incorporation and subcellular distribution. *Biochem Pharmacol.* 2004, 68, 1443-52.

Bonneau, S; Vever-Bizet, C; Morlière, P; Mazière, JC; Brault, D. Equilibrium and kinetic studies of the interactions of a porphyrin with low-density lipoproteins. *Biophys J.* 2002, 83, 3470-81.

Boyle, RW; Dolphin, D. Structure and biodistribution relationships of photodynamic sensitizers. *Photochem Photobiol.* 1996, 64, 469-85.

Brault, D. Physical chemistry of porphyrins and their interactions with membranes: the importance of pH. *J Photochem Photobiol B.* 1990, 6, 79-86.

Brevet, D; Gary-Bobo, M; Raehm, L; Richeter, S; Hocine, O; Amro, K; Loock, B; Couleaud, P; Frochot, C; Morère, A; Maillard, P; Garcia, M; Durand, JO. Mannose-targeted mesoporous silica nanoparticles for photodynamic therapy. *Chem Commun (Camb).* 2009, 28, 1475-7.

Brown, LF; Berse, B; Jackman, RW; Tognazzi, K; Guidi, AJ; Dvorak, HF; Senger, DR; Connolly, JL; Schnitt, SJ. Expression of vascular permeability factor (vascular endothelial growth factor) and its receptors in breast cancer. *Hum Pathol.* 1995, 6, 86-91.

Brown, SB; Brown, EA; Walker, I. The present and future role of photodynamic therapy in cancer treatment. *Lancet Oncol.* 2004, 5, 497-508.

Byrne, JD; Betancourt, T; Brannon-Peppas, L. Active targeting schemes for nanoparticle systems in cancer therapeutics. *Adv Drug Deliv Rev.* 2008, 60, 1615-26.

Candide, C; Morlière, P; Mazière, JC; Goldstein, S; Santus, R; Dubertret, L; Reyftmann, JP; Polonovski, J. In vitro interaction of the photoactive anticancer porphyrin derivative photofrin II with low density lipoprotein, and its delivery to cultured human fibroblasts. *FEBS Lett.* 1986, 207, 133-8.

Cavanaugh, PG. Synthesis of chlorin e6-transferrin and demonstration of its light-dependent in vitro breast cancer cell killing ability. *Breast Cancer Res Treat.* 2002, 72, 117-30.

Chaleix, V; Sol, V; Guilloton, M; Granet, R; Krausz, P. Efficient synthesis of RGD-containing cyclic peptide–porphyrin conjugates by ring-closing metathesis on solid support. *Tetrahedron Letters.* 2004, 45, 5295-5299.

Chan, WS; Marshall, JF; Svensen, R; Phillips, D; Hart, IR. Photosensitising activity of phthalocyanine dyes screened against tissue culture cells. *Photochem Photobiol.* 1987, 45, 757-61.

Chatterjee, DK; Fong, LS; Zhang, Y. Nanoparticles in photodynamic therapy: an emerging paradigm. *Adv Drug Deliv Rev.* 2008, 60, 1627-37.

Chen, J; Bierhaus, A; Schiekofer, S; Andrassy, M; Chen, B; Stern, DM; Nawroth, PP. Tissue factor--a receptor involved in the control of cellular properties, including angiogenesis. *Thromb Haemost.* 2001, 86, 334-45.

Chen, W; Zhang, J. Using nanoparticles to enable simultaneous radiation and photodynamic therapies for cancer treatment. *J Nanosci Nanotechnol.* 2006, 6, 1159-66.

Chen, X; Hui, L; Foster, DA; Drain, CM. Efficient synthesis and photodynamic activity of porphyrin-saccharide conjugates: targeting and incapacitating cancer cells. *Biochemistry.* 2004, 43, 10918-29.

Chen, ZL; Sun, Y; Huang, P; Yang, XX; Zhou, XP. Studies on Preparation of Photosensitizer Loaded Magnetic Silica Nanoparticles and Their Anti-Tumor Effects for Targeting Photodynamic Therapy. *Nanoscale Res Lett.* 2009, 4, 400-408.

Chetty, C; Lakka, SS; Bhoopathi, P; Rao, JS. MMP-2 alters VEGF expression via alphaVbeta3 integrin-mediated PI3K/AKT signaling in A549 lung cancer cells. *Int J Cancer.* 2010, 127, 1081-95.

Choi, Y; Weissleder, R; Tung, CH. Selective antitumor effect of novel protease-mediated photodynamic agent. *Cancer Res.* 2006, 66, 7225-9.

Clauss, M; Gerlach, M; Gerlach, H; Brett, J; Wang, F; Familletti, PC; Pan, YC; Olander, JV; Connolly, DT; Stern, D. Vascular permeability factor: a tumor-derived polypeptide that induces endothelial cell and monocyte procoagulant activity, and promotes monocyte migration. *J Exp Med.* 1990, 172, 1535-45.

Collaud, S; Juzeniene, A; Moan, J; Lange, N. On the selectivity of 5-aminolevulinic acid-induced protoporphyrin IX formation. *Curr Med Chem Anticancer Agents.* 2004, 4, 301-16.

Contrino, J; Hair, G; Kreutzer, DL; Rickles, FR. In situ detection of tissue factor in vascular endothelial cells: correlation with the malignant phenotype of human breast disease. *Nat Med.* 1996, 2, 209-15.

Cornia, M; Binacchi, S; Soldato, TD; Zanardi, F; Casiraghi, G. *J Org Chem*, 1995, 60, 4964-4965.

Couffinhal, T; Kearney, M; Witzenbichler, B; Chen, D; Murohara, T; Losordo, DW; Symes, J; Isner, JM. Vascular endothelial growth factor/vascular permeability factor

(VEGF/VPF) in normal and atherosclerotic human arteries. *Am J Pathol.* 1997, 150, 1673-85.

Cuomo, V; Jori, G; Rihter, B; Kenney, ME; Rodgers, MA. Tumour-localising and - photosensitizing properties of liposome-delivered Ge(IV)-octabutoxy-phthalocyanine. *Br J Cancer.* 1991, 64, 93-5.

D'Auria, S; Petrova, L; John, C; Russev, G; Varriale, A; Bogoeva, V. Tumor-specific protein human galectin-1 interacts with anticancer agents. *Mol Biosyst.* 2009, 5, 1331-6.

De Luca, S; Tesauro, D; Di Lello, P; Fattorusso, R; Saviano, M; Pedone, C; Morelli, G. Synthesis and solution characterization of a porphyrin-CCK8 conjugate. *J Pept Sci.* 2001, 7, 386-94.

Derycke, AS ; Kamuhabwa, A; Gijsens, A; Roskams, T; De Vos, D; Kasran, A; Huwyler, J; Missiaen, L; de Witte, PA. Transferrin-conjugated liposome targeting of photosensitizer AlPcS4 to rat bladder carcinoma cells. *J Natl Cancer Inst.* 2004, 96, 1620-30.

Derycke, AS; De Witte, PA. Transferrin-mediated targeting of hypericin embedded in sterically stabilized PEG-liposomes. *Int J Oncol.* 2002, 20, 181-7.

Di Stasio, B; Frochot, C; Dumas, D; Even, P; Zwier, J; Müller, A; Didelon, J; Guillemin, F; Viriot, ML; Barberi-Heyob, M. The 2-aminoglucosamide motif improves cellular uptake and photodynamic activity of tetraphenylporphyrin. *Eur J Med Chem.* 2005, 40, 1111-22.

Dixit, V; Van den Bossche, J; Sherman, DM; Thompson, DH; Andres, RP. Synthesis and grafting of thioctic acid-PEG-folate conjugates onto Au nanoparticles for selective targeting of folate receptor-positive tumor cells. *Bioconjug Chem.* 2006, 17, 603-9.

Drake, TA; Morrissey, JH; Edgington, TS. Selective cellular expression of tissue factor in human tissues. Implications for disorders of hemostasis and thrombosis. *Am J Pathol.* 1989, 134, 1087-97.

Egeblad, M; Werb, Z. New functions for the matrix metalloproteinases in cancer progression. *Nat Rev Cancer.* 2002, 2, 161-74.

Eidne, KA; Flanagan, CA; Harris, NS; Millar, RP. Gonadotropin-releasing hormone (GnRH)-binding sites in human breast cancer cell lines and inhibitory effects of GnRH antagonists. *J Clin Endocrinol Metab.* 1987, 64, 425-32.

Eisenberg, C; Seta, N; Appel, M; Feldmann, G; Durand, G; Feger, J. Asialoglycoprotein receptor in human isolated hepatocytes from normal liver and its apparent increase in liver with histological alterations. *J Hepatol.* 1991, 13, 305-9.

Emons, G; Schally, AV. The use of luteinizing hormone releasing hormone agonists and antagonists in gynaecological cancers. *Hum Reprod.* 1994, 9, 1364-79.

Fallon, RJ; Schwartz, AL. Receptor-mediated endocytosis and targeted drug delivery. *Hepatology.* 1985, 5, 899-901.

Ferrara, N. Vascular endothelial growth factor: basic science and clinical progress. *Endocr Rev.* 2004, 25, 581-611.

Ferreira, CS; Cheung, MC; Missailidis, S; Bisland, S; Gariépy, J. Phototoxic aptamers selectively enter and kill epithelial cancer cells. *Nucleic Acids Res.* 2009, 37, 866-76.

Fiore, E; Fusco, C; Romero, P; Stamenkovic, I. Matrix metalloproteinase 9 (MMP-9/gelatinase B) proteolytically cleaves ICAM-1 and participates in tumor cell resistance to natural killer cell-mediated cytotoxicity. *Oncogene.* 2002, 21, 5213-23.

Firestone, RA. Low-density lipoprotein as a vehicle for targeting antitumor compounds to cancer cells. *Bioconjug Chem.* 1994, 5, 105-13.

Fister, S; Günthert, AR; Emons, G; Gründker, C. Gonadotropin-releasing hormone type II antagonists induce apoptotic cell death in human endometrial and ovarian cancer cells in vitro and in vivo. *Cancer Res.* 2007, 67, 1750-6.

Folkman, J. Fighting cancer by attacking its blood supply. *Sci Am.* 1996, 275, 150-4.

Frochot, C; Di Stasio, B; Vanderesse, R; Belgy, MJ; Dodeller, M; Guillemin, F; Viriot, ML; Barberi-Heyob, M. Interest of RGD-containing linear or cyclic peptide targeted tetraphenylchlorin as novel photosensitizers for selective photodynamic activity. *Bioorg Chem.* 2007, 35, 205-20.

Funovics, M; Weissleder, R; Tung, CH. Protease sensors for bioimaging. *Anal Bioanal Chem.* 2003, 377, 956-63.

Garin-Chesa, P; Campbell, I; Saigo, PE; Lewis, JL Jr; Old, LJ; Rettig, WJ. Trophoblast and ovarian cancer antigen LK26. Sensitivity and specificity in immunopathology and molecular identification as a folate-binding protein. *Am J Pathol.* 1993, 142, 557-67.

Gijsens, A; Derycke, A; Missiaen, L; De Vos, D; Huwyler, J; Eberle, A; de Witte, P. Targeting of the photocytotoxic compound AlPcS4 to Hela cells by transferrin conjugated PEG-liposomes. *Int J Cancer.* 2002, 101, 78-85.

Gijsens, A; Missiaen, L; Merlevede, W; de Witte, P. Epidermal growth factor-mediated targeting of chlorin e6 selectively potentiates its photodynamic activity. *Cancer Res.* 2000, 60, 2197-202.

Grant, WE; Speight, PM; Hopper, C; Bown, SG. Photodynamic therapy: an effective, but non-selective treatment for superficial cancers of the oral cavity. *Int J Cancer.* 1997, 71, 937-42.

Gravier, J; Schneider, R; Frochot, C; Bastogne, T; Schmitt, F; Didelon, J; Guillemin, F; Barberi-Heyob, M. Improvement of meta-tetra(hydroxyphenyl)chlorin-like photosensitizer selectivity with folate-based targeted delivery. synthesis and in vivo delivery studies. *J Med Chem.* 2008, 51, 3867-77.

Gründker, C; Günthert, AR; Westphalen, S; Emons, G. Biology of the gonadotropin-releasing hormone system in gynecological cancers. *Eur J Endocrinol.* 2002, 146, 1-14.

Hamblin, MR; Newman, EL. Photosensitizer targeting in photodynamic therapy. I. Conjugates of haematoporphyrin with albumin and transferrin. *J Photochem Photobiol B.* 1994, 26, 45-56.

Hammer, RP; Owens, CV; Hwang, SH; Sayes, CM; Soper, SA. Asymmetrical, water-soluble phthalocyanine dyes for covalent labeling of oligonucleotides. *Bioconjug Chem.* 2002, 13, 1244-52.

Hazum, E; Meidan, R; Liscovitch, M; Keinan, D; Lindner, HR; Koch, Y. Receptor-mediated internalization of LHRH antagonists by pituitary cells. *Mol Cell Endocrinol.* 1983, 30, 291-301.

Hong, TM; Chen, YL; Wu, YY; Yuan, A; Chao, YC; Chung, YC; Wu, MH; Yang, SC; Pan, SH; Shih, JY; Chan, WK; Yang, PC. Targeting neuropilin 1 as an antitumor strategy in lung cancer. *Clin Cancer Res.* 2007, 13, 4759-68.

Hu, Z; Rao, B; Chen, S; Duanmu, J. Targeting tissue factor on tumour cells and angiogenic vascular endothelial cells by factor VII-targeted verteporfin photodynamic therapy for breast cancer in vitro and in vivo in mice. *BMC Cancer.* 2010, 26, 10:235.

Hu, Z; Sun, Y; Garen, A. Targeting tumor vasculature endothelial cells and tumor cells for immunotherapy of human melanoma in a mouse xenograft model. *Proc Natl Acad Sci U S A.* 1999, 96, 8161-6.

Ishigami, SI; Arii, S; Furutani, M; Niwano, M; Harada, T; Mizumoto, M; Mori, A; Onodera, H; Imamura, M. Predictive value of vascular endothelial growth factor (VEGF) in metastasis and prognosis of human colorectal cancer. *Br J Cancer*. 1998, 78, 1379-84.

Jans, DA; Xiao, CY; Lam, MH. Nuclear targeting signal recognition: a key control point in nuclear transport? *Bioessays*. 2000, 22, 532-44.

Jans, DA; Briggs, LJ; Gustin, SE; Jans, P; Ford, S; Young, IG. A functional bipartite nuclear localisation signal in the cytokine interleukin-5. *FEBS Lett*. 1997, 40, 315-20.

Kamen, BA; Capdevila, A. Receptor-mediated folate accumulation is regulated by the cellular folate content. *Proc Natl Acad Sci U S A*. 1986, 83, 5983-7.

Kessel, D. The role of low-density lipoprotein in the biodistribution of photosensitizing agents. *J Photochem Photobiol B*. 1992, 14, 261-2.

Khan, EH; Ali, H; Tian, H; Rousseau, J; Tessier, G; Shafiullah van Lier, JE. Synthesis and biological activities of phthalocyanine-estradiol conjugates. *Bioorg Med Chem Lett*. 2003, 13, 1287-90.

Konan, YN; Gurny, R; Allémann, E. State of the art in the delivery of photosensitizers for photodynamic therapy. *J Photochem Photobiol B*. 2002, 66, 89-106.

Konan-Kouakou, YN; Boch, R; Gurny, R; Allémann, E. In vitro and in vivo activities of verteporfin-loaded nanoparticles. *J Control Release*. 2005, 103, 83-91.

Koval, VV; Chernonosov, AA; Abramova, TV; Ivanova, TM; Fedorova, OS; Derkacheva, VM; Lukyanets, EA. Photosensitized and catalytic oxidation of DNA by metallophthalocyanine-oligonucleotide conjugates. *Nucleosides Nucleotides Nucleic Acids*. 2001, 20, 1259-62.

Kremer C, Breier G, Risau W, Plate KH. Up-regulation of flk-1/vascular endothelial growth factor receptor 2 by its ligand in a cerebral slice culture system. Cancer Res. 1997 Sep 1;57(17):3852-9. PubMed PMID: 9288799.

Laptev, R; Nisnevitch, M; Siboni, G; Malik, Z; Firer, MA. Intracellular chemiluminescence activates targeted photodynamic destruction of leukaemic cells. *Br J Cancer*. 2006, 95, 189-96.

Law, B; Tung, CH. Proteolysis: a biological process adapted in drug delivery, therapy, and imaging. *Bioconjug Chem*. 2009, 20, 1683-95.

Leamon, CP; Low, PS. Delivery of macromolecules into living cells: a method that exploits folate receptor endocytosis. *Proc Natl Acad Sci U S A*. 1991, 88, 5572-6.

Leinonen, T; Pirinen, R; Böhm, J; Johansson, R; Ropponen, K; Kosma, VM. Expression of matrix metalloproteinases 7 and 9 in non-small cell lung cancer. Relation to clinicopathological factors, beta-catenin and prognosis. *Lung Cancer*. 2006, 51, 313-21.

Levy, JG; Obochi, M. New applications in photodynamic therapy. Introduction. *Photochem Photobiol*. 1996, 64, 737-9.

Li, H; Marotta, DE; Kim, S; Busch, TM; Wileyto, EP; Zheng, G. High payload delivery of optical imaging and photodynamic therapy agents to tumors using phthalocyanine-reconstituted low-density lipoprotein nanoparticles. *J Biomed Opt*. 2005, 10, 41203.

Li, H; Qian, ZM. Transferrin/transferrin receptor-mediated drug delivery. *Med Res Rev*. 2002, 22, 225-50.

Li, X; Ng, DKP. Synthesis and spectroscopic properties of the first phthalocyanine–nucleobase conjugates. *Tetrahedron Letters*. 2001, 42, 305-309.

Liu, FT; Rabinovich, GA. Galectins as modulators of tumour progression. *Nat Rev Cancer*. 2005, 5, 29-41.

Lovell, J F; Zheng, G. Activatable smart probes for molecular optical imaging and therapy. *J. Innov. Opt. Health. Sci.* 2008, 1, 45-61

Lu, Y; Low, PS. Folate-mediated delivery of macromolecular anticancer therapeutic agents. *Adv Drug Deliv Rev.* 2002, 54, 675-93.

Lutsenko, SV; Feldman, NB; Finakova, GV; Posypanova, GA; Severin, SE; Skryabin, KG; Kirpichnikov, MP; Lukyanets, EA; Vorozhtsov, GN. Targeting phthalocyanines to tumor cells using epidermal growth factor conjugates. *Tumour Biol.* 1999, 20, 218-24.

Makky, A; Michel, JP; Kasselouri, A; Briand, E; Maillard, P; Rosilio, V. Evaluation of the specific interactions between glycodendrimeric porphyrins, free or incorporated into liposomes, and concanavalin A by fluorescence spectroscopy, surface pressure, and QCM-D measurements. *Langmuir.* 2010a, 26, 12761-8.

Makky, A; Michel, JP; Ballut, S; Kasselouri, A; Maillard, P; Rosilio, V. Effect of cholesterol and sugar on the penetration of glycodendrimeric phenylporphyrins into biomimetic models of retinoblastoma cells membranes. *Langmuir.* 2010b, 26, 11145-56.

Maree, SE; Nyokong, T. *J Porph Phthalo*, 2001, 5, 782-792.

Maree, S; Phillips, D; Nyokong, T. *J. Porphyrins Phthalocyanines,* 2002, 5, 17-25.

Miao, HQ; Lee, P; Lin, H; Soker, S; Klagsbrun, M. Neuropilin-1 expression by tumor cells promotes tumor angiogenesis and progression. *FASEB J.* 2000, 14, 2532-9.

Milgrom, LR. Towards recombinant antibody-fragment targeted photodynamic therapy. *Sci Prog.* 2008, 91, 241-63.

Mordon, S; Devoisselle, JM; Soulié, S. Fluorescence spectroscopy of pH in vivo using a dual-emission fluorophore (C-SNAFL-1). *J Photochem Photobiol B.* 1995, 28, 19-23.

Moriwaki, SI; Misawa, J; Yoshinari, Y; Yamada, I; Takigawa, M; Tokura, Y. Analysis of photosensitivity in Japanese cancer-bearing patients receiving photodynamic therapy with porfimer sodium (Photofrin). *Photodermatol Photoimmunol Photomed.* 2001, 17, 241-3.

Myc, A; Majoros, IJ; Thomas, TP; Baker, JR Jr. Dendrimer-based targeted delivery of an apoptotic sensor in cancer cells. *Biomacromolecules.* 2007, 8, 13-8.

Nasarre, C; Roth, M; Jacob, L; Roth, L; Koncina, E; Thien, A; Labourdette, G; Poulet, P; Hubert, P; Crémel, G; Roussel, G; Aunis, D; Bagnard, D. Peptide-based interference of the transmembrane domain of neuropilin-1 inhibits glioma growth in vivo. *Oncogene.* 2010, 29, 2381-92.

Nemerson, Y. Tissue factor and hemostasis. *Blood.* 1988, 71(1):1-8.

O-charoenrat, P; Rhys-Evans, P; Eccles, P. Correlations between MMPs and SA. MMPs and TIMP-1 with invasion and metastasis in head and neck squamous cell carcinomas. *Clin Exp Metastasis*, 1999,17,773.

Okarvi, SM; Jammaz, IA. Preparation and in vitro and in vivo evaluation of technetium-99m-labeled folate and methotrexate conjugates as tumor imaging agents. *Cancer Biother Radiopharm.* 2006, 21, 49-60.

Pandey, SK; Zheng, X; Morgan, J; Missert, JR; Liu, TH; Shibata, M; Bellnier, DA; Oseroff, AR; Henderson, BW; Dougherty, TJ; Pandey, RK. Purpurinimide carbohydrate conjugates: effect of the position of the carbohydrate moiety in photosensitizing efficacy. *Mol Pharm.* 2007, 4, 448-64.

Parker, N; Turk, MJ; Westrick, E; Lewis, JD; Low, PS; Leamon, CP. Folate receptor expression in carcinomas and normal tissues determined by a quantitative radioligand binding assay. *Anal Biochem.* 2005, 338, 284-93.

Paulos, CM; Reddy, JA; Leamon, CP; Turk, MJ; Low, PS. Ligand binding and kinetics of folate receptor recycling in vivo: impact on receptor-mediated drug delivery. *Mol Pharmacol.* 2004, 66, 1406-14.

Perret, GY; Starzec, A; Hauet, N; Vergote, J; Le Pecheur, M; Vassy, R; Léger, G; Verbeke, KA; Bormans, G; Nicolas, P; Verbruggen, AM; Moretti, JL. In vitro evaluation and biodistribution of a 99mTc-labeled anti-VEGF peptide targeting neuropilin-1. *Nucl Med Biol.* 2004, 31, 575-81.

Pham, W; Choi, Y; Weissleder, R; Tung, CH. Developing a peptide-based near-infrared molecular probe for protease sensing. *Bioconjug Chem.* 2004, 15, 1403-7.

Plate, KH; Breier, G; Weich, HA; Risau, W. Vascular endothelial growth factor is a potential tumour angiogenesis factor in human gliomas in vivo. *Nature.* 1992, 359, :845-8.

Ponka, P; Beaumont, C; Richardson, DR. Function and regulation of transferrin and ferritin. *Semin Hematol.* 1998, 35, 35-54.

Rahimipour, S; Ben-Aroya, N; Ziv, K; Chen, A; Fridkin, M; Koch, Y. Receptor-mediated targeting of a photosensitizer by its conjugation to gonadotropin-releasing hormone analogues. *J Med Chem.* 2003, 46, 3965-74.

Redmond, RW; Kochevar, IE. Spatially resolved cellular responses to singlet oxygen. *Photochem Photobiol.* 2006, 82, 1178-86.

Renno, RZ ; Terada, Y; Haddadin, MJ; Michaud, NA; Gragoudas, ES; Miller, JW. Selective photodynamic therapy by targeted verteporfin delivery to experimental choroidal neovascularization mediated by a homing peptide to vascular endothelial growth factor receptor-2. *Arch Ophthalmol.* 2004, 122, 1002-11.

Reubi, JC. Peptide receptors as molecular targets for cancer diagnosis and therapy. *Endocr Rev.* 2003, 24, 389-427.

Ricchelli, F. Photophysical properties of porphyrins in biological membranes. *J Photochem Photobiol B.* 1995, 29, 109-18.

Rifkin, RM; Gregory, SA; Mohrbacher, A; Hussein, MA. Pegylated liposomal doxorubicin, vincristine, and dexamethasone provide significant reduction in toxicity compared with doxorubicin, vincristine, and dexamethasone in patients with newly diagnosed multiple myeloma: a Phase III multicenter randomized trial. *Cancer.* 2006, 106, 848-58.

Rittenhouse-Diakun, K; Van Leengoed, H; Morgan, J; Hryhorenko, E; Paszkiewicz, G; Whitaker, JE; Oseroff, AR. The role of transferrin receptor (CD71) in photodynamic therapy of activated and malignant lymphocytes using the heme precursor delta-aminolevulinic acid (ALA). *Photochem Photobiol.* 1995, 61, 523-8.

Rosenkranz, AA; Jans, DA; Sobolev, AS. Targeted intracellular delivery of photosensitizers to enhance photodynamic efficiency. *Immunol Cell Biol.* 2000, 78,452-64.

Sasnouski, S; Kachatkou, D; Zorin, V; Guillemin, F; Bezdetnaya, L. Redistribution of Foscan from plasma proteins to model membranes. *Photochem Photobiol Sci.* 2006, 5, 770-7.

Savellano, MD; Pogue, BW; Hoopes, PJ; Vitetta, ES; Paulsen, KD. Multiepitope HER2 targeting enhances photoimmunotherapy of HER2-overexpressing cancer cells with pyropheophorbide-a immunoconjugates. *Cancer Res.* 2005, 65, 6371-9.

Schally, AV. Luteinizing hormone-releasing hormone analogs: their impact on the control of tumorigenesis. *Peptides.* 1999, 20, 1247-62.

Schneider, R; Tirand, L; Frochot, C; Vanderesse, R; Thomas, N; Gravier, J; Guillemin, F; Barberi-Heyob, M. Recent improvements in the use of synthetic peptides for a selective photodynamic therapy. *Anticancer Agents Med Chem.* 2006, 6, 469-88.

Schneider, R; Schmitt, F; Frochot, C; Fort, Y; Lourette, N; Guillemin, F; Müller, JF; Barberi-Heyob, M. Design, synthesis, and biological evaluation of folic acid targeted tetraphenylporphyrin as novel photosensitizers for selective photodynamic therapy. *Bioorg Med Chem.* 2005, 13, 2799-808.

Semeraro, N; Colucci, M. Tissue factor in health and disease. *Thromb Haemost.* 1997, 78, 759-64.

Sessler, JL ; Jayawickramaradjah, J; Gouloumis, A ; Pantos, GD ; Torres, T ; Guldi, DM. *Tetrahydron,* 2006, 62, 2123-2131.

Sharwani, A; Jerjes, W; Hopper, C; Lewis, MP; El-Maaytah, M; Khalil, HS; Macrobert, AJ; Upile, T; Salih, V. Photodynamic therapy down-regulates the invasion promoting factors in human oral cancer. *Arch Oral Biol.* 2006, 51, 1104-11.

Sheldon, K; Liu, D; Ferguson, J; Gariépy, J. Loligomers: design of de novo peptide-based intracellular vehicles. *Proc Natl Acad Sci U S A.* 1995, 92, 2056-60.

Shiomi, T; Okada, Y. MT1-MMP and MMP-7 in invasion and metastasis of human cancers. *Cancer Metastasis Rev.* 2003, 22, 145-52.

Shoji, M; Sun, A; Kisiel, W; Lu, YJ; Shim, H; McCarey, BE; Nichols, C; Parker, ET; Pohl, J; Mosley, CA; Alizadeh, AR; Liotta, DC; Snyder JP. Targeting tissue factor-expressing tumor angiogenesis and tumors with EF24 conjugated to factor VIIa. *J Drug Target.* 2008, 16, 185-97.

Sibani, SA; McCarron, PA; Woolfson, AD; Donnelly, RF. Photosensitiser delivery for photodynamic therapy. Part 2: systemic carrier platforms. *Expert Opin Drug Deliv.* 2008, 5, 1241-54.

Singh, D; Bisland, SK; Kawamura, K; Gariépy, J. Peptide-based intracellular shuttle able to facilitate gene transfer in mammalian cells. *Bioconjug Chem.* 1999a, 10, 745-54.

Singh, D; Kiarash, R; Kawamura, K; LaCasse, EC; Gariépy, J. Penetration and intracellular routing of nucleus-directed peptide-based shuttles (loligomers) in eukaryotic cells. *Biochemistry.* 1998, 37, 5798-809.

Singh, M. Transferrin As A targeting ligand for liposomes and anticancer drugs. *Curr Pharm Des.* 1999b, 5, 443-51.

Sobolev, AS; Jans, DA; Rosenkranz, AA. Targeted intracellular delivery of photosensitizers. *Prog Biophys Mol Biol.* 2000, 73, 51-90.

Sobolev, AS; Akhlynina, TV; Yachmenev, SV; Rosenkranz, AA; Severin, ES. Internalizable insulin-BSA-chlorin E6 conjugate is a more effective photosensitizer than chlorin E6 alone. *Biochem Int.* 1992, 26, 445-50.

Soker, S; Miao, HQ; Nomi, M; Takashima, S; Klagsbrun, M. VEGF165 mediates formation of complexes containing VEGFR-2 and neuropilin-1 that enhance VEGF165-receptor binding. *J Cell Biochem.* 2002, 85, 357-68.

Soker, S; Takashima, S; Miao, HQ; Neufeld, G; Klagsbrun, M. Neuropilin-1 is expressed by endothelial and tumor cells as an isoform-specific receptor for vascular endothelial growth factor. *Cell.* 1998, 20, 735-45.

Solban, N; Rizvi, I; Hasan, T. Targeted photodynamic therapy. *Lasers Surg Med.* 2006, 38, 522-31.

Song, L; Li, H; Sunar, U; Chen, J; Corbin, I; Yodh, AG; Zheng, G. Naphthalocyanine-reconstituted LDL nanoparticles for in vivo cancer imaging and treatment. *Int J Nanomedicine.* 2007, 2, 767-74.

Soppimath, KS; Liu, LH; Yang Seow, W; Liu, SQ; Powell, R; Chan, P; Yang, YY. Multifunctional Core/Shell Nanoparticles Self-Assembled from pH-Induced Thermosensitive Polymers for Targeted Intracellular Anticancer Drug Delivery. *Adv. Funct. Mater.* 2007, 17, 355–362.

Soukos, NS; Hamblin, MR; Keel, S; Fabian, RL; Deutsch, TF; Hasan, T. Epidermal growth factor receptor-targeted immunophotodiagnosis and photoimmunotherapy of oral precancer in vivo. *Cancer Res.* 2001, 61, 4490-6.

Staneloudi, C; Smith, KA; Hudson, R; Malatesti, N; Savoie, H; Boyle, RW; Greenman, J. Development and characterization of novel photosensitizer : scFv conjugates for use in photodynamic therapy of cancer. *Immunology.* 2007, 120, 512-7.

Stefflova, K; Li, H; Chen, J; Zheng, G. Peptide-based pharmacomodulation of a cancer-targeted optical imaging and photodynamic therapy agent. *Bioconjug Chem.* 2007, 18, 379-88.

Sun, Y ; Chen, ZL ; Yang, XX ; Huang, P; Zhou, XP; Du, XX. Magnetic chitosan nanoparticles as a drug delivery system for targeting photodynamic therapy. *Nanotechnology.* 2009, 20, 135102.

Thijssen, VL; Barkan, B; Shoji, H; Aries, IM; Mathieu, V; Deltour, L; Hackeng, TM; Kiss, R; Kloog, Y; Poirier, F; Griffioen, AW. Tumor cells secrete galectin-1 to enhance endothelial cell activity. *Cancer Res.* 2010, 70, 6216-24.

Thomas, N; Bechet, D; Becuwe, P; Tirand, L; Vanderesse, R; Frochot, C; Guillemin, F; Barberi-Heyob, M. Peptide-conjugated chlorin-type photosensitizer binds neuropilin-1 in vitro and in vivo. *J Photochem Photobiol B.* 2009, 96, 101-8.

Thomas, N; Tirand, L; Chatelut, E; Plénat, F; Frochot, C; Dodeller, M; Guillemin, F; Barberi-Heyob, M. Tissue distribution and pharmacokinetics of an ATWLPPR-conjugated chlorin-type photosensitizer targeting neuropilin-1 in glioma-bearing nude mice. *Photochem Photobiol Sci.* 2008, 7, 433-41.

Tirand, L; Thomas, N; Dodeller, M; Dumas, D; Frochot, C; Maunit, B; Guillemin, F; Barberi-Heyob, M. Metabolic profile of a peptide-conjugated chlorin-type photosensitizer targeting neuropilin-1: an in vivo and in vitro study. *Drug Metab Dispos.* 2007, 35, 806-13.

Tirand, L; Frochot, C; Vanderesse, R; Thomas, N; Trinquet, E; Pinel, S; Viriot, ML; Guillemin, F; Barberi-Heyob, M. A peptide competing with VEGF165 binding on neuropilin-1 mediates targeting of a chlorin-type photosensitizer and potentiates its photodynamic activity in human endothelial cells. *J Control Release.* 2006, 111, 153-64.

Tozer, GM; Kanthou, C; Baguley, BC. Disrupting tumour blood vessels. *Nat Rev Cancer.* 2005, 5, 423-35.

Trail, PA; King, HD; Dubowchik, GM. Monoclonal antibody drug immunoconjugates for targeted treatment of cancer. *Cancer Immunol Immunother.* 2003, 52, 328-37.

Turek, JJ; Leamon, CP; Low, PS. Endocytosis of folate-protein conjugates: ultrastructural localization in KB cells. *J Cell Sci.* 1993, 106, 423-30.

Van den Brûle, F; Califice, S; Castronovo, V. Expression of galectins in cancer: a critical review. *Glycoconj J.* 2004, 19, 537-42.

Van Kempen, LC; de Visser, KE; Coussens, LM. Inflammation, proteases and cancer. *Eur J Cancer.* 2006, 42, 728-34.

Verma, S; Watt, GM; Mai, Z; Hasan, T. Strategies for enhanced photodynamic therapy effects. *Photochem Photobiol.* 2007, 83, 996-1005.

Vihinen, P; Ala-aho, R; Kähäri, VM. Matrix metalloproteinases as therapeutic targets in cancer. *Curr Cancer Drug Targets.* 2005, 5, 203-20.

Vitols, S. Uptake of low-density lipoprotein by malignant cells--possible therapeutic applications. *Cancer Cells.* 1991, 3, 488-95.

Wang, CH; Hsiue, GH. Polymer-DNA hybrid nanoparticles based on folate-polyethylenimine-block-poly(L-lactide). *Bioconjug Chem.* 2005, 16, 391-6.

Whitaker, GB; Limberg, BJ; Rosenbaum, JS. Vascular endothelial growth factor receptor-2 and neuropilin-1 form a receptor complex that is responsible for the differential signaling potency of VEGF(165) and VEGF(121). *J Biol Chem.* 2001, 276, 25520-31.

Wormald, R; Evans, J; Smeeth, L; Henshaw, K. Photodynamic therapy for neovascular age-related macular degeneration. *Cochrane Database Syst Rev.* 2007, CD002030.

Wu, DQ; Li, ZY; Li, C; Fan, JJ; Lu, B; Chang, C; Cheng, SX; Zhang, XZ; Zhuo, RX. Porphyrin and galactosyl conjugated micelles for targeting photodynamic therapy. *Pharm Res.* 2010, 27, 187-99.

Yano, T; Pinski, J; Radulovic, S; Schally, AV. Inhibition of human epithelial ovarian cancer cell growth in vitro by agonistic and antagonistic analogues of luteinizing hormone-releasing hormone. *Proc Natl Acad Sci U S A.* 1994, 91, 1701-5.

Zhang, M; Zhang, Z; Blessington, D; Li, H; Busch, TM; Madrak, V; Miles, J; Chance, B; Glickson, JD; Zheng, G. Pyropheophorbide 2-deoxyglucosamide: a new photosensitizer targeting glucose transporters. *Bioconjug Chem.* 2003, 14, 709-14.

Zheng, G; Chen, J; Stefflova, K; Jarvi, M; Li, H; Wilson, BC. Photodynamic molecular beacon as an activatable photosensitizer based on protease-controlled singlet oxygen quenching and activation. *Proc Natl Acad Sci U S A.* 2007, 104, 8989-94.

Zheng, G; Chen, J; Li, H; Glickson, JD. Rerouting lipoprotein nanoparticles to selected alternate receptors for the targeted delivery of cancer diagnostic and therapeutic agents. *Proc Natl Acad Sci U S A.* 2005, 102, 17757-62.

Zheng, G; Li, H; Zhang, M; Lund-Katz, S; Chance, B; Glickson, JD. Low-density lipoprotein reconstituted by pyropheophorbide cholesteryl oleate as target-specific photosensitizer. *Bioconjug Chem.* 2002, 13, 392-6.

Zucker, S; Mirza, H; Conner, CE; Lorenz, AF; Drews, MH; Bahou, WF; Jesty, J. Vascular endothelial growth factor induces tissue factor and matrix metalloproteinase production in endothelial cells: conversion of prothrombin to thrombin results in progelatinase A activation and cell proliferation. *Int J Cancer.* 1998, 75, 780-6.

In: Brain Cancer, Tumor Targeting and Cervical Cancer
Editor: Elena K. Salvatti

ISBN: 978-1-61122-738-3

Chapter II

Emerging Targets and Novel Nanotechnology-based Strategies: Changing the Treatment Paradigm in Oncology

Vera Moura[a,b]***, Lígia C. Gomes da Silva***[a,b]***, Nuno A. Fonseca***[a,b]***, Maria C. Pedroso de Lima***[b,c]***, Sérgio Simões***[a,b] ***and João Nuno Moreira***[a,b,*]

[a]Laboratory of Pharmaceutical Technology, Faculty of Pharmacy,
[b]Center for Neuroscience and Cell Biology, University of Coimbra, Portugal,
[c]Department of Life Sciences, Faculty of Sciences and Technology,
University of Coimbra, Portugal

Abstract

The entwined mechanisms of survival that allow tumors to grow and spread in the human body have been intensively studied, in the attempt to recognize specific patterns and create effective clinical approaches. New insights into the molecular basis of tumor development, progression and metastases gave rise to new targets in cancer therapy and opened new opportunities for science to create and identify novel targeting strategies.

Tumors start to grow and survive along with the supportive role of the microenvironment, sustained by processes such as angiogenesis (formation of new blood vessels from preexisting ones). Cancer cells interact with different players and promote the organization of an immature vessel network to access nutrients and oxygen, but also to colonize distant areas from the tumor primary site. The thorough knowledge of this process has opened a portal for understanding cancer associated mechanisms and design

* Corresponding author: jmoreira@ff.uc.pt

treatment strategies, accounting for the interaction of therapeutic agents directly with the unique features of tumors and their supporting cells. Moreover, the discovery of the regulatory machinery behind tumor progression provided a rationale for the development of therapies concerning cancer stem cells, dormant cells, circulating tumor cells, metastases suppressor genes and non-coding RNAs. The understanding of tumor cell interaction with the stroma, inflammatory infiltrates, cell-cell interaction and even mechanical forces also gave rise to the integration of new key regulators and possible targets for interference with tumor progression and metastases.

Alongside with the discovery of efficient and selective molecular targets for the development of novel ligands with high specificity towards cancer cells and/or their microenvironment are the means for reaching such targets. Many non-invasive methods for delivering therapeutic agents have been designed and improved to minimize inherent risk association. Nowadays, delivery systems are engineered to comprise a multitude of features, such as safety, efficacy, target accumulation, cargo delivery, remote action and, more importantly, therapeutic and diagnostic features. As a consequence of the urgency in creating novel treatment solutions within cancer therapy, the number of targeted nanoparticles in clinical trials has increased over the last years and is expected to present major benefits in cancer therapeutic approaches, as it leads to a more personalized medical care.

In this chapter, we discuss emerging targets in cancer therapy and novel nanotechnology-based therapeutic strategies that have been changing the paradigm in cancer treatment.

Abreviations List

AIDS	*Acquired Immune Deficiency Syndrome*
Bcl-2	B-cell lymphoma/leukemia 2
Bcl-XL	Bcl2-like protein 1
COX	Cyclooxigenase
CSC	Cancer Stem Cell
DNA	Deoxyribonucleic Acid
EC	Endothelial Cell
EGFR	Endothelial Growth Factor Receptor
erbB-2	V-erb-b2 erythroblastic leukemia viral oncogene homolog 2 (Human Epidermal Growth Factor Receptor 2)
ERK	Extracellular Signal-Regulated Kinase
FDA	Food And Drug Administration
FGFR	Fibroblast Growth Factor Receptor
HIF	Hypoxia Inducible Factor
IFP	Interstitial Fluid Pressure
IL	Interleucin
miRNA	microRNA
MMP	Matrix Metalloproteinase
NSAID	Nonsteroidal Anti-Inflammatory Drug
MRI	*Magnetic Resonance Imaging*
MT-MMP	Transmembrane MMP
PDGFR	Platelet Derived Growth Factor Receptor
PE	Phosphatidylethanolamine
PECAM	Platelet Endothelial Cell Adhesion Molecule

PEG	Poly(ethylene glycol)
PI3K	Phosphoinositide 3-Kinase
RNA	Ribonucleic Acid
siRNA	Small Interfering RNA
TGF-β	Transforming Growth Factor B
TLR	Toll-Like Receptor
TNF	Tumor Necrosis Factor
VDA	Vascular Disrupting Agent
VEGF	Vascular Endothelial Growth Factor
VEGFR	Vascular Endothelial Growth Factor Receptor

1. Understanding Tumor Development, Progression and Metastases: The Hallmarks of Cancer

The knowledge generated in the field of oncobiology over the last decades has revealed cancer as a disease unlike any other. An important property of tumor cells is their immortality (the ability to divide indefinitely), in opposition to normal cells, which become senescent during their lifespan. Also, they are not dependent on external growth factors to initiate their proliferation, since they are able to produce them on their own (autocrine signalling), resulting in permanent activation of signalling pathways that stimulate cellular proliferation. Another feature common to cancer is the ability to evade apoptosis (programmed cell death), which allows survival of cells with genetic defects and therefore indulges tumor progression.

In order to survive, tumor cells proliferating in an uncontrolled manner engage a metabolic reprogramming process to invade the surrounding microenvironment, which is poor in oxygen and nutrients. Moreover, tumor cells are able to survive, invade and colonize distant sites from the primary tumor, taking advantage of a dedicated vessel network, architected by a controlled process called angiogenesis.

These and other specific features of cancer were systematically described by Hanahan and Weinberg, in a landmark review with strong acceptance in the scientific community. These authors described six fundamental and common hallmarks in cancer physiology: (I) self-sufficiency in growth signal; (II) insensitivity to antigrowth signals; (III) evasion to apoptosis; (IV) limitless replicative potential; (V) sustained angiogenesis and (VI) tissue invasion and metastases [1]. However, in the last decade, additional hallmarks of cancer have been proposed such as: (VII) avoidance of immunosurveillance [2,3]; (VIII) cancer-related inflammation [4]; (IX) and stress phenotype of cancer cells (metabolic stress, proteotoxic stress, mitotic stress, oxidative stress and DNA damage stress) [5]. The mentioned hallmarks take into consideration not only the tumor cells *per se,* but also surrounding cells with a strong contribution to tumor growth and progression, such as endothelial cells (ECs), fibroblasts, pericytes, cancer stem cells (CSCs) and dormant cells, among others [6].

From the above, it is clear that a successful therapeutic approach for cancer treatment should target simultaneously different sets of cells and, within these, different pathways and processes fundamental for cancer progression. In this regard, the following discussion will envision the successes and challenges of targeted technologies to tumor cells and their microenvironment.

2. Successes and Challenges of Targeted Technologies

2.1. The Problem of Conventional Therapeutic Approaches for Cancer Treatment

Traditional chemotherapy constitutes one of the most important treatment modalities against cancer, in the current clinical practice. Chemotherapeutic agents, upon systemic administration, are generally characterized by a high volume of distribution which leads to a poor selectivity towards tumor cells and accumulation in healthy tissues. Such pattern of distribution can lead to increased toxicities against normal tissues that also show enhanced proliferative rates, such as the bone marrow, gastrointestinal tract and hair follicles. Myelosuppression, alopecia or mucositis are examples of some of the most undesired consequences of fighting cancer with conventional therapy [7]. Side effects that occur as a result of toxicity to normal tissues are the main reason for administration of sub-optimal doses of anticancer chemotherapeutics, resulting in the eventual failure of therapy. This is often accompanied by the development of drug resistance and metastatic disease, which hinders patient's overall survival. Hence, there is an evident need for a more specific and personalized medical care.

Targeted drug delivery towards tumor cells offers the possibility of directing and concentrating the therapeutic agent only at the desired target site, improving therapeutic efficacy through increased tumor cell death and decreasing the incidence of side effects in healthy tissues [8]. In general terms, this treatment selectivity can be achieved by designing a system specifically directed to the target site, taking advantage of one or more distinct features of the pathological cell and/or microenvironment. In this regard, one of the most important strategies in molecularly guided cancer pharmacology is the development of techniques that can modify the kinetic features of drugs by encapsulating them in nanosystems, and the recognition of specific molecular and cellular targets. This strategy can be planned from the point of view of either attacking a cellular element, a deregulated pathway or the tumor microenvironment.

2.2. The Quest for Specificity - Emerging Targets in Cancer Therapy

2.2.1. Targeting Tumor Cells and their Deregulated Signaling Pathways

The ability of tumor cells to reprogram their metabolism and adapt to changes through the interconnection between the different hallmarks of cancer, involves complex processes that depend on a large number of players and factors [9,10]. Fortunately, this dependence provides an opportunity for designing strategies to hinder tumor progression and invasion, upon selection of one or more targets from the pool of intervenients.

The following section intends to approach several pathways already subject of therapeutic intervention as well as a few of the major concerns in oncology: the role of cancer stem cells, dormant cells and non-coding RNAs in cancer progression.

2.2.1.1. Signaling Pathways

The accumulation of genetic modifications allows cancer cells to acquire specific competences to adapt to internal and environmental changes. These modifications can be controlled or inhibited once a molecular target, relevant to cell viability, is identified. Some of the most recent examples of molecular targets within the different hallmarks of cancer, previously discussed, and the corresponding drugs and agents used for specific molecular targeting are listed in Table 1.

2.2.1.2. Cancer Stem Cells

Several models have been proposed to explain why tumors never heal, but the concept of an existing cell that can differentiate into a rapidly dividing tumor cell is currently gaining increased acceptance in the oncology field. In opposition to the stochastic model (which states that all tumor cells are mutated somatic cells dividing in a non-controlled manner capable of forming new tumor niches) [35], the hierarchic model assumes that a tumor can be originated from one single stem cell. This last approach deviates cancer from a merely proliferative illness, and explains why conventional chemotherapy is incapable of completely destroying cancer. This theory provides the answer to the unlimited self-renewal and differentiation ability of tumors, since a stem cell is capable of duplicating with minimal damage to its DNA, differentiating into distinct cells according to the encountered microenvironment. Whether these cells are located in all known niches (bone marrow, mesoderme, digestive tract) or the malignant transformation occurs in the totipotent or multipotent mother cell, remains uncertain, despite all the efforts to reproduce the assumptions of this theory in an adequate *in vivo* model.

Chann H. Lagadec *et al.*, when studying stem cells, assumed that proteasome activity can be a means to track and target cancer initiating cells *in vivo* and *in vitro*, and therefore, it might have an impact on tumor therapeutics. In their work, human glioma and breast cancer cells were engineered to stably express a fluorescent protein (ZsGreen) fused to the carboxyl-terminal of the murine ornithine decarboxylase (cODC) degron, resulting in a fluorescent fusion protein that accumulated in cells with low 26S proteasome activity. This small three aminoacid sequence construct was directly recognized by the proteasome, and accumulated in CSCs with low proteasome degradation. The authors correlated ZsGreen positive transfected cells with CSCs markers ($CD44^{high}/CD24^{-/low}$ expression) and with increased sphere forming capacity and metastases. CSCs were then selectively targeted via a proteasome-dependent suicide gene, and their elimination in vivo led to tumor regression [36]. In fact, proteasome inhibitors have been seen as promising anticancer agents in breast cancer (Bortezomib, in clinical trials, July 2008).

But, do all CSCs repopulate equally? How can one interfere with tumor progression, given its heterogeneity? The work of Phillips *et al.* reinforced the unpredictable behavior of CSCs when they showed that breast cancer initiating cells increase in number when radiotherapy is interrupted, which may be connected with repopulation of tumors during therapeutic gaps [37].

Table 1. Examples of drugs used for approaching different molecular targets within the proposed hallmarks of cancer

Hallmark of cancer	Drug		Molecular targets	Refs
	Name	Class		
(I) Self-sufficiency in growth signal	Sorafenib Sunitinib	small inhibitors	Raf, PDGFR, VEGFR2/3, c-Kit	[11,12]
	LErafAODN	antisense oligodeoxynucleotide	Raf-1	[13]
	LErafsiRNA	small interfering RNA		
	Trastuzumab	monoclonal antibody	HER2	[11,12]
	Lapatinib	small inhibitor	HER2, EGFR	[14]
	Imatinib Dasatinib Nilotinib	small inhibitors	BCR-ABL, PDGFR, c-Kit	[15,16]
(II) Insensitivity to antigrowth signals	Trabedersen	antisense oligodeoxynucleotide	TGFβ	[17]
	Lerdelimubab Metelimubab	monoclonal antibodies		
	ALK-5	small inhibitor		
(III) Evasion to apoptosis	Oblimersen	antisense oligodeoxynucleotide	Bcl-2	[18]
	ABT-263 ABT-737	small inhibitors	Bcl-2, Bcl-XL, Bcl-w	[19]
	Mapatumumab	monoclonal antibody	TRAIL receptor	[20]
	CAL-101	small inhibitor	PI3K-delta	[21]
	LY2181308	antisense oligodeoxynucleotide	Survivin	[22]
(IV) Limitless replicative potential	GRN163L	antisense oligodeoxynucleotide	hTERT	[23]
	Nutlin	small inhibitor	MDM2- p53 interaction	[24]
(V) Angiogenesis	See table 3			
(VI) Tissue invasion and metastases	Denosumab	monoclonal antibody	RANKL	[25]
(VII) Avoidance of immunosurveillance	Trastuzumab	monoclonal antibody	HER2	[26]
	Lenalidomide, Pomalidomide	small inhibitors	TNF-α, IFN-γ, ILs	[27]
(VIII) Cancer-related inflammation	Celecoxib Etoricoxib Rofecoxib	small inhibitors	COX2	[28]
	Lovastatin	small inhibitor	HMGCoA	[29,30]
(IX) Stress phenotype of cancer cells	Rapamycin	small inhibitor	mTOR	[31]
	CALAA-01	small interfering RNA in a targeted nanoparticle	RRM2	[32]
	Bortezomib	small inhibitor	Proteasome	[33,34]

Abbreviations: Akt, v-akt murine thymoma viral oncogene homolog; Bcl-2, B-cell lymphoma/leukemia 2; Bcl-XL, Bcl2-like protein 1; Bcl-W, Bcl2-like protein 2; c-Kit, cluster of differentiation 117;COX, cyclooxigenase; EGFR, endothelial growth factor receptor; HMGCoA, 3-hydroxy-3-methyl-glutaryl-CoA reductase; HER2, Human Epidermal Growth Factor Receptor 2; hTERT, human telomerase reverse transcriptase; IFN, interferon; IL, interleukin; MDM2, murine double minute 2; mTOR, *mammalian target of rapamycin;* PDGFR, platelet derived growth factor receptor; PI3K, phosphoinositide 3-kinase; Raf-1, v-raf-1 murine leukemia viral oncogene homolog 1; RANKL, receptor activator of NF-Kb ligand; RRM2, ribonucleotide reductase small subunit; TGF, transforming growth factor; TRAIL, TNF-related apoptosis-inducing ligand; VEGFR, vascular endothelium growth factor receptor.

2.2.1.3. Dormant Cells

Dormant cells are associated with the colonization of a secondary site and recurrence. They are extremely difficult to eradicate with conventional therapy, since most cytotoxic agents are only effective against actively dividing cells. Imbalances in the activity of ERK/p38 [38] and inactivation of oncogenes Myc [39] and Wnt [40] have been reported to induce and maintain dormancy, as well as the multifaceted action of the immune system. The conditions under which these dormant cells acquire an invasive and proliferating phenotype are not yet understood, nor the strategies to induce their elimination, although they have been related with poor overall survival [41].

In the attempt to gather more pieces of the puzzle, additional questions have emerged. Does it matter to determine the tumor's origin? Can the phenotype of the tumor be defined if the involved oncogenes and the surrounding environment are known? Are malignancy, dormant and stem cells involved in independent mechanisms? It is further needed to understand which molecular pathways are involved in tumor proliferation and the influence of the microenvironment; which biomarkers and imaging techniques can be more accurate to perform early diagnosis; which is the cell of origin and where is it located [42,43], and what are the best molecular targets for dormant and CSCs.

2.2.1.4. Non-coding RNAs

Small molecules that regulate gene expression, such as microRNAS (miRNAs), appear to have a marked effect on tumor progression [44] and angiogenesis. miR-126, miR-221 and miR-222 are constitutively expressed in ECs from tumor blood vessels, while miR-296 and 130a are induced by angiogenic growth factors and miR-210 by hypoxia. Others, such as miR-17-92, act in a non-cell autonomous manner on angiogenesis. Therefore, silencing these or other pro-angiogenic miRNAs is a potential strategy for targeted anticancer therapies. miR-296, which regulates growth factor receptors overexpression in ECs, has been downregulated by antagomirs, small synthetic RNA molecules, resulting in reduction of angiogenesis in tumor xenografts *in vivo* [45]. Nevertheless, miRNA effects can be controversial. As an example, miR-126 expressed in ECs, blocks angiogenesis and decreases tumor growth [46], while loss of expression in tumor cells increases the incidence of metastases [47].

From what has been demonstrated so far, more efficient and sophisticated strategies are required to control tumor progression. Taking into consideration the versatility of tumor cells described throughout the nine hallmarks of cancer, the new rationale for anti-cancer therapy contemplates targets that are not as prone to mutations as tumor cells and that are fundamental for cancer survival and progression, such as tumor-stroma interactions and mechanical forces, the *seed and soil* theory in metastases development, the inflammatory process, the immune system, and the angiogenic process. Therefore, targeting the tumor cell through the tumor microenvironment appears as a more effective option that is changing the paradigm in cancer treatment.

2.3. Targeting Tumor Cells and the Tumor Microenvironment - The New Paradigm in Cancer Treatment

2.3.1. Tumor-stroma Interactions and Mechanical Forces

It is widely known and accepted that stromal-epithelial interactions promote tumor progression and the role of mechanical forces in tumor invasive behaviour is being progressively recognized. The most representative example is breast cancer.

In normal breast cells, integrins and growth factors interact with the extracellular matrix and with the cytoskeleton forming large complexes, but no stiffness is observed. Matrix stiffness disrupts acinar morphogenesis of breast cells and induces cancer progression. In fact, palpation of stiffness is helpful in diagnosis with MRI and ultrasound, and already reports an alteration of integrins behaviour in breast cancer. Stiffness increase occurs first than tumor growth, regulating growth factor-dependent ERK and PI3K activity [48]. Therefore, stromal tension must be considered in tumor progression and metastases, especially in relation to lysil oxidase. This enzyme is implicated in collagen homeostasis and increases in circulation before evidence of a tumor mass. Inhibiting the enzyme activity induces alterations in breast histology, increases cytokeratin-4 expression and decreases collagen cross-linking, reducing extracellular matrix stiffness. Moreover, it accumulates at premetastatic sites and is essential for CD11b$^+$ myeloid cell recruitment. Therefore, targeting lysil oxidase can inhibit tumor growth and prevent metastatic disease [48,49]. Matrix stiffness also cooperates with erbB2 oncogene to promote invasion [50].

Overall, stroma composition, degradation and physical forces must also be seen as key regulators of tumor progression. However, not only *in situ* tumor proliferation is dependent on the microenvironment. The metastatic process is intimately reliant on the secondary site's particular features.

2.3.2. The "Seed and Soil" Theory in Metastases Development

Tumors are able to invade distant organs of the body through three major routes: lymphatic vessels, blood vessels and serosal surfaces [51]. In order to accomplish such a task, cells acquire a different phenotype, becoming mobile and invasive. The extracellular matrix and bone marrow, as well as the surrounding vascular network, undergo disruption and proteolysis, creating a scaffold for tumor cells migration [52,53]. However, once in the blood stream or in the lymphatic vessels, these cells are under the effect of velocity-induced shear forces, lack of a substratum and the presence of immune cells, facing the physical challenge of arresting and extravasating from the vessel. This last task is performed mainly by nonspecific binding to coagulation factors including tissue factor, fibrinogen, fibrin and thrombin. In the presence of an embolus, ECs retract, exposing the extracellular matrix, whereas in lymphatic vessels, specific adhesive interactions with E- and P-selectins and cadherins or immunoglobulin-like cell adhesion molecules promote firm endothelial–tumor cell adhesion and extravasation from the vessel. The site of tumor arrest can be identified with the use of peptides that selective bind to proteins on the surface of the endothelium. As an example, the protein metadherin mediates tumor homing to the lung, taking advantage of the differences between the endothelia of the lung and the skin, kidney and other organs. Approaches to interfere with this kind of proteins can have an impact in the metastatic process and enhance the efficacy of chemotherapeutic treatments [52,53]. Brown and

Ruoslahti showed that metadherin is overexpressed in breast cancer tissue and xenografts and mediates localization at the metastatic site, since siRNA-mediated knockdown of the protein in breast cancer cells inhibited experimental lung metastases [54].

However, once in the secondary site, only a small fraction of tumor cells is capable of surviving and developing colonies. The stromal composition is usually different from the original location and cells are challenged in two ways: adjust or induce changes in the microenvironment. In fact, the colonization of distant sites can be the result of a careful preparation involving the recruitment of endothelial and hematopoietic progenitor cells [55,56], such as evidenced in the 1889 "seed and soil" theory of Stephen Paget. The assumption that tumors are able to prepare the secondary site is also reinforced by a recently identified gene prognosis signature in early-stage primary tumors that metastasize [57,58], and by identification of more than twenty metastases suppressor genes [59,60,61,62], discarding the original perception that the metastatic potential is undissociated from tumor size. Moreover, tumor growth, development and metastases have been extensively paired with numerous players in the stroma, with the associated matrix and the immune system, supporting Virchow's theory that *tumors are wounds that never heal.*

2.3.3. The Inflammatory Process and the Immune System

Carcinogenesis and chronic unresolved inflammation have many players in common and never reach equilibrium. They progress in a chaotic manner, as a consequence of cell proliferation and mutation, and share common pathways.

Examples of mutual intervenients include ECs and their precursors, mural cells (pericytes and smooth-muscle cells), fibroblasts of various phenotypes, neutrophils and other granulocytes (eosinophils and basophils), T-, B- and mast cells, natural killer lymphocytes, antigen-presenting cells (macrophages and dendritic cells), the blood and lymphatic vascular networks, and the extracellular matrix [63].

When an injury is made, the inflammatory process follows a series of sequenced steps in order to restore homeostasis: activation of resident cells, recruitment of granulocytes and macrophages, infiltration of lymphocytes and activation of mesenchymal cells to form new blood vessels and a collagenous matrix, ending with tissue remodeling. Similarly, tumor cells can produce cytokines and chemokines, attracting macrophages, dendritic cells, mast cells, T-cells and haematopoietic progenitors [64], and trigger the angiogenic switch, promoting survival and replication of dysfunctional epithelial cells.

Michael Karin's group correlated inflammatory factors secreted by cancer cells with immune cell receptors from the host and metastases. In this study, animals bearing Lewis lung carcinoma lacking inflammatory cytokines, like Tumor Necrosis Factor-α (TNF-α), showed increased survival and decreased number and size of metastases. In addition, they have observed that tumor cells in mice depleted of Toll-like Inflammatory Receptor 2 (TLR2) failed to disseminate, with reduction in the number and size of metastases in the lungs, liver and adrenal glands. These effects were corroborated and yet improved in irradiated wild type mice receiving bone marrow transplants from TLR2 knockout mice. Therefore, macrophages can be related with the rate of tumor progression and metastases, strengthening the intimate connection between tumor development and the inflammatory process [65], as well as reinforcing the importance of the microenvironment in anti-cancer therapies.

Hu *et al.* investigated TLR2 signaling on tumor metastases in a B16-melanoma mouse model, reporting reduced pulmonary metastases and increased survival by reversing B16 cells immunosuppressive microenvironment and restoring tumor-killing cells such as CD8$^+$ T-cells and M1 macrophages with an anti-TLR2 antibody [66]. Moreover, TLR4 has been identified in several types of cancer, while prostate cancer has also been linked to TLR2 and TLR10 [67,68]. Therefore, TLRs can be useful therapeutic targets.

Reisfeld *et al.* studied the modulation of immune polarization in the tumor microenvironment by carcinoma-activated fibroblasts, in a 4T1 murine model of metastatic breast cancer. The elimination of such activated fibroblasts, upon targeting the fibroblast activation protein with a DNA vaccine, resulted in improved anti-metastatic effects of doxorubicin chemotherapy and increased protein expression of IL-2 and IL-7, enhanced suppression of IL-6 and IL-4, while increasing recruitment of dendritic cells and CD8$^+$ T-cells. Moreover, recruitment of tumor-associated macrophages, myeloid derived suppressor cells and T regulatory cells was suppressed [69]. Therefore, protection against inflammatory states and host-derived factors is critical in triggering the disease and controlling its progression. In fact, chronic nonsteroidal anti-inflammatory drugs (NSAIDs), such as COX2 inhibitors, have been related to a low incidence of cancer in patients, especially colon cancer [70]. In a breast cancer study, a 50% reduction in breast cancer risk was reported with aspirin use, and relative risk dropped to 0.29 with specific COX2 inhibitors [71].

Statins have also been related to reduction of cancer incidence, possibly for their involvement in inflammation and angiogenesis suppression [72]. Endostatin is known to promote EC apoptosis, by downregulation of Bcl-2 and Bcl-XL [73] and impairs their invasive behaviour by inhibiting the activation of matrix metalloproteinases (MMP), such as MMP2 and MT1-MMP [74], essential for the first steps of the angiogenic process.

These studies, among others, have raised the possibility of preventing carcinogenesis by interfering with the inflammatory process and the immune system, since progression towards lethal cancer is due to the accumulation of a critical number of genetic defects in cancer cells, recruitment of the endothelium, involvement of the extracellular matrix, fibroblasts and pericytes, and to the engagement of the immune system.

The challenge now is to specifically inhibit the protumorigenic roles of normal cells in cancer, while maintaining their homeostatic functions in normal tissues. Examples of such cells are myofibroblasts and adipocytes, which represent a risk factor for several types of cancer, due to their role in the inflammatory process. Myofibroblasts abut on epithelial and glandular cells and produce growth factors, cytokines, chemokines and extracellular matrix components, being related to liver and colon cancer. Adipocytes also secrete cytokines, such as the adipokines leptin, adiponectin, resistin and visfatin, as well as matrix metalloproteinases that contribute to the inflammatory process [75].

Overall, there is still room for new approaches for cancer treatment and for pursuing novel cellular and molecular targets, within tumor's dependence on the surrounding microenvironment. One important process crucial for tumor sustainability is angiogenesis.

2.3.4. Angiogenesis

In order to overcome the obstacles inherent to lack of an effective blood supply, early solid tumors start to grow towards pre-existing nearby blood vessels. This process, known as 'vessel cooption', allows tumors to survive in a restrained form for months or even years, since it only serves the periphery of the cell mass. To grow and expand exponentially to other tissues, a functional vascular network is required. Tumors establish their blood supply via a number of processes in addition to angiogenesis, the process of new vessel formation from preexisting ones. These include vasculogenesis (blood vessel formation occurring by a *de novo* production of ECs), vascular remodelling, intussusceptions (new blood vessel creation by the splitting of an existing blood vessel in two) and possibly, in certain tumors, vascular mimicry (conduits lined by tumor cells for the transport of nutrients).

2.3.4.1. The History

The vascular nature of tumors has been recognized for many centuries, since Fallopio noted dilated vessels in 1600 and Hunter documented an increase in the number of tumor vessels (1828, published posthumously). The systematic studies done in the first half of the twentieth century [76,77,78,79] revealed the structured nature of tumor vascularization and in 1971, Folkman raised the hypothesis that tumor growth is dependent on angiogenesis [80]. This specific research field was greatly accelerated by the establishment of long-term EC cultures achieved by Gimbrone and colleagues in the early 1970s [81] and by the subsequent development of *in vitro* models of capillary network formation [82]. Meanwhile, in 1991, Noel Weidner, Folkman and colleagues highlighted human tumor blood vessels immunohistochemically using antibodies to factor VIII-related antigen, allowing the quantification of microvessels in the tumor [83]. The first endogenous tissue angiogenesis inhibitor associated with negative tumor regulation was thrombospondin-1 [84] and a large number has followed, such as angiostatin, endostatin, vasostatin. Although, it was the identification of Vascular Endothelial Growth Factor (VEGF) [85] and Hypoxia-Induced Factor (HIF) [86] that raised the veil covering a vast number of entwined pathways capable of promoting vessel destabilization, maturation and specialization.

With the rise of new technologies to identify and correlate the players and pathways in the new vessel formation process, the number of publications regarding angiogenesis increased significantly in the past decades, reflecting the efforts to explain the unique features of tumor blood vessel architecture and their influence on tumor growth and metastases.

2.3.4.2. The Process

Angiogenesis is considered a hallmark of cancer, because it ensures tumor survival and proliferation through a better access to nutrients and oxygen, and allows migration and expansion of the solid tumor to other tissues. Physiologic angiogenesis is usually focal and temporary, while the pathologic process can persist for years and gives rise to bleeding, vascular leakage and tissue destruction [87]. The latest is not exclusive of advanced tumors, but it is an integral part of tumor development, which becomes activated during the "angiogenic switch" and is required for early pre-malignant lesions to progress to malignant tumors [88,89].

The best-known conditions in which angiogenesis is triggered are malignant, ocular and inflammatory disorders, but many additional processes are affected, such as obesity, asthma, diabetes, cirrhosis, multiple sclerosis, endometriosis, AIDS (Acquired Immune Deficiency Syndrome), bacterial infections and autoimmune disease [64]. Vascular development can be divided into the following processes, which are sequenced and entangled: migration, formation, stabilization, maturation (branching, remodeling and pruning), and specialization.

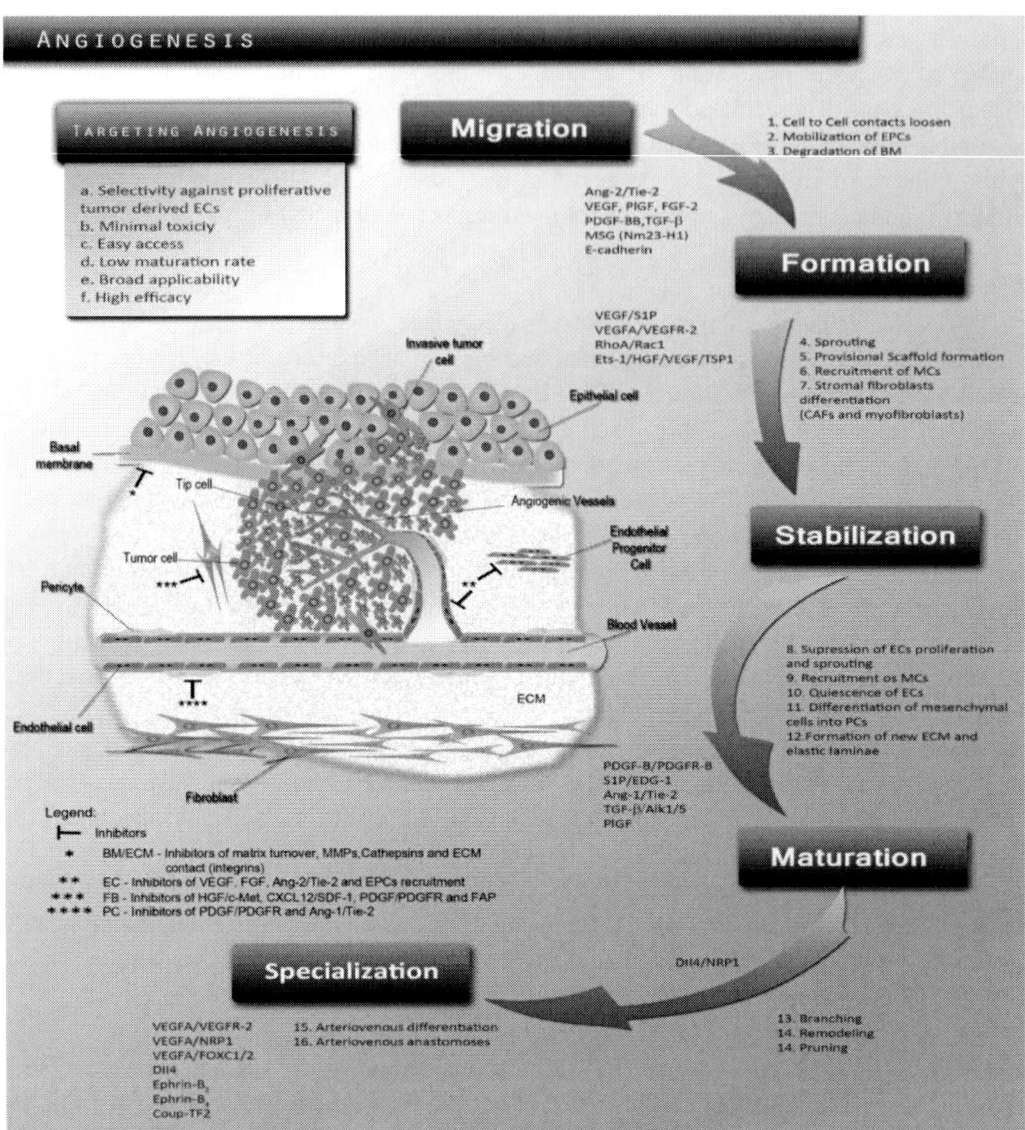

Abbreviations: Alk1/5, activin-like kinase receptor 1/5; Ang, angiopoietins; BM, basement membrane; CAFs, carcinoma-activated fibroblasts; Coup-TF, chicken ovalbumin upstream promoter-transcription factor; CXCL, chemokine (C-X-C motif) ligand; Dll4, Notch ligand Delta-like 4; EC, endothelial cell; ECM, extracellular matrix; EDG, endoglin; EPC, endothelial progenitor cell; Ets-1, erythroblastosis virus E26 oncogene homolog 1; FAP, fibroblasts activated protein; FB, fibroblasts; FGF, fibroblast growth factor; FOX, forkhead transcription factor; HGF, hepatocyte growth factor; MC, mural cell; MMP, matrix metalloproteinase; MSG, metastases suppressor gene; NRP1, neuropilin 1; PC, pericytes; PDGF, platelet derived growth factor; PDGFR, platelet derived growth factor receptor; PlGF, placental growth factor; S1P, sphingosine-1-phosphatase; TGF, transforming growth factor; Tie, tyrosine kinase receptor; TSP1, thrombospondin 1; VEGF, vascular endothelium growth factor; VEGFR, vascular endothelium growth factor receptor.

Figure 1. Designing cancer therapeutics based on the angiogenic process. The main steps, the most important players and signaling pathways in the angiogenic process are represented on the right side of the figure. The main advantages of targeting angiogenesis and targeting opportunities in cancer therapeutics are displayed on the top left and central areas of the figure.

The timing of most of these processes overlaps [90] and requires a careful orchestration between ECs, mural cells (pericytes and smooth-muscle cells), basement membrane and a wide number of signalling molecules (either stimulatory or inhibitory), such as integrins, angiopoietins, chemokines, junctional molecules, oxygen sensors, endogenous inhibitors and many others [91].

When a pool of angiogenic factors is released and cell-cell contacts loosen, mobilization of cells occurs through a provisional scaffold of fibronectin and fibrin, and migration takes place [91]. This is the first step of angiogenesis.

Maturation and stabilization of the nascent vessel follow, but these processes are compromised in pathological conditions. They can be divided into different steps, although they are entwined: suppression of ECs proliferation by establishment of their quiescence; suppression of sprouting; protection against VEGF withdrawal; stabilization of existing vascular tubes; recruitment of mural cells; proliferation, migration and differentiation of mesenchymal cell into pericytes; investment of new vessels with pericytes and smooth muscle cells; development of new surrounding matrix and elastic laminae; formation of new basal membrane; organ-specific specialization of ECs, mural cells and matrix by strengthening of cell-cell contacts (interendothelial junctions, fenestrations, surface receptors); optimal patterning of the network by branching, expanding and pruning. Finally, arteriovenous differentiation takes place to give rise to arteries, capillaries, veins and lymphatics and also arterio-venous anastomoses [91].

The major steps during the angiogenic process, along with the most important players and signaling pathways are described in Figure 1.

2.3.4.2. *Why Target Tumor Angiogenesis in Cancer Treatment?*

Lymphatic and blood vessels structure is abnormal in the tumor environment, when compared to healthy tissues. Endothelial lining is seldom continuous and lacks adequate mural cells coverage, presenting large fenestrations. Blood flow is normally irregular and regions with high, poor or no perfusion are often encountered [92]. Elevated vascular permeability and impaired lymphatic drainage are therefore two of the main typical features of tumors [93,94], which increases interstitial fluid pressure (IFP). Therapeutic agents administered intravenously can hence easily permeate the fenestrated endothelium by diffusion, convection or transcytosis (depending on their size, shape, charge, flexibility and water or lipid solubility) [95] and accumulate in the extracellular matrix, which functions as a reservoir and promotes a sustained release, increasing therapeutic effect [96,97]. However, the high interstitial pressure gradient towards the tumor core impairs penetration of therapeutic agents into the tumor mass. In addition, as a result of hydrostatic pressure increase throughout time, even extravasation of therapeutic drugs or engineered delivery systems from the blood stream can be hindered, which decreases the probability of anticancer therapies to succeed. Thus, normalizing the blood vessel network, by pruning immature and leaky blood vessels, can significantly improve blood flow and reduce IFP, which potentiates extravasation of therapeutic agents to the tumor interstitial space [94,98,99,100]. Moreover, targeting tumor angiogenesis in cancer treatment, as opposed to attacking tumor cells directly, can also deprive tumor cells from nutrient supply, which impairs their survival and proliferation [101].

Such strategy presents several advantages: selectivity of the treatment against proliferative tumor-derived ECs; minimal toxicity because angiogenesis in the adult is limited to wound healing, ovulation, pregnancy and atherosclerosis; easy access for drugs from the blood to tumor vascular ECs (compared with drugs that have to penetrate large bulky tumor masses); low mutation rate within the vasculature and independence of tumor-cell resistance mechanisms; broad applicability (most solid tumors and leukemia, are dependent on angiogenesis for their survival); and high efficacy, because each tumor capillary supplies hundreds of tumor cells [102].

Furthermore, being a portal for tumor cells to colonize distant areas of the body, angiogenesis can also be targeted to prevent metastases. In a breast xenograft model with established lung metastases, the administration of a potent VEGF inhibitior (the VEGF-Trap) resulted in the regression of both the primary and secondary tumors [103]. Similarly, interfering with tumor lymphangiogenesis, using a soluble VEGFR3 inhibitor [104] or a VEGF-C antibody [105], resulted in decreased lymph node metastases [106].

Moreover, antiangiogenic therapies can be beneficial for radiotherapy and other therapeutics associated with radiation, since pruning immature and inefficient vessels and promoting maturation of the remaining ones, increases perfusion and oxygen levels. As an example, bevacizumab (anti-VEGF monoclonal antibody) administered with chemotherapeutics showed an improved effect, by inhibiting angiogenesis. In a Phase 3 clinical trial, 12 days after combined treatment, tumor perfusion and tumor microvessel densities decreased, tumor interstitial fluid pressure was reduced, and it appeared that pre-existing vessels had increased pericyte coverage [107]. Moreover, bevacizumab, DC101 (anti-VEGFR-2 antibody) and SU11657 (small molecule inhibitor of VEGFRs and PDGFR) decreased IFP in breast, colon cancers, and gliomas [94,108,109], respectively.

As for incidence of adverse side effects with antiangiogenic therapeutic approaches, only a few cases were reported. These include hypertension, intratumoral bleeding and bowel perforation for treatment with VEGF inhibitors (particularly in tumors of the intestine), and thrombosis. This complication was documented in the case of combined administration of angiogenesis inhibitors with conventional chemotherapy, such as bevacizumab and thalidomide, an immunomodulatory drug (Thalomid®; Celgene) [110,111]. Lenalidomide (Revlimid®; Celgene), a derivative of thalidomide with less significant side effects, was also placed under investigation for potential safety problems, in March 2008. However, the incidence of bone-marrow suppression, weakness, alopecia, severe vomiting and diarrhea adverse effects are less common in antiangiogenic therapies than with conventional chemotherapy.

Angiogenesis is, to the date, a promising therapeutic target and has been extensively studied to unfold the mechanisms responsible for tumor progression. However, there are important aspects to consider when targeting angiogenesis, mainly affinity to the right molecular target.

2.3.4. Important Aspects for Successful Vascular Targeting

2.3.4.1. Target Selection

Angiogenesis blockade has become a significant process to consider in therapeutic interventions, because it can contribute to hinder tumor cell survival and to a more effective delivery of bioactive molecules (including, but not limited to, drugs, cytokines, procoagulant factors, radionuclides, etc.) to the tumor microenvironment. Specificity, local accumulation of an effective concentration of the therapeutic agent or interference with signaling pathways can be achieved, depending on the selection of a suitable molecular and/or cellular target candidate.

An ideal receptor or molecular target must be exclusively expressed or overexpressed on easily accessible vascular cells, preferentially ECs. Target molecules selectively up-regulated in ECs from angiogenic blood vessels include: cell adhesion molecules (such as integrins, cadherins, selectins, vascular cell adhesion molecules, molecules associated with procoagulant changes on angiogenic endothelium (such as phosphatidylserine) [112,113], endothelial ectopeptidases [114], caveolar proteins [115], growth factor receptors (such as VEGR [116,117], EGFR [118], FGFR [119], PDGFR [120]), transport proteins [121] and other proteins (such as CD31 (PECAM-1) [122] and angiotensin-converting enzyme [123].

To unmask the molecular topography of the cell surface, various techniques have been developed. Proteomics of EC plasma membrane is a good example that allows micro- and nano-scale range "cartography" of EC surface determinants in organs of interest and identification of "zip codes" characteristic of vascular areas [124,125]. Once targets are selected, the process of identifying or designing suitable targeting ligands for therapeutic purposes can thus be initiated.

2.3.4.2. Ligand Selection - Phage Display Libraries

Peptides with targeting abilities can be derived by different methods. They can be identified using the structure-activity relationship for known receptors or even by molecular modeling of the known X-Ray structure of the receptor [126,127]. However, these approaches may be limited, since they provide only small subsets of possible targeting ligands [127]. To overcome this limitation, one can use combinatorial library chemistry (a powerful resource in drug discovery), being phage display library the most common methodology. A combinatorial library is a collection of molecules, chemical or biologically synthesized, that can be screened for a function or affinity [127]. Phage display libraries are displayed by modified bacteriophages (or phage), which are single stranded viruses with the ability to infect bacteria. The process is accomplished by inserting random olignucleotides into the N-terminus of a filamentous phage coating proteins [127]. The inserts are positioned in such a way that capsid protein functions only as an anchor and does not influence the activity of the displayed peptide [128]. This allows the fusion protein to be introduced to different molecular targets. The affinity selection is accomplished *in vitro* or *in vivo* by passing the library over molecular targets, using cell cultures and tumor xenografts. The unbound phages are washed away and the bound phages maintain their infectious properties and can proliferate by infecting new host bacterial cells. The identification of the inserted peptide is then performed by sequencing the viral DNA [127].

The advantage of this method in angiogenesis lays in the built-in recognition (and selection) of sites accessible only to the blood [129]. This technique allows generation and screening of 10^8 to 10^9 different phages, representing a powerful method to routinely develop and test millions of structures [126,127]. Different evolutions of the technology have been accomplished and are extensively reviewed [128].

Many peptides for different targets have been discovered using this technology and are reviewed elsewhere [126]. Arap, Pentz, Zurita and co-workers [130,131,132] have used phage display technologies to map human vascular diversity. The RGD-4C peptide was introduced in 1998 by Arap and co-workers [133] and consists in the Arg-Gly-Asp motif that selectively binds to $\alpha v\beta 3$ integrin expressed in ECs of the tumor vasculature.

Pasqualini *et al.* identified Aminopeptidase N/CD13 as the receptor for the NGR motif [134]. Marchió *et al.* recognized Aminopeptidase A as the receptor for the CPRECESIC motif and demonstrated *in vivo* that it is able to home to the tumor vasculature and inhibit tumor growth [135]. Recently, Loi *et al.* used doxorubicin-loaded liposomes coupled with CPRECESA cyclic motif to demonstrate specific cell association in cells transfected with aminopeptidase A, *in vitro*, and also tumor accumulation, in a murine model of neuroblastoma. A combination of these liposomes with the CNGRCGV motif to target the tumor vasculature marker aminopeptidase N, was also tested *in vivo* by the authors. This strategy revealed an increase in tumor cell apoptosis and animal survival when compared to separate administration of each formulation [136]. These and other angiogenic homing motifs, such as signal sequence-based peptides, cell penetrating peptides, synthetic polyarginines and DNA-binding peptides are described in Table 2.

Table 2. Angiogenic homing motifs used for targeting purposes

Signal sequence-based peptides	**Refs**	**Cell penetrating peptides**	**Refs**
WIFPWIQL	[137]	VIP	[138]
RGD	[139,140,141]	antennapedia – Antp (RQIKIYFQNRRMKWKK)	[142,143]
CNGRC	[136,144]	model amphipathic peptide (MAP)	[145]
CPRECES	[135]	TAT peptide (TATp)	[142,146]
GSL	[133]	Transportan	[147]
CTTHWGFTLC	[148]		
CGRRAGGSC	[132]	**Synthetic polyarginines**	
CKGGRAKDC	[149]	R9 analog of TAT (R9-TAT); HIV-1 Rev	[150]
CREKA	[151]	flock house virus coat peptide (RRRRNRTRRNRRRVR)	[152]
CVPELGHEC	[153]		
F3	[154,155]	**DNA-binding peptides**	
LyP-1	[156]	c-Fos; c-Jun; yeast GCN4	[157,158,159]

2.3.4.5. Anti-angiogenic and Vascular Disrupting Agents

A current goal in the clinical use of antiangiogenic therapies is to convert cancer into a chronic manageable disease. Anti-angiogenic agents can be divided into two major groups: angiogenesis inhibitors and vascular disrupting agents (VDAs).

Angiogenesis inhibitors inhibit the early steps of tumor angiogenesis with the purpose of preventing new blood vessel formation, namely by interrupting signaling pathways between tumor, endothelial and stromal cells. For the design of these inhibitors, four main strategies have been used:

1. Block the ability of ECs to break down their surrounding tissue barriers;
2. Inhibit endothelial progenitor cells recruitment;
3. Block the chemical signals that stimulate angiogenesis or induce the production of angiogenesis inhibitors;
4. Block the action of integrins on the surface of ECs, responsible for their continued survival during angiogenesis.

On the other hand, the destruction of an established tumor vessel network is the purpose of VDAs, causing a rapid and selective vascular shutdown and secondary tumor cell death by ischemia [160,161]. VDAs induce a dose-dependent suppression of functional blood vessels, resulting in widespread tumor necrosis, although a viable rim remains at the periphery of the tumor due to nutritional support from the adjacent normal tissue vasculature [162].

The combined strategy of VDAs and angiogenesis inhibitors is capable of working in complementary fronts since the former destroy the tumor mass and prevent progression, while the latter impair tumor regrowth and limit the metastatic potential. However, such therapy requires chronicle administration over months or even years in an intermittent fashion, as only repeated multiple-dose treatments show impact on tumor growth. Angiogenesis inhibitors targeting ECs rather than tumor cells may not necessarily kill tumors, but hold them in check indefinitely. It is then beneficial to design combinations with standard anticancer therapies, such as radio or chemotherapy, since the viable rim is composed of rapid dividing cells.

Several antiangiogenic agents are currently in clinical trials and some have already received FDA approval.

Marimastat is an orally administered, synthetic broad spectrum inhibitor of matrix metalloproteinases activity that has been shown to inhibit tumor growth and metastatic spread, but has not proven to induce tumor regression [163,164].

COX-2 inhibitors, such as NS-398 and celecoxib, have been used to suppress tumor angiogenesis [165].

Anti-VEGF agents have been extensively explored over the last decades, since VEGF is over-produced in approximately 60% of human tumors [87]. The most advanced antiangiogenic agents in the clinic targeting VEGF family members are the anti-VEGF antibody bevacizumab (Avastin®, Genentech/Roche), a VEGF165 aptamer, pegaptanib (Macugen®, Eyetech/Pfizer) and various RTKIs, which target VEGFRs and other receptors: sorafenib (Nexavar®, Bayer Pharmaceuticals Corp.), sunitinib malate (Sutent®, Pfizer, Inc.), erlotinib (Tarceva®, OSI Pharmaceuticals Inc.). Additional compounds currently in development include VEGF trap-eye (Regeneron/Bayer) and antibodies against VEGFR-2 or VEGFR-1 in advanced hepatocellular carcinoma (IMC-1121B and IMC-18F1, ImClone Systems Incorporated) and antibodies against the VEGFR-1 ligand PlGF (TB-403, Thrombogenics Ltd and BioInvent International). This class of antiangiogenic agents not only arrests EC proliferation and prevents vessel growth, but also induces regression of existing vessels by increasing EC death. Immature vessels, devoid of pericytes, are most susceptible to this approach. In addition, VEGF inhibitors suppress the mobilization of EPCs from the bone marrow. However, VEGF inhibitors are cytotoxic for some malignant cells, activate the anti-tumor immune attack and suppress the pro-angiogenic activity of haematopoietic cells [64].

Other FDA-approved drugs revealed antiangiogenic activity in addition to anticancer activity directed against tumor cells. For example, bortezomib (Velcade®, Millennium Pharmaceuticals Inc., Takeda Oncology Company, Johnson & Johnson Pharmaceutical Research and Development), approved as a proteasome inhibitor for the treatment of mantle cell lymphoma in 2006 and multiple myeloma in 2008, was subsequently demonstrated to also have potent antiangiogenic activity.

Anti-angiogenic agents' approval does not limit their field of application. In fact, if a disease or a different type of cancer than the one originally approved is also dependent on angiogenesis, the same agent is susceptible of being applied or even used off-label. Administration with other anticancer agents has also added major benefits in patients treatment modalities, as is the example of co-administration of Avastin® with Tarceva® (approved by the FDA for the treatment of non-small cell lung cancer), which blocks production of VEGF, bFGF and TGF-α; or with Caplostatin®, which has a larger spectrum than Tarceva's [166].

Table 3. Angiogenesis inhibitors currently in clinical trials or FDA approved

Drug	Mechanism of action	FDA approval
ABT-518	MMP inhibitor	
Angiostatin	Inhibitior of EC proliferation, migration and survival (αvβ3 integrin and ATP synthase ligand)	
Angiozyme	Ribozyme targeting VEGF receptor-1 mRNA	
Avastin® (Bevacizumab)	Anti-VEGF antibody	2009 (second-line treatment of glioblastoma) 2008 (HER2-negative breast cancer in association with Taxol®) 2006 (first-line treatment of NSCLC and second-line treatment of metastatic CRC) 2004 (first-line treatment of metastatic CRC)
BMS-275291	MMP inhibitor	
CAI	Calcium influx inhibitor	
Celebrex® (Celecoxib)	COX-2 inhibitor; promotes EC proliferation and survival	
Combretastatin A4	Tubulin-binding agent; induction of EC apoptosis	
COL-3	MMP inhibitor	
EMD-121974 (Cilengitide)	Integrin αv inhibitor	
Endostatin	Inhibitior of EC proliferation, migration and survival (α5β1 integrin ligand)	
Fragmin® (Dalteparin)	Inhibitior of invasion promoted by enzymatic degradation of the stroma and BM	
Gleevec® (Imatinib mesylate)	Inhibitor of Bcr-Abl kinase-, PDGF receptor- and Klt-mediated effects	Different types of leukemia, GIST, Dermatofibrosarcoma protuberans, myelodysplastic/mieloproliferative disorders, systemic mastocytosis.
Herceptin® (Trastuzumab)	Inhibitor of TGFB, PAI-1, angiopoietin and VEGF production; stimulation of TSP-1 production	2006 (HER2-overexpressing breast cancer)
IMC-1221B	Anti-VEGFR-2 antibody	
IMC-18FI	Anti-VEGFR-1 antibody	
IM862	Inhibitor of grow factor-receptor binding	
Interferon-α	Inhibition of bFGF and VEGF production	
Interleukin 12	Induces interferonα and IP-10 expression	
Iressa® (Gefitinib)	Inhibitor of VEGF, bFGF, TGFα and IL-8 production	2003 (NSCLC)
Macugen® (Pegaptanib)	Anti-VEGF antibody	2004 (wet AMD)

Table 3 (Continued)

Drug	Mechanism of action	FDA approval
Marimastat	MMP inhibitor	
2-ME	Unknown	
Metastat	MMP inhibitor	
Neovastat® (AE941)	Inhibitor of VEGF and MMP, stimulator of tPA	
Nexavar® (Sorafenib)	RTKI	2007 (liver cancer); 2005 (kidney cancer)
NM-3	Unknown	
PI-88	Heparan sulfate antagonist	
Prinomast® (AG3340)	MMP inhibitor	
PTK-787 (ZK222584)	Inhibitor of VEGFR-2, PDGFR- and Klt-mediated effects	
S-3304	MMP inhibitor	
Squalamine	NHE3 sodium-hydrogen exchanger inhibitor	
Suramin	Inhibitior of invasion promoted by enzymatic degradation of the stroma and BM	
Sutent® (Sunitinib)	RTKI	2006 (GIST and kidney cancer)
SU11248	Inhibitor of VEGFR-, PDGFR-, KIT- and FLT3-mediated effects	
SU6668	Inhibitor of VEGF-, FGF-, and PDGF-receptor mediated effects	
SU5416 (Semaxanib)	VEGF receptor-2 inhibitor	
Tarceva® (Erlotinib)	Receptor Tyrosine Kinase inhibitor; inhibitor of VEGF, bFGF, TGFα and IL-8 production	2005 (pancreatic cancer in association with gemcitabine) 2004 (local or metastatic NSCLC)
TB-403	Anti-PlGF antibody	
Thalidomid® (Thalidomide)	Inhibitor of TNFα and other cytokine effects	2006 (multiple myeloma)
Tumstatin	Inhibitor of EC proliferation, migration and survival ($\alpha v\beta 3$ ligand)	
TNP-470	Inhibitor of EC proliferation	
Velcade® (Bortezomib)	Proteasome inhibitor	2008 (multiple myeloma) 2006 (mantle cell lymphoma)
Vitaxin® (hLM609)	Anti-vitronectin receptor (antibody against integrin $\alpha v\beta 3$) antibody; inhibition of EC proliferation, migration and survival	
Vioxx® (Rofecoxib)	COX-2 inhibitor; promotes EC proliferation and survival	
WX-UK-1	MMP inhibitor	
ZD6474	Inhibitor of VEGFR-2 mediated effects	
ZD4190	Inhibitor of EC signal transduction	

Abbreviations: AMD, age-related macular degeneration; bFGF, basic fibroblast growth factor; BM, basement membrane; CRC, colorectal cancer; GIST, gastrointestinal stromal tumor; Klt, stem cell factor receptor; MMP, matrix metalloproteinase; RTKI, receptor tyrosine kinase inhibitor; VEGF, vascular endothelial growth factor; FGF, fibroblast growth factor; FLT3, FMS-like tyrosine kinase 3 receptor; NSCLC, non-small cell lung cancer; PDGF, Platelet-derived growth factor; PlGF, placental growth factor; tPA, plasminogen activator; TSP-1, thrombospondin-1; VEGF, vascular endothelial growth factor.

Therapy with antiangiogenic agents however, has implied a continued regimen of application, since it reduces cancer to a chronic disease state. Studies performed with thalidomide, approved in Australia in the year of 2003 for the treatment of advanced multiple myeloma, showed no evidence of resistance to therapy in 3-5 years therapeutic regimens [167] and low dose interferon-α has also been used successfully to treat life-threatening angioblastomas over the years [168].

These and other angiogenesis inhibitors, their mechanisms of action and FDA approval are summarized in Table 3.

The above-mentioned examples demonstrate that choosing a promising target within the tumor and its microenvironment and selecting high affinity ligand(s) are key elements in a therapeutic strategy that intends to be specific and effective towards cancer regression and/or inhibition. However, when considering therapeutic drugs, effectiveness depends on efficient delivery, therefore reinforcing the importance of adequate delivery platforms in the broad field of cancer therapeutics. In this regard, it is important to emphasize that delivery does not only rely on specificity towards a defined target cellular population. It is also of crucial importance to trigger efficient intracellular delivery of the cargo.

2.4. Effective Strategies for Targeting the Tumor and its Microenvironment

2.4.1. Drug Delivery Systems

Rational design of drug delivery systems (nanocarriers) is essential and includes [169]:

- selection of proper target determinants;
- production of ligands with high receptor affinity for targeting (peptides, antibodies, or their fragments);
- selection of the most adequate nanocarrier (depending on size, composition and pharmacokinetic properties).

Invasive methods for delivering membrane-impermeable molecules, such as microinjection or electroporation, can damage the cellular membrane [170], and the use of viral vectors for DNA delivery suffer from lack of specificity and inherent risk of virus-induced complications [171]. However, many more types of nanocarriers have been designed, including polymer conjugates, polymeric nanoparticles, lipid-based carriers (liposomes and micelles), quantum dots, polyplexes, dendrimers, carbon nanotubes, gold nanoparticles (including nanoshells and nanocages).

Regardless the nature of the nanocarrier, there are a number of different features they must incorporate aiming at their systemic application for cancer therapy [172]:

- biocompatibility and functionality;
- high differential uptake efficiency into the target cells over the healthy ones;
- soluble or colloidal under aqueous conditions;

- low rate of aggregation;
- physical and chemical stability in biological fluids;
- extended circulating half-life.

The use of drug delivery systems can provide the following advantages [173]:

- optimize pharmacokinetics properties of a drug in the bloodstream and offer protection against degradation, inactivation and premature activity en route to the target;
- provide fine control of drug-release kinetics;
- enhance absorption of the drug into a selected tissue (such as a solid tumor);
- provide a template for multivalent affinity binding sites, enhancing effectiveness of anchoring on the target cells;
- modulate subcellular delivery of drugs.

Key controllable parameters of nanocarriers that define their utility for drug delivery include structural materials, plasticity, morphology, size, shape, charge, permeability, biocompatibility, and biodegradability. These unique properties, especially size, make them suitable for deep penetration into tumor tissues. Moreover, specificity and selectivity can be achieved by coupling targeting moieties at their interactive surface [174,175].

The potential of targeted and non-targeted nanoparticles has been exploited for drug delivery, imaging, photothermal ablation of tumors, radiation sensitizers, detection of apoptosis, sentinel lymph node mapping and manipulation and tracking of stem cells [176,177,178]. Among the different types of nanocarriers, liposomes are the ones that have been studied the most, being nowadays a reality in a clinical setting. Many of the lessons learnt from the use of lipossomes on drug delivery will be important for the development of nanocarriers. The following section illustrates some of the main features of liposomes on the field of drug delivery.

2.4.2. Lipid-based Targeted Nanocarriers

Liposomes conjugated with affinity moieties have proven to be ideal for encapsulating and delivering a specific payload to the intracellular compartment. These vehicles have been developed for more than fifty years and have been subject to progressive improvements, taking into consideration distinct applications in clinic. They can decrease elimination by the reticuloendothelial system and enhance circulation time through a stealth effect, dependent on the introduction of a hydrophilic molecule, such as PEG, on the liposome's surface [179]. They can selectively accumulate inside the affected organ or tissue and increase the concentration of the liposomal drug in the desired site of action [180]. This process can even be controlled through a triggered-release mechanism, thus improving therapeutic efficacy [181].

2.4.2.1. Controlled-release

Efficiency of delivery can be optimized by engineering nanocarriers in a way to control the release of their cargo, in a specific site and at a determined time schedule, taking advantage of tumor cells metabolism and tumor microenvironment special features. Certain stimuli intrinsically characteristic of the pathological area (or applied externally to this zone) can beneficially modify the properties of the drug-nanocarrier system, providing enhanced or controlled drug release, improvement of the cellular drug uptake, control of the intracellular drug fate or even allowing for certain physical action on the surrounding pathological tissue. The stimuli can include temperature increase within the tissue (hyperthermia from inflammation) or induced externally (ultrasound, electromagnetic field) [182]; redox potential alteration [183] and pH variations [184].

Intratumoral pH value in solid tumors may decrease to 6.5, i.e. one pH unit lower than in normal blood (7.4), due to hypoxia and massive cell death inside the tumor [185]. pH decreases even more inside the cell, especially in the endosome (5.5 or even below). Different pH-sensitive formulations have been described, and are reviewed elsewhere, but most of them are based on phosphatidylethanolamine (PE) containing acidic moieties (like carboxylic groups) [181]. However, very recent work emphasizes the use of pH-sensitive polymers, more specifically the backbone structure of the polymer and its importance in the interaction of liposomes and biological membranes. Yuba *et al.* demonstrated a significant increase in efficacy of payload delivery into the cytosol of dendritic cells DC2.4 by liposomes modified with hyperbranched poly(glycidol) derivatives, when compared to linear polymer-grafted liposomes [186].

2.4.2.2. Active Targeting – Ligand Functionalized Surface

Tumor highly permeable microvasculature and poorly developed lymphatic system are unique properties of tumors responsible for the enhanced permeability and retention effect (EPR) [187]. This effect alters the distribution of drugs, but also of nanocarriers, like liposomes, leading to an increased accumulation in the tumor tissue. This type of targeting strategy is called passive targeting, as it depends only on tumor pathophysiological properties and liposome size for accumulation. Particle size is important, since pore dimensions in the endothelial lining of peripheral tumor vessels ranges from 100 to 600 nm, depending on the type of tumor [188]· However, accumulation in the interstitial space is not all that is required for therapeutic success. Treatment efficacy is somewhat compromised by the increasing pressure gradient towards the core of the tumor that hinders access to cancer cells. Despite the advantages of PEG-grafted nanocarriers, also known as Stealth®, to reach the pathologic site, passive targeting fails to provide significant selectivity for enhanced therapeutic effect. Therefore, novel and improved strategies are required for overcoming therapeutic drawbacks such as the random nature of delivery and the heterogeneity of solid tumors [172].

To improve selectivity, the use of nanocarriers with a ligand functionalized surface allows active targeting to tumor cells by specific ligand-receptor interactions, leaving the surrounding healthy tissues unaffected or minimally affected by the drug load. For maximal selectivity, one has to target a cell surface marker exclusively or abnormally overexpressed in tumor cells or in other cells of the surrounding microenvironment, such as ECs of angiogenic vessels [172].

Different types of ligands can be attached to the nanocarrier surface: antibodies and their fragments, nucleic acids (aptamers) or even peptides or compounds such as vitamins and carbohydrates, without affecting their integrity and binding properties. Ligand attachment can be accomplished with different methodologies, such as direct adsorption to the nanocarrier surface or covalent coupling. In nanocarriers, the most common ligands (antibodies or peptides) are normally covalently coupled to the end of functionalized PEG-derivatized lipids, like DSPE-PEG^{2000kD} (1,2-distearoyl-sn-glycero-3-phosphoethanolamine-N- [amino (polyethylene glycol)-2000]). Different groups can be used to functionalize the PEG-end, like amino, thiol or carboxyl groups. To allow a single step coupling reaction between PEG and an amino group of a peptide, for instance, a lipid with a PEG-modified end by p-nitrophenylcarbonyl group can be introduced onto the nanocarrier [189]. Another method uses maleimide modified PEG-lipid to promote a coupling reaction with an amino-carrying moiety, like a peptide, functionalized with a thiol group [8].

The use of a spacer between the modified PEG and the ligand can also be beneficial for the extension of the latter outside the PEG brush, minimizing steric hindrance during the ligand-receptor interaction [175].

Regarding the different types of targeting moieties, antibodies are customarily used as targeting agents, either as whole or only fragments, like antigen binding fragment (Fab) or single chain fragment variable (scFv) regions. The use of monoclonal antibodies can increase the affinity of a nanocarrier to the target site, since two binding sites are available. However, given that the Fc region also binds to its receptor in normal cells, such as macrophages, immunogenicity of nanocarriers targeted with antibodies can increase, thus leading to enhanced removal from circulation by the reticuloendothelial system in the liver [172]. Moreover, antibodies' size (160kD) renders nanocarriers as less suitable particles for infiltrating the tumor cell mass [126], which can impair the efficacy of drug delivery. Some of these drawbacks are likely to be overcome using only antibody fragments, like Fab or scFv regions, despite being less stable in vivo [172]. For example, anti-HER2 antibody fragments were used to target doxorubicin-loaded liposomes in murine tumor models. The therapeutic efficacy was higher for the immunoliposomes when compared with free doxorubicin or the non-targeted counterpart. Equal efficacy was observed when immunoliposomes containing Fab or scFv regions of mAb-HER2 were compared to each other [190]. More recently, Gao *et al.* used PEGylated liposomes conjugated with anti-HER2 Fab' to deliver Pseudomonas derived exotoxin PE38KDL (antitumor agent). As expected, anti-HER2 Fab'-coupled liposomes exhibited receptor-specific binding and superior cytotoxic effect to HER2-overexpressing breast cancer cells when compared to non-targeted liposomes [191].

Besides targeted drug delivery, antibodies or antibodies fragments can be used to deliver agents for radiological treatment or cancer imaging techniques. Boron neutron capture therapy is a good example, being based on the release of lethal radiation particles in proliferating and non-proliferating cells, after external irradiation of boron with low energy neutrons. Pan *et al.* demonstrated the success of this therapy through specific delivery of boron compounds to malignant cells. They used cetuximab immunoliposomes targeting EGFR for targeted delivery of a boron derived agent. An 8-fold increase in the uptake of boron by F98 glioma cells, modified to express EGFR, was observed when compared to non-targeted liposomes (human IgG liposomes) [192].

However, the problems imposed with antibodies or their fragments associated with different types of nanoparticles cannot be neglected. As an alternative, other agents can be used, namely, peptides, as they are small, easy to synthesize and very stable molecules [126]. They can also be engineered to confer suitable pharmacokinetic properties, i.e., specificity, high affinity and efficient binding and cellular association [127].

With peptides as targeting moieties, multivalency is an important strategy for active targeting, since peptides typically bind to their targets with relatively modest (low micromolar) affinity [193]. The recent work of Jayanna *et al.* is one such example. They showed the use of a phage coating protein displaying a homing peptide for prostate cancer cells to self integrate in membranes when separated from phages. The authors used this property to prepare and modify Doxil® liposomes by post-inserting the phage coating protein fused with peptides with affinity to PC3 prostate cancer cells. High association of targeted liposomes by PC3 cancer cells and also superior cytotoxicity over the control non-targeted liposomes was demonstrated [194].

The use of specifically targeted nanocarriers, such as liposomes, has and will continue to be extended to other applications besides the ones described herein and documented in other scientific papers. The potential rendered to them by targeting strategies is only modestly known.

Conclusion

Tumor growth, progression and metastases are more complex processes than what was originally anticipated. Knowledge of mechanisms, interactions and players involved in this process is the key for acting on tumor progression. However, tumors are live entities and heterogeneity is their main feature, alongside with adaptability and versatility.

The genetic revolution allowed a deepest understanding of the signaling pathways that are deregulated in tumor cells, resulting in the identification of several oncogenes that theoretically, upon inhibition, can bring a positive impact in treatment of human cancer. However, even this case is not of clear understanding, since tumor cells also depend on the cellular function of normal genes acting in the oncogenic pathway.

It is not likely that inhibition of just one pathway in tumor cells will be enough. The contribution of angiogenesis, stromal interactions, mechanical forces, the immune system and the inflammatory process, stem/progenitor and dormant cells, among others, for tumor progression should not be disregarded. Therefore, targeting different cells, different signaling pathways and distinct process simultaneously seems a more reasonable approach for promoting tumor regression and/or inhibition.

It is hard to foresee how therapeutic agents will be effective in the long run. Fortunately, overall survival has been increased in many types of cancer and actively targeted nanocarriers with controlled release of drugs in the intracellular compartment hold a very promising outcome. The ability of nanotechnology-based strategies, namely lipid-based carriers, to be tailor-made to different targets related to the tumor microenvironment element(s) and/or signaling pathway(s), confer them a great potential for bridging the gap between fundamental knowledge and efficient therapies for cancer patients. The search for better targets and more efficient strategies is still urgently required. However, science is growing exponentially and new tools are discovered and engineered every day, enabling us to take giant steps towards what will be the full understanding of how tumors survive and how their Achilles tendon can be targeted.

Acknowledgments

Vera Moura and Nuno Fonseca are recipients of a fellowship from the Portuguese Foundation for Science and Technology (FCT) (SFRH/BD/21648/2005 and SFRH/BD/ 64243/2009, respectively). L.C. Gomes da Silva is a student of the international PhD program on Biomedicine and Experimental Biology, from the Center for Neuroscience and Cell Biology, and recipient of a fellowship from FCT (ref.: SFRH/BD/33184/2007). The work in the authors' laboratory was supported by a Portuguese grant from FCT, POCTI and FEDER (ref.: POCI/SAU-OBS/57831/2004) and by the Portugal-Spain capacitation program in nanoscience and nanotechnology (ref.: NANO/NMed-AT/0042/2007).

References

[1] Hanahan D, Weinberg RA (2000) The hallmarks of cancer. Cell 100: 57-70.

[2] Dunn GP, Old LJ, Schreiber RD (2004) The three Es of cancer immunoediting. Annu Rev Immunol 22: 329-360.

[3] Zitvogel L, Tesniere A, Kroemer G (2006) Cancer despite immunosurveillance: immunoselection and immunosubversion. Nat Rev Immunol 6: 715-727.

[4] Colotta F, Allavena P, Sica A, Garlanda C, Mantovani A (2009) Cancer-related inflammation, the seventh hallmark of cancer: links to genetic instability. Carcinogenesis 30: 1073-1081.

[5] Luo J, Solimini NL, Elledge SJ (2009) Principles of cancer therapy: oncogene and non-oncogene addiction. Cell 136: 823-837.

[6] Pietras K, Ostman A (2010) Hallmarks of cancer: interactions with the tumor stroma. Exp Cell Res 316: 1324-1331.

[7] Ferrara N, Hillan KJ, Novotny W (2005) Bevacizumab (Avastin), a humanized anti-VEGF monoclonal antibody for cancer therapy. Biochem Biophys Res Commun 333: 328-335.

[8] Moreira JN, Ishida T, Gaspar R, Allen TM (2002) Use of the post-insertion technique to insert peptide ligands into pre-formed stealth liposomes with retention of binding activity and cytotoxicity. Pharm Res 19: 265-269.

[9] Kroemer G, Pouyssegur J (2008) Tumor cell metabolism: cancer's Achilles' heel. Cancer Cell 13: 472-482.

[10] Seyfried TN, Shelton LM (2010) Cancer as a metabolic disease. Nutr Metab (Lond) 7: 7.

[11] Hynes NE, Lane HA (2005) ERBB receptors and cancer: the complexity of targeted inhibitors. Nat Rev Cancer 5: 341-354.

[12] Hynes NE, MacDonald G (2009) ErbB receptors and signaling pathways in cancer. Curr Opin Cell Biol 21: 177-184.

[13] Zhang C, Newsome JT, Mewani R, Pei J, Gokhale PC, et al. (2009) Systemic delivery and pre-clinical evaluation of nanoparticles containing antisense oligonucleotides and siRNAs. Methods Mol Biol 480: 65-83.

[14] Medina PJ, Goodin S (2008) Lapatinib: a dual inhibitor of human epidermal growth factor receptor tyrosine kinases. Clin Ther 30: 1426-1447.

[15] Agrawal M, Garg RJ, Cortes J, Quintas-Cardama A (2010) Tyrosine kinase inhibitors: the first decade. Curr Hematol Malig Rep 5: 70-80.

[16] Agrawal M, Garg RJ, Kantarjian H, Cortes J (2010) Chronic Myeloid Leukemia in the Tyrosine Kinase Inhibitor Era: What Is the "Best" Therapy? Curr Oncol Rep 12(5):302-313.

[17] Bonafoux D, Lee WC (2009) Strategies for TGF-beta modulation: a review of recent patents. Expert Opin Ther Pat 19: 1759-1769.

[18] Moreira JN, Santos A, Simoes S (2006) Bcl-2-targeted antisense therapy (Oblimersen sodium): towards clinical reality. Rev Recent Clin Trials 1: 217-235.

[19] Vogler M, Dinsdale D, Dyer MJ, Cohen GM (2009) Bcl-2 inhibitors: small molecules with a big impact on cancer therapy. Cell Death Differ 16: 360-367.

[20] Carlo-Stella C, Lavazza C, Locatelli A, Vigano L, Gianni AM, et al. (2007) Targeting TRAIL agonistic receptors for cancer therapy. Clin Cancer Res 13: 2313-2317.

[21] Herman SE, Gordon AL, Wagner AJ, Heerema NA, Zhao W, et al. (2010) The phosphatidylinositol 3-kinase-{delta} inhibitor CAL-101 demonstrates promising pre-clinical activity in chronic lymphocytic leukemia by antagonizing intrinsic and extrinsic cellular survival signals. Blood 116(12):2078-2088.

[22] Ryan BM, O'Donovan N, Duffy MJ (2009) Survivin: a new target for anti-cancer therapy. Cancer Treat Rev 35: 553-562.

[23] Kelland L (2007) Targeting the limitless replicative potential of cancer: the telomerase/telomere pathway. Clin Cancer Res 13: 4960-4963.

[24] Shangary S, Wang S (2009) Small-molecule inhibitors of the MDM2-p53 protein-protein interaction to reactivate p53 function: a novel approach for cancer therapy. Annu Rev Pharmacol Toxicol 49: 223-241.

[25] Lee RJ, Saylor PJ, Smith MR (2010) Treatment and prevention of bone complications from prostate cancer. Bone.

[26] Beano A, Signorino E, Evangelista A, Brusa D, Mistrangelo M, et al. (2008) Correlation between NK function and response to trastuzumab in metastatic breast cancer patients. J Transl Med 6: 25.

[27] De Sanctis JB, Mijares M, Suarez A, Compagnone R, Garmendia J, et al. (2010) Pharmacological properties of thalidomide and its analogues. Recent Pat Inflamm Allergy Drug Discov 4: 144-148.

[28] Harris RE (2009) Cyclooxygenase-2 (cox-2) blockade in the chemoprevention of cancers of the colon, breast, prostate, and lung. Inflammopharmacology 17: 55-67.

[29] Bjorkhem-Bergman L, Acimovic J, Torndal UB, Parini P, Eriksson LC (2010) Lovastatin prevents carcinogenesis in a rat model for liver cancer. Effects of ubiquinone supplementation. Anticancer Res 30: 1105-1112.

[30] Martirosyan A, Clendening JW, Goard CA, Penn LZ (2010) Lovastatin induces apoptosis of ovarian cancer cells and synergizes with doxorubicin: potential therapeutic relevance. BMC Cancer 10: 103.

[31] Faivre S, Kroemer G, Raymond E (2006) Current development of mTOR inhibitors as anticancer agents. Nat Rev Drug Discov 5: 671-688.

[32] Davis ME, Zuckerman JE, Choi CH, Seligson D, Tolcher A, et al. (2010) Evidence of RNAi in humans from systemically administered siRNA via targeted nanoparticles. Nature 464: 1067-1070.

[33] Wu WK, Cho CH, Lee CW, Wu K, Fan D, et al. (2010) Proteasome inhibition: a new therapeutic strategy to cancer treatment. Cancer Lett 293: 15-22.

[34] Dick LR, Fleming PE (2010) Building on bortezomib: second-generation proteasome inhibitors as anti-cancer therapy. Drug Discov Today 15: 243-249.

[35] Quintana E, Shackleton M, Sabel MS, Fullen DR, Johnson TM, et al. (2008) Efficient tumour formation by single human melanoma cells. Nature 456: 593-598.

[36] Vlashi E, Kim K, Lagadec C, Donna LD, McDonald JT, et al. (2009) In vivo imaging, tracking, and targeting of cancer stem cells. J Natl Cancer Inst 101: 350-359.

[37] Phillips TM, McBride WH, Pajonk F (2006) The response of CD24(-/low)/CD44+ breast cancer-initiating cells to radiation. J Natl Cancer Inst 98: 1777-1785.

[38] Aguirre-Ghiso JA, Ossowski L, Rosenbaum SK (2004) Green fluorescent protein tagging of extracellular signal-regulated kinase and p38 pathways reveals novel dynamics of pathway activation during primary and metastatic growth. Cancer Res 64: 7336-7345.

[39] Giuriato S, Ryeom S, Fan AC, Bachireddy P, Lynch RC, et al. (2006) Sustained regression of tumors upon MYC inactivation requires p53 or thrombospondin-1 to reverse the angiogenic switch. Proc Natl Acad Sci U S A 103: 16266-16271.

[40] Gattelli A, Cirio MC, Quaglino A, Schere-Levy C, Martinez N, et al. (2004) Progression of pregnancy-dependent mouse mammary tumors after long dormancy periods. Involvement of Wnt pathway activation. Cancer Res 64: 5193-5199.

[41] Cameron MD, Schmidt EE, Kerkvliet N, Nadkarni KV, Morris VL, et al. (2000) Temporal progression of metastasis in lung: cell survival, dormancy, and location dependence of metastatic inefficiency. Cancer Res 60: 2541-2546.

[42] Joseph NM, Mosher JT, Buchstaller J, Snider P, McKeever PE, et al. (2008) The loss of Nf1 transiently promotes self-renewal but not tumorigenesis by neural crest stem cells. Cancer Cell 13: 129-140.

[43] Kim CF, Dirks PB (2008) Cancer and stem cell biology: how tightly intertwined? Cell Stem Cell 3: 147-150.

[44] Ventura A, Young AG, Winslow MM, Lintault L, Meissner A, et al. (2008) Targeted deletion reveals essential and overlapping functions of the miR-17 through 92 family of miRNA clusters. Cell 132: 875-886.

[45] Wurdinger T, Tannous BA, Saydam O, Skog J, Grau S, et al. (2008) miR-296 regulates growth factor receptor overexpression in angiogenic endothelial cells. Cancer Cell 14: 382-393.

[46] van Solingen C, Seghers L, Bijkerk R, Duijs JM, Roeten MK, et al. (2008) Antagomir-Mediated Silencing of Endothelial Cell Specific MicroRNA-126 Impairs Ischemia-Induced Angiogenesis. J Cell Mol Med 13(8A):1577-1585.

[47] Tavazoie SF, Alarcon C, Oskarsson T, Padua D, Wang Q, et al. (2008) Endogenous human microRNAs that suppress breast cancer metastasis. Nature 451: 147-152.

[48] Erler JT, Weaver VM (2009) Three-dimensional context regulation of metastasis. Clin Exp Metastasis 26: 35-49.

[49] Erler JT, Bennewith KL, Cox TR, Lang G, Bird D, et al. (2009) Hypoxia-induced lysyl oxidase is a critical mediator of bone marrow cell recruitment to form the premetastatic niche. Cancer Cell 15: 35-44.

[50] Levental KR, Yu H, Kass L, Lakins JN, Egeblad M, et al. (2009) Matrix crosslinking forces tumor progression by enhancing integrin signaling. Cell 139: 891-906.

[51] Bacac M, Stamenkovic I (2008) Metastatic cancer cell. Annu Rev Pathol 3: 221-247.
[52] Hu G, Wei Y, Kang Y (2009) The multifaceted role of MTDH/AEG-1 in cancer progression. Clin Cancer Res 15: 5615-5620.
[53] Steeg PS (2006) Tumor metastasis: mechanistic insights and clinical challenges. Nat Med 12: 895-904.
[54] Brown DM, Ruoslahti E (2004) Metadherin, a cell surface protein in breast tumors that mediates lung metastasis. Cancer Cell 5: 365-374.
[55] Kaplan RN, Rafii S, Lyden D (2006) Preparing the "soil": the premetastatic niche. Cancer Res 66: 11089-11093.
[56] Kaplan RN, Riba RD, Zacharoulis S, Bramley AH, Vincent L, et al. (2005) VEGFR1-positive haematopoietic bone marrow progenitors initiate the pre-metastatic niche. Nature 438: 820-827.
[57] Mook S, Schmidt MK, Weigelt B, Kreike B, Eekhout I, et al. (2010) The 70-gene prognosis signature predicts early metastasis in breast cancer patients between 55 and 70 years of age. Ann Oncol 21: 717-722.
[58] van't Veer LJ, Paik S, Hayes DF (2005) Gene expression profiling of breast cancer: a new tumor marker. J Clin Oncol 23: 1631-1635.
[59] Shevde LA, Welch DR (2003) Metastasis suppressor pathways--an evolving paradigm. Cancer Lett 198: 1-20.
[60] Felgner PL, Gadek TR, Holm M, Roman R, Chan HW, et al. (1987) Lipofection: a highly efficient, lipid-mediated DNA-transfection procedure. Proc Natl Acad Sci U S A 84: 7413-7417.
[61] Steeg PS (2003) Metastasis suppressors alter the signal transduction of cancer cells. Nat Rev Cancer 3: 55-63.
[62] Horak CE, Lee JH, Marshall JC, Shreeve SM, Steeg PS (2008) The role of metastasis suppressor genes in metastatic dormancy. APMIS 116: 586-601.
[63] Albini A, Sporn MB (2007) The tumour microenvironment as a target for chemoprevention. Nat Rev Cancer 7: 139-147.
[64] Carmeliet P (2005) Angiogenesis in life, disease and medicine. Nature 438: 932-936.
[65] Kim S, Takahashi H, Lin WW, Descargues P, Grivennikov S, et al. (2009) Carcinoma-produced factors activate myeloid cells through TLR2 to stimulate metastasis. Nature 457: 102-106.
[66] Yang HZ, Cui B, Liu HZ, Mi S, Yan J, et al. (2009) Blocking TLR2 activity attenuates pulmonary metastases of tumor. PLoS One 4: e6520.
[67] Kormann MS, Depner M, Hartl D, Klopp N, Illig T, et al. (2008) Toll-like receptor heterodimer variants protect from childhood asthma. J Allergy Clin Immunol 122: 86-92, 92 e81-88.
[68] Stevens VL, Hsing AW, Talbot JT, Zheng SL, Sun J, et al. (2008) Genetic variation in the toll-like receptor gene cluster (TLR10-TLR1-TLR6) and prostate cancer risk. Int J Cancer 123: 2644-2650.
[69] Liao D, Luo Y, Markowitz D, Xiang R, Reisfeld RA (2009) Cancer associated fibroblasts promote tumor growth and metastasis by modulating the tumor immune microenvironment in a 4T1 murine breast cancer model. PLoS One 4: e7965.
[70] Mann JR, Backlund MG, DuBois RN (2005) Mechanisms of disease: Inflammatory mediators and cancer prevention. Nat Clin Pract Oncol 2: 202-210.

[71] Harris RE, Beebe-Donk J, Alshafie GA (2006) Reduction in the risk of human breast cancer by selective cyclooxygenase-2 (COX-2) inhibitors. BMC Cancer 6: 27.
[72] Demierre MF, Higgins PD, Gruber SB, Hawk E, Lippman SM (2005) Statins and cancer prevention. Nat Rev Cancer 5: 930-942.
[73] Dhanabal M, Ramchandran R, Waterman MJ, Lu H, Knebelmann B, et al. (1999) Endostatin induces endothelial cell apoptosis. J Biol Chem 274: 11721-11726.
[74] Kim HJ, Park CI, Park BW, Lee HD, Jung WH (2006) Expression of MT-1 MMP, MMP2, MMP9 and TIMP2 mRNAs in ductal carcinoma in situ and invasive ductal carcinoma of the breast. Yonsei Med J 47: 333-342.
[75] Wellen KE, Hotamisligil GS (2005) Inflammation, stress, and diabetes. J Clin Invest 115: 1111-1119.
[76] Algire G, Chalkley H, Legallais F, Park H (1945) Vascular reactions to normal and malignant tissue in vivo. I Vascular reactions of mice to wound and to normal and neoplastic transplants. J Natl Cancer Inst 6.
[77] Goldman E (1907) The growth of malignant disease in man and the lower animals with special reference to the vascular system. Lancet 2: 1236.
[78] Ide A, Baker N, Warren S (1939) Vascularization of the BrownPierce rabbit epithelioma transplant as seen in the transparent ear chamber. Am J Roentgen Radiother 42.
[79] Lewis W (1927) The vascular pattern of tumors. Johns Hopkins Hosp Bull 41: 156-162.
[80] Folkman J, Merler E, Abernathy C, Williams G (1971) Isolation of a tumor factor responsible for angiogenesis. J Exp Med 133: 275-288.
[81] Gimbrone MA, Jr., Cotran RS, Folkman J (1973) Endothelial regeneration: studies with human endothelial cells in culture. Ser Haematol 6: 453-455.
[82] Folkman J, Haudenschild C (1980) Angiogenesis in vitro. Nature 288: 551-556.
[83] Weidner N, Semple JP, Welch WR, Folkman J (1991) Tumor angiogenesis and metastasis--correlation in invasive breast carcinoma. N Engl J Med 324: 1-8.
[84] Dameron KM, Volpert OV, Tainsky MA, Bouck N (1994) Control of angiogenesis in fibroblasts by p53 regulation of thrombospondin-1. Science 265: 1582-1584.
[85] Senger DR, Perruzzi CA, Feder J, Dvorak HF (1986) A highly conserved vascular permeability factor secreted by a variety of human and rodent tumor cell lines. Cancer Res 46: 5629-5632.
[86] Wang GL, Jiang BH, Rue EA, Semenza GL (1995) Hypoxia-inducible factor 1 is a basic-helix-loop-helix-PAS heterodimer regulated by cellular O2 tension. Proc Natl Acad Sci U S A 92: 5510-5514.
[87] Folkman J (2006) Angiogenesis. Annu Rev Med 57: 1-18.
[88] Atiqur Rahman M, Toi M (2003) Anti-angiogenic therapy in breast cancer. Biomed Pharmacother 57: 463-470.
[89] Gasparini G, Longo R, Fanelli M, Teicher BA (2005) Combination of antiangiogenic therapy with other anticancer therapies: results, challenges, and open questions. J Clin Oncol 23: 1295-1311.
[90] Jain RK (2003) Molecular regulation of vessel maturation. Nat Med 9: 685-693.
[91] Carmeliet P (2003) Angiogenesis in health and disease. Nat Med 9: 653-660.
[92] Fukumura D, Jain RK (2007) Tumor microenvironment abnormalities: causes, consequences, and strategies to normalize. J Cell Biochem 101: 937-949.

[93] Boucher Y, Jain RK (1992) Microvascular pressure is the principal driving force for interstitial hypertension in solid tumors: implications for vascular collapse. Cancer Res 52: 5110-5114.

[94] Tong RT, Boucher Y, Kozin SV, Winkler F, Hicklin DJ, et al. (2004) Vascular normalization by vascular endothelial growth factor receptor 2 blockade induces a pressure gradient across the vasculature and improves drug penetration in tumors. Cancer Res 64: 3731-3736.

[95] Jain RK (1987) Transport of molecules across tumor vasculature. Cancer Metastasis Rev 6: 559-593.

[96] Vaage J, Mayhew E, Lasic D, Martin F (1992) Therapy of primary and metastatic mouse mammary carcinomas with doxorubicin encapsulated in long circulating liposomes. Int J Cancer 51: 942-948.

[97] Williams SS, Alosco TR, Mayhew E, Lasic DD, Martin FJ, et al. (1993) Arrest of human lung tumor xenograft growth in severe combined immunodeficient mice using doxorubicin encapsulated in sterically stabilized liposomes. Cancer Res 53: 3964-3967.

[98] Jain RK, Tong RT, Munn LL (2007) Effect of vascular normalization by antiangiogenic therapy on interstitial hypertension, peritumor edema, and lymphatic metastasis: insights from a mathematical model. Cancer Res 67: 2729-2735.

[99] Wildiers H, Guetens G, De Boeck G, Verbeken E, Landuyt B, et al. (2003) Effect of antivascular endothelial growth factor treatment on the intratumoral uptake of CPT-11. Br J Cancer 88: 1979-1986.

[100] Jain RK (2005) Normalization of tumor vasculature: an emerging concept in antiangiogenic therapy. Science 307: 58-62.

[101] Folkman J (1971) Tumor angiogenesis: therapeutic implications. N Engl J Med 285: 1182-1186.

[102] Folkman J (1997) Addressing tumor blood vessels. Nat Biotechnol 15: 510.

[103] Huang J, Frischer JS, Serur A, Kadenhe A, Yokoi A, et al. (2003) Regression of established tumors and metastases by potent vascular endothelial growth factor blockade. Proc Natl Acad Sci U S A 100: 7785-7790.

[104] He Y, Kozaki K, Karpanen T, Koshikawa K, Yla-Herttuala S, et al. (2002) Suppression of tumor lymphangiogenesis and lymph node metastasis by blocking vascular endothelial growth factor receptor 3 signaling. J Natl Cancer Inst 94: 819-825.

[105] Makinen T, Veikkola T, Mustjoki S, Karpanen T, Catimel B, et al. (2001) Isolated lymphatic endothelial cells transduce growth, survival and migratory signals via the VEGF-C/D receptor VEGFR-3. EMBO J 20: 4762-4773.

[106] Von Marschall Z, Scholz A, Stacker SA, Achen MG, Jackson DG, et al. (2005) Vascular endothelial growth factor-D induces lymphangiogenesis and lymphatic metastasis in models of ductal pancreatic cancer. Int J Oncol 27: 669-679.

[107] Willett CG, Boucher Y, di Tomaso E, Duda DG, Munn LL, et al. (2004) Direct evidence that the VEGF-specific antibody bevacizumab has antivascular effects in human rectal cancer. Nat Med 10: 145-147.

[108] Huber PE, Bischof M, Jenne J, Heiland S, Peschke P, et al. (2005) Trimodal cancer treatment: beneficial effects of combined antiangiogenesis, radiation, and chemotherapy. Cancer Res 65: 3643-3655.

[109] Lee CG, Heijn M, di Tomaso E, Griffon-Etienne G, Ancukiewicz M, et al. (2000) Anti-Vascular endothelial growth factor treatment augments tumor radiation response under normoxic or hypoxic conditions. Cancer Res 60: 5565-5570.

[110] Zangari M, Anaissie E, Barlogie B, Badros A, Desikan R, et al. (2001) Increased risk of deep-vein thrombosis in patients with multiple myeloma receiving thalidomide and chemotherapy. Blood 98: 1614-1615.

[111] Ozen A, Cicin I, Sezer A, Uzunoglu S, Saynak M, et al. (2009) Dural sinus vein thrombosis in a patient with colon cancer treated with FOLFIRI/bevacizumab. J Cancer Res Ther 5: 130-132.

[112] Brekken RA, Li C, Kumar S (2002) Strategies for vascular targeting in tumors. Int J Cancer 100: 123-130.

[113] Brekken RA, Thorpe PE (2001) VEGF-VEGF receptor complexes as markers of tumor vascular endothelium. J Control Release 74: 173-181.

[114] Danilov SM, Muzykantov VR, Martynov AV, Atochina EN, Sakharov I, et al. (1991) Lung is the target organ for a monoclonal antibody to angiotensin-converting enzyme. Lab Invest 64: 118-124.

[115] Schnitzer JE (2001) Caveolae: from basic trafficking mechanisms to targeting transcytosis for tissue-specific drug and gene delivery in vivo. Adv Drug Deliv Rev 49: 265-280.

[116] Carmeliet P, Collen D (2000) Molecular basis of angiogenesis. Role of VEGF and VE-cadherin. Ann N Y Acad Sci 902: 249-262; discussion 262-244.

[117] Ferrara N (1999) Molecular and biological properties of vascular endothelial growth factor. J Mol Med 77: 527-543.

[118] Amin DN, Hida K, Bielenberg DR, Klagsbrun M (2006) Tumor endothelial cells express epidermal growth factor receptor (EGFR) but not ErbB3 and are responsive to EGF and to EGFR kinase inhibitors. Cancer Res 66: 2173-2180.

[119] Winter SF, Acevedo VD, Gangula RD, Freeman KW, Spencer DM, et al. (2007) Conditional activation of FGFR1 in the prostate epithelium induces angiogenesis with concomitant differential regulation of Ang-1 and Ang-2. Oncogene 26: 4897-4907.

[120] Sennino B, Kuhnert F, Tabruyn SP, Mancuso MR, Hu-Lowe DD, et al. (2009) Cellular source and amount of vascular endothelial growth factor and platelet-derived growth factor in tumors determine response to angiogenesis inhibitors. Cancer Res 69: 4527-4536.

[121] Lee HJ, Engelhardt B, Lesley J, Bickel U, Pardridge WM (2000) Targeting rat anti-mouse transferrin receptor monoclonal antibodies through blood-brain barrier in mouse. J Pharmacol Exp Ther 292: 1048-1052.

[122] Muro S, Muzykantov VR (2005) Targeting of antioxidant and anti-thrombotic drugs to endothelial cell adhesion molecules. Curr Pharm Des 11: 2383-2401.

[123] Danilov SM, Gavrilyuk VD, Franke FE, Pauls K, Harshaw DW, et al. (2001) Lung uptake of antibodies to endothelial antigens: key determinants of vascular immunotargeting. Am J Physiol Lung Cell Mol Physiol 280: L1335-1347.

[124] Durr E, Yu J, Krasinska KM, Carver LA, Yates JR, et al. (2004) Direct proteomic mapping of the lung microvascular endothelial cell surface in vivo and in cell culture. Nat Biotechnol 22: 985-992.

[125] Oh P, Li Y, Yu J, Durr E, Krasinska KM, et al. (2004) Subtractive proteomic mapping of the endothelial surface in lung and solid tumours for tissue-specific therapy. Nature 429: 629-635.

[126] Aina OH, Liu R, Sutcliffe JL, Marik J, Pan CX, et al. (2007) From combinatorial chemistry to cancer-targeting peptides. Mol Pharm 4: 631-651.

[127] Landon LA, Deutscher SL (2003) Combinatorial discovery of tumor targeting peptides using phage display. J Cell Biochem 90: 509-517.

[128] Bratkovic T (2010) Progress in phage display: evolution of the technique and its application. Cell Mol Life Sci 67: 749-767.

[129] Pasqualini R, McDonald DM, Arap W (2001) Vascular targeting and antigen presentation. Nat Immunol 2: 567-568.

[130] Arap W, Kolonin MG, Trepel M, Lahdenranta J, Cardo-Vila M, et al. (2002) Steps toward mapping the human vasculature by phage display. Nat Med 8: 121-127.

[131] Pentz RD, Flamm AL, Pasqualini R, Logothetis CJ, Arap W (2003) Revisiting ethical guidelines for research with terminal wean and brain-dead participants. Hastings Cent Rep 33: 20-26.

[132] Zurita AJ, Troncoso P, Cardo-Vila M, Logothetis CJ, Pasqualini R, et al. (2004) Combinatorial screenings in patients: the interleukin-11 receptor alpha as a candidate target in the progression of human prostate cancer. Cancer Res 64: 435-439.

[133] Arap W, Pasqualini R, Ruoslahti E (1998) Cancer treatment by targeted drug delivery to tumor vasculature in a mouse model. Science 279: 377-380.

[134] Pasqualini R, Koivunen E, Kain R, Lahdenranta J, Sakamoto M, et al. (2000) Aminopeptidase N is a receptor for tumor-homing peptides and a target for inhibiting angiogenesis. Cancer Res 60: 722-727.

[135] Marchio S, Lahdenranta J, Schlingemann RO, Valdembri D, Wesseling P, et al. (2004) Aminopeptidase A is a functional target in angiogenic blood vessels. Cancer Cell 5: 151-162.

[136] Loi M, Marchio S, Becherini P, Di Paolo D, Soster M, et al. (2010) Combined targeting of perivascular and endothelial tumor cells enhances anti-tumor efficacy of liposomal chemotherapy in neuroblastoma. J Control Release 145(1):66-73.

[137] Arap MA, Lahdenranta J, Mintz PJ, Hajitou A, Sarkis AS, et al. (2004) Cell surface expression of the stress response chaperone GRP78 enables tumor targeting by circulating ligands. Cancer Cell 6: 275-284.

[138] Collado B, Sanchez MG, Diaz-Laviada I, Prieto JC, Carmena MJ (2005) Vasoactive intestinal peptide (VIP) induces c-fos expression in LNCaP prostate cancer cells through a mechanism that involves Ca2+ signalling. Implications in angiogenesis and neuroendocrine differentiation. Biochim Biophys Acta 1744: 224-233.

[139] Kok RJ, Schraa AJ, Bos EJ, Moorlag HE, Asgeirsdottir SA, et al. (2002) Preparation and functional evaluation of RGD-modified proteins as alpha(v)beta(3) integrin directed therapeutics. Bioconjug Chem 13: 128-135.

[140] Schraa AJ, Kok RJ, Botter SM, Withoff S, Meijer DK, et al. (2004) RGD-modified anti-CD3 antibodies redirect cytolytic capacity of cytotoxic T lymphocytes toward alphavbeta3-expressing endothelial cells. Int J Cancer 112: 279-285.

[141] Mulder WJ, Strijkers GJ, Habets JW, Bleeker EJ, van der Schaft DW, et al. (2005) MR molecular imaging and fluorescence microscopy for identification of activated tumor endothelium using a bimodal lipidic nanoparticle. FASEB J 19: 2008-2010.

[142] Cantelmo AR, Cammarota R, Noonan DM, Focaccetti C, Comoglio PM, et al. (2010) Cell delivery of Met docking site peptides inhibit angiogenesis and vascular tumor growth. Oncogene 29: 5286–5298.
[143] Bernatchez PN, Bauer PM, Yu J, Prendergast JS, He P, et al. (2005) Dissecting the molecular control of endothelial NO synthase by caveolin-1 using cell-permeable peptides. Proc Natl Acad Sci U S A 102: 761-766.
[144] Pastorino F, Di Paolo D, Piccardi F, Nico B, Ribatti D, et al. (2008) Enhanced antitumor efficacy of clinical-grade vasculature-targeted liposomal doxorubicin. Clin Cancer Res 14: 7320-7329.
[145] Torchilin VP (2008) Cell penetrating peptide-modified pharmaceutical nanocarriers for intracellular drug and gene delivery. Biopolymers 90: 604-610.
[146] Torchilin VP (2008) Tat peptide-mediated intracellular delivery of pharmaceutical nanocarriers. Adv Drug Deliv Rev 60: 548-558.
[147] Muratovska A, Eccles MR (2004) Conjugate for efficient delivery of short interfering RNA (siRNA) into mammalian cells. FEBS Lett 558: 63-68.
[148] Stefanidakis M, Bjorklund M, Ihanus E, Gahmberg CG, Koivunen E (2003) Identification of a negatively charged peptide motif within the catalytic domain of progelatinases that mediates binding to leukocyte beta 2 integrins. J Biol Chem 278: 34674-34684.
[149] Kolonin MG, Saha PK, Chan L, Pasqualini R, Arap W (2004) Reversal of obesity by targeted ablation of adipose tissue. Nat Med 10: 625-632.
[150] Torchilin VP, Levchenko TS, Rammohan R, Volodina N, Papahadjopoulos-Sternberg B, et al. (2003) Cell transfection in vitro and in vivo with nontoxic TAT peptide-liposome-DNA complexes. Proc Natl Acad Sci U S A 100: 1972-1977.
[151] Simberg D, Duza T, Park JH, Essler M, Pilch J, et al. (2007) Biomimetic amplification of nanoparticle homing to tumors. Proc Natl Acad Sci U S A 104: 932-936.
[152] Ho A, Schwarze SR, Mermelstein SJ, Waksman G, Dowdy SF (2001) Synthetic protein transduction domains: enhanced transduction potential in vitro and in vivo. Cancer Res 61: 474-477.
[153] Vidal CI, Mintz PJ, Lu K, Ellis LM, Manenti L, et al. (2004) An HSP90-mimic peptide revealed by fingerprinting the pool of antibodies from ovarian cancer patients. Oncogene 23: 8859-8867.
[154] Porkka K, Laakkonen P, Hoffman JA, Bernasconi M, Ruoslahti E (2002) A fragment of the HMGN2 protein homes to the nuclei of tumor cells and tumor endothelial cells in vivo. Proc Natl Acad Sci U S A 99: 7444-7449.
[155] Reddy GR, Bhojani MS, McConville P, Moody J, Moffat BA, et al. (2006) Vascular targeted nanoparticles for imaging and treatment of brain tumors. Clin Cancer Res 12: 6677-6686.
[156] Laakkonen P, Porkka K, Hoffman JA, Ruoslahti E (2002) A tumor-homing peptide with a targeting specificity related to lymphatic vessels. Nat Med 8: 751-755.
[157] Marconcini L, Marchio S, Morbidelli L, Cartocci E, Albini A, et al. (1999) c-fos-induced growth factor/vascular endothelial growth factor D induces angiogenesis in vivo and in vitro. Proc Natl Acad Sci U S A 96: 9671-9676.
[158] Chang L, Karin M (2001) Mammalian MAP kinase signalling cascades. Nature 410: 37-40.

[159] Cho CH, Kammerer RA, Lee HJ, Steinmetz MO, Ryu YS, et al. (2004) COMP-Ang1: a designed angiopoietin-1 variant with nonleaky angiogenic activity. Proc Natl Acad Sci U S A 101: 5547-5552.

[160] Thorpe PE (2004) Vascular targeting agents as cancer therapeutics. Clin Cancer Res 10: 415-427.

[161] Siemann DW, Horsman MR (2009) Vascular targeted therapies in oncology. Cell Tissue Res 335: 241-248.

[162] Siemann DW (2004) Therapeutic strategies that selectively target and disrupt established tumor vasculature. Hematol Oncol Clin North Am 18: 1023-1037, viii.

[163] Bramhall SR, Schulz J, Nemunaitis J, Brown PD, Baillet M, et al. (2002) A double-blind placebo-controlled, randomised study comparing gemcitabine and marimastat with gemcitabine and placebo as first line therapy in patients with advanced pancreatic cancer. Br J Cancer 87: 161-167.

[164] Bramhall SR, Hallissey MT, Whiting J, Scholefield J, Tierney G, et al. (2002) Marimastat as maintenance therapy for patients with advanced gastric cancer: a randomised trial. Br J Cancer 86: 1864-1870.

[165] Milkiewicz M, Ispanovic E, Doyle JL, Haas TL (2006) Regulators of angiogenesis and strategies for their therapeutic manipulation. Int J Biochem Cell Biol 38: 333-357.

[166] Satchi-Fainaro R, Mamluk R, Wang L, Short SM, Nagy JA, et al. (2005) Inhibition of vessel permeability by TNP-470 and its polymer conjugate, caplostatin. Cancer Cell 7: 251-261.

[167] Hutchins LF, Moon J, Clark JI, Thompson JA, Lange MK, et al. (2007) Evaluation of interferon alpha-2B and thalidomide in patients with disseminated malignant melanoma, phase 2, SWOG 0026. Cancer 110: 2269-2275.

[168] Marler JJ, Rubin JB, Trede NS, Connors S, Grier H, et al. (2002) Successful antiangiogenic therapy of giant cell angioblastoma with interferon alfa 2b: report of 2 cases. Pediatrics 109: E37.

[169] Ding BS, Dziubla T, Shuvaev VV, Muro S, Muzykantov VR (2006) Advanced drug delivery systems that target the vascular endothelium. Mol Interv 6: 98-112.

[170] Chang DC, Reese TS (1990) Changes in membrane structure induced by electroporation as revealed by rapid-freezing electron microscopy. Biophys J 58: 1-12.

[171] Torchilin VP (2006) Recent approaches to intracellular delivery of drugs and DNA and organelle targeting. Annu Rev Biomed Eng 8: 343-375.

[172] Peer D, Karp JM, Hong S, Farokhzad OC, Margalit R, et al. (2007) Nanocarriers as an emerging platform for cancer therapy. Nat Nanotechnol 2: 751-760.

[173] Vasir JK, Labhasetwar V (2008) Quantification of the force of nanoparticle-cell membrane interactions and its influence on intracellular trafficking of nanoparticles. Biomaterials 29: 4244-4252.

[174] Cuenca AG, Jiang H, Hochwald SN, Delano M, Cance WG, et al. (2006) Emerging implications of nanotechnology on cancer diagnostics and therapeutics. Cancer 107: 459-466.

[175] Torchilin VP (2007) Targeted pharmaceutical nanocarriers for cancer therapy and imaging. AAPS J 9: E128-147.

[176] Ferrari M (2005) Cancer nanotechnology: opportunities and challenges. Nat Rev Cancer 5: 161-171.

[177] LaVan DA, McGuire T, Langer R (2003) Small-scale systems for in vivo drug delivery. Nat Biotechnol 21: 1184-1191.

[178] Ferreira L, Karp JM, Nobre L, Langer R (2008) New opportunities: the use of nanotechnologies to manipulate and track stem cells. Cell Stem Cell 3: 136-146.

[179] Gabizon A, Papahadjopoulos D (1988) Liposome formulations with prolonged circulation time in blood and enhanced uptake by tumors. Proc Natl Acad Sci U S A 85: 6949-6953.

[180] Gabizon A, Catane R, Uziely B, Kaufman B, Safra T, et al. (1994) Prolonged circulation time and enhanced accumulation in malignant exudates of doxorubicin encapsulated in polyethylene-glycol coated liposomes. Cancer Res 54: 987-992.

[181] Simoes S, Moreira JN, Fonseca C, Duzgunes N, de Lima MC (2004) On the formulation of pH-sensitive liposomes with long circulation times. Adv Drug Deliv Rev 56: 947-965.

[182] Dromi S, Frenkel V, Luk A, Traughber B, Angstadt M, et al. (2007) Pulsed-high intensity focused ultrasound and low temperature-sensitive liposomes for enhanced targeted drug delivery and antitumor effect. Clin Cancer Res 13: 2722-2727.

[183] Meister A, Anderson ME (1983) Glutathione. Annu Rev Biochem 52: 711-760.

[184] Torchilin VP (2005) Recent advances with liposomes as pharmaceutical carriers. Nat Rev Drug Discov 4: 145-160.

[185] Vaupel P, Kallinowski F, Okunieff P (1989) Blood flow, oxygen and nutrient supply, and metabolic microenvironment of human tumors: a review. Cancer Res 49: 6449-6465.

[186] Yuba E, Harada A, Sakanishi Y, Kono K (2010) Carboxylated hyperbranched poly(glycidol)s for preparation of pH-sensitive liposomes. J Control Release.

[187] Maeda H, Sawa T, Konno T (2001) Mechanism of tumor-targeted delivery of macromolecular drugs, including the EPR effect in solid tumor and clinical overview of the prototype polymeric drug SMANCS. J Control Release 74: 47-61.

[188] Yuan F, Dellian M, Fukumura D, Leunig M, Berk DA, et al. (1995) Vascular permeability in a human tumor xenograft: molecular size dependence and cutoff size. Cancer Res 55: 3752-3756.

[189] Torchilin VP, Levchenko TS, Lukyanov AN, Khaw BA, Klibanov AL, et al. (2001) p-Nitrophenylcarbonyl-PEG-PE-liposomes: fast and simple attachment of specific ligands, including monoclonal antibodies, to distal ends of PEG chains via p-nitrophenylcarbonyl groups. Biochim Biophys Acta 1511: 397-411.

[190] Park JW, Hong K, Kirpotin DB, Colbern G, Shalaby R, et al. (2002) Anti-HER2 immunoliposomes: enhanced efficacy attributable to targeted delivery. Clin Cancer Res 8: 1172-1181.

[191] Gao J, Zhong W, He J, Li H, Zhang H, et al. (2009) Tumor-targeted PE38KDEL delivery via PEGylated anti-HER2 immunoliposomes. Int J Pharm 374: 145-152.

[192] Pan X, Wu G, Yang W, Barth RF, Tjarks W, et al. (2007) Synthesis of cetuximab-immunoliposomes via a cholesterol-based membrane anchor for targeting of EGFR. Bioconjug Chem 18: 101-108.

[193] Ruoslahti E, Bhatia SN, Sailor MJ (2010) Targeting of drugs and nanoparticles to tumors. J Cell Biol 188: 759-768.

[194] Jayanna PK, Bedi D, Gillespie JW, Deinnocentes P, Wang T, et al. (2010) Landscape phage fusion protein-mediated targeting of nanomedicines enhances their prostate tumor cell association and cytotoxic efficiency. Nanomedicine 6(4):538-546.

In: Brain Cancer, Tumor Targeting and Cervical Cancer ISBN: 978-1-61122-738-3
Editor: Elena K. Salvatti

Chapter III

Biological and Clinical Relevance of Neural Stem Cells in the Pathogenesis of Childhood Medulloblastoma

***Cornelia M. Bertram*[1], *Nicholas G. Gottardo*[1,2], *Ursula R. Kees*[1], *Kanti D. Bhoola*[3] *and Peter B. Dallas*[1]**

[1] Brain Tumour Research Program, The Telethon Institute for Child Health Research and Centre for Child Health Research, The University of Western Australia, Perth Australia

[2] Department of Paediatric Oncology and Haematology, Princess Margaret Hospital for Children, Perth, Australia

[3] Lung Institute of Western Australia, Centre for Asthma, Allergy and Respiratory Research, University of Western Australia, Sir Charles Gairdner Hospital, Perth, Australia

Abstract

Paediatric brain tumours have overtaken leukaemias as the leading cause of cancer-related death in children despite being only half as common. Medulloblastoma (MB), the most common type of malignant paediatric brain tumour, is a heterogeneous group with at least five histological and four molecular subtypes. Despite major progress in the understanding of the genetic and molecular pathogenesis of MB, relapse rate and side effects remain major concerns. The existence of multiple MB subtypes suggests that the current clinical categorisation of two risk stratification groups is inadequate and highlights the need for more tailored therapeutic strategies. The best understood subtype comprises sonic hedgehog (SHH)-dependent MB. Several components of the SHH pathway are viable drug targets that may translate to effective subtype-specific treatment options in the near future. Similarly, subtype-specific targets for clinical intervention are being investigated in other molecular variants of MB. Some MB may rely on a small population of CD133+ tumour cells termed brain tumour stem cells (BTSCs) that phenotypically and functionally resemble normal CD133+ neural stem cells (NSCs) in the developing foetal brain. The resemblance suggests that BTSCs could arise from transformed NSCs residing in the cerebellum. BTSCs are more resistant to therapy and are not effectively targeted by current treatments. New methodologies allow the generation, maintenance and manipulation of human NSCs and BTSCs, providing valuable tools and reagents for studying the mechanisms of neoplastic deregulation. The development of NSC- and BTSC-based models may be of benefit for elucidating the underlying molecular mechanisms of MB pathogenesis and facilitate transition of new knowledge into clinical setting in the future.

Abbreviations

AAV	Adeno-associated virus
AdV	Adenovirus
APC	Adenomatous polyposis coli
AT/RT	Atypical teratoid/rhabdoid tumour
bFGF	Fibroblast growth factor - beta
BMI1	B lymphoma Mo-MLV insertion region 1
BMP	Bone morphogenic protein
BTSC	Brain tumour stem cell
CAR	Coxackie-adenovirus receptor
CBF1	C-repeat binding factor 1
CNS	Central nervous system
CNS-PNET	Central nervous system-primitive neuroectodermal tumours
CSC	Cancer stem cell
CSL	CBF1, Suppressor of Hairless, Lag-1
DHH	Desert hedgehog
DLL	Delta-like
ECM	Extracellular matrix
EGF	Epidermal growth factor
EGL	External granular layer
ESC	Embryonic stem cells
FZD	Frizzled
GBM	Glioblastoma multiforme
GEMM	Genetically engineered mouse model
GFAP	Glial fibrillary acidic protein
GLI	Glioma-associated
GPC	Granule precursor cell
GSK1β	Glycogen synthase kinase 1beta
HES	Hairy enhancer of split
IGF	Insulin-like growth factor
IHH	Indian hedgehog
JAG	Jagged
MB	Medulloblastoma
MBEN	Medulloblastoma with extensive nodularity
MELK	Maternal embryonic leucine zipper kinase

MSI1	Musashi-1
MYC	v-myc myelocytomatosis viral oncogene homolog (avian)
MYCC	v-myc myelocytomatosis viral oncogene homolog C
MYCN	v-myc myelocytomatosis viral oncogene homolog N
NICD	NOTCH intracellular domain
NPC	Neural precursor cells
NSC	Neural stem cells
PAX	Paired box
PDGF	Platelet derived growth factor
PTCH	Patched
SHH	Sonic hedgehog
SMOH	Smoothened
SOX	SRY (sex determining region Y)-box
SSEA-1	Stage-specific embryonic antigen 1
SUFU	Suppressor of fused
SVZ	Subventricular zone
TSG	Tumour suppressor gene
VEGF	Vascular endothelial growth factor
vWF	von Willebrand factor
WHO	World health organisation
WNT	Wingless

Introduction

Incidence of Childhood Brain Tumours

Childhood cancer is a significant burden on the health systems of developed as well as developing countries. Childhood cancer related mortality is twice as prevalent as death from infectious diseases in the western world, and second only to infections in developing countries [1]. Central nervous system (CNS) tumours are the most common solid tumours occurring in children and the second most common of all malignancies after leukaemia [1, 2]. In 2001 in Australia, 2.9 per 100,000 children per year were newly diagnosed with brain cancer and 5.6 per 100,000 with leukaemia [3], which is similar to global incidence data [2, 4-8]. Overall, CNS tumours account for approximately 20% of paediatric neoplasms [3, 6, 8], with males being 1.3 times more likely to suffer from brain tumours than females [9]. The most common type of malignant paediatric brain tumour is medulloblastoma (MB), which occurs in the cerebellum, and accounts for one fifth of all paediatric CNS malignancies [10]. About 60% of cancers of the cerebellum develop in children under 14 years with the incidence rate peaking around the age of seven [11, 12].

Aetiology

Despite a long history of epidemiological research, the aetiology of paediatric brain tumours remains largely unknown. The evenly distributed geographic incidence rates provide little indication of genetic and environmental influences. The only hereditary syndromes linked to a higher incidence of brain tumours include Li-Fraumeni, Gorlin and Turcot syndrome, which are autosomal genetic conditions [13]. The Li-Fraumeni syndrome arises through *TP53* germline mutations or during embryogenesis, and patients with this mutation show greatly increased susceptibility to malignancies during childhood, in particular breast cancer, brain tumours and leukaemia [14]. Patients with Gorlin syndrome carry mutations in the *Patched* (*PTCH*) gene, which is associated with the development of basal cell carcinoma and other malignancies. In addition, a higher susceptibility to malignancies of the neuroepithelial tissues including gliomas and MB has been observed [15]. Similarly, Turcot syndrome patients exhibit mutations in the familial *Adenomatous Polyposis Coli* (*APC*) genes that are associated with a high risk of MB and glioma [16].

Many physical, chemical and infectious agents have been investigated for their potential to cause brain tumours after pre- and post-natal exposure. Also, low-frequency electromagnetic fields, as emitted from mobile phones, have been a great concern, but an explicit link between exposure and brain tumour pathogenesis has yet to be demonstrated [17, 18]. Small cohort studies of the effects of vitamins, maternal diet, medication, pesticides or smoking have produced contradictory results and potential associations have not yet been assessed in larger cohorts [19].

The common denominator for elevated risk of brain tumours may be derived from an insult affecting neurological development during pregnancy. This is supported by extrapolations from animal models where exposure to carcinogens during critical developmental windows led specifically to neural impairments [20, 21]. The vulnerability of neural cells to such disruption at different stages of development may be augmented by individual genetic polymorphisms. This is consistent with the idea of a multi-step process of tumorigenesis involving environmental as well as genetic factors *in utero* and postnatally.

More recent efforts have been made to connect epidemiological data such as birth weight and accelerated foetal growth with underlying mechanisms involved in childhood cancers [22, 23]. One hypothesis is that abnormal levels of regulators such as Insulin-like growth factors (IGFs) that normally contribute to general foetal growth may also be linked to developmental pathology [24-26]. The importance of IGF in brain development and brain pathology highlights the potential of epidemiological studies to help elucidate the molecular mechanisms driving brain tumorigenesis [27]. However, even though the expression level of IGF has been associated with some paediatric brain tumours and leukaemias, there was no association detected with MB [23, 28-30]. This may be a reflection of study limitations due to the relatively small patient numbers, the heterogeneity of this tumour group and the need for better parameters for measuring embryonal development. Nevertheless, epidemiological studies that assess environmental and genetic factors may be valuable in the future in order to increase our understanding of susceptibility to brain tumour development.

Brain Tumour Classification an Historical Perspective

Originally, CNS tumours were named after the cell type they most resembled histologically. In neuroepithelial tissue, the main cell types and corresponding cancers are glial cells (gliomas), oligodendrocytes (oligodendrocytomas), neurons (gangiogliomas) or ependymal cells (ependymomas) [31]. However, tumours which could not be readily classified according to this system, either due to the lack of lineage specific morphology or their relatively undifferentiated phenotype, were grouped together and categorised as "embryonal" tumours. These tumours were generally referred to as central nervous system primitive neuroectodermal tumours (CNS-PNETs). At the time, CNS-PNETs were defined as tumours consisting of cells that resemble the primitive neuroepithelial precursor cell types present during embryonal brain development [32]. Common features of CNS-PNETs included patternless sheets of blast-like cells with high nuclear-cytoplasmic ratio and high mitotic count [33]. CNS-PNETs could display a variety of neuronal, glial or epithelial lineage-specific protein expression patterns but lacked mature lineage-specific morphology.

In 1983, Rorke [34] proposed that a variety of high-grade undifferentiated tumours of the neuroepithelium from different regions of the CNS should be classified as CNS-PNETs based on the assumption that they were derived from the same early neuroepithelial cell type. This cell was assumed to be a neural progenitor cell (NPC) with multi-lineage differentiation potential derived from the subependymal matrix located in the floor of the lateral ventricle, above the caudate nucleus and the caudothalamic groove. The histological variation of subtypes was thought to be partially derived from the distinct underlying patterns of genetic deregulation [34-36]. This concept caused controversy amongst pathologists and clinicians, who could not agree on one category for such a heterogeneous population of tumours and found such a generalised classification scheme a limitation to the elucidation of their molecular pathogenesis [37]. Consequently, it was proposed that CNS-PNETs located in the cerebellum be termed MB, representing embryonal tumours unique to this region. In contrast, CNS-PNETs with similar histological features located in other regions of the CNS should be named differently. Thus, tumours occurring in the cerebrum were termed cerebral CNS-PNETs, whereas tumours located in the pineal region were called pineoblastomas. Interestingly, the latter were removed completely from the embryonal tumours and assigned to a separate category - tumours of the pineal region - despite the histological resemblance to MB and high prevalence in children [38, 39].

This system became widely accepted as MB, pineal and cerebral PNETs exhibit different immunohistochemical features and clinical behaviour [40, 41]. Subsequently, improved cytogenetic and molecular methods have identified expression signatures exclusive to subgroups of embryonal tumours from different locations which facilitated the development of a more refined subclassification scheme for MB variants [41-43]. Since then, new insights into genetic signatures suggest the existence of at least four or five molecular subgroups of MB [44-46] that will require consideration for the next edition of the World Health Organisation (WHO) classification of brain tumours. The ongoing periodic adjustment of brain tumour classification schemes highlights the importance of re-evaluating the existing paradigms on a regular basis [31, 38, 41] as additional data will facilitate the gradual clarification of the relationships between MB subtypes and other embryonal tumours.

According to the current WHO classification scheme, embryonal tumours of the CNS include the three main clinical subgroups MB, CNS-PNETs, and Atypical teratoid/rhabdoid tumours (AT/RTs) [31]. MB are classified as tumours of the cerebellum with signs of neuronal differentiation that can be further sub-classified according to their cyto-architectural appearance (Fig. 1). In contrast, CNS-PNETs mainly develop in the cerebrum, contain almost exclusively poorly differentiated neuroepithelial cells and are particularly aggressive. They include supratentorial CNS-PNETs (CNS neuroblastomas and CNS ganglioblastomas), medulloepitheliomas and ependymoblastomas [47]. AT/RTs are slightly more common than CNS-PNETS and generally consist of rhabdoid cells with or without areas of epithelial tissue, neoplastic mesenchyme and primitive morphology. A robust system for distinguishing AT/RTs, from other embryonal tumours was suggested in 2002 based on the loss, deletion or mutation of the tumour suppressor gene (TSG) *hSNF5/INI1* on chromosome 22q11.23 [48-51]. Mutations affecting *hSNF5/INI1* are present in over 90% of AT/RTs clearly delineating them from other embryonal tumours.

Embryonal tumours

Medulloblastoma
Classic medulloblastoma
Desmoplastic/nodular medulloblastoma
Medulloblastoma with extensive nodularity
Anaplastic medulloblastoma
Large cell medulloblastoma

CNS-primitive neuroectodermal tumours
CNS Neuroblastoma
CNS Ganglioblastoma
Medulloepithelioma
Ependymoblastoma

Atyical teratoid/rhabdoid tumour

Figure 1. Classification of embryonal tumours according to the 2007 WHO.

Current Therapy for Medulloblastoma

Children diagnosed with MB receive risk-adapted therapy, based on clinical characteristics and histopathological features at presentation [52]. Paediatric brain tumours are treated first with surgery, in order to provide a histological diagnosis and at the same time reduce the tumour burden [4]. Depending on the extent of surgical resection (gross total or near-total vs. sub-total), the presence of leptomeningeal dissemination and a histological diagnosis of anaplastic MB, children with MB are classified into average (also known as standard) risk or high risk treatment groups, based on the risk of treatment failure. Patients with gross total or near-total resection and without signs of tumour dissemination are classified as average risk, whilst the remaining patients are classified as high risk. More recently, some cooperative groups have expanded this classification to include the anaplastic MB variant in the high risk group, based on evidence revealing a worse outcome for these patients [41, 53]. Infants and young children under the age of three years are also considered high risk, but due to the unacceptable long-term effects of radiotherapy on the developing CNS, they are managed on separate therapeutic regimes which exclude radiotherapy [54]. Over the past four decades substantial improvements in survival have been achieved for children with MB, with cure rates exceeding 80% for average risk patients and up to 70% for high-risk patients in some centres [55, 56]. However, patients with MB presenting with macro-metastatic disease continue to have poor outcomes, despite contemporary treatment approaches [57-59]. Moreover, effective regimens remain elusive for the more than 30% of children with relapsed or refractory MB, with little chance of cure for these patients [10]. Of particular concern is that survivors are frequently left with a multitude of significant long-term sequelae, including impaired neurocognitive, neuroendocrine, and social functioning, auditory deficits and the development of second malignancies [4, 60-62], which adversely impact on quality of life.

Current therapy for MB consists of a multimodal approach utilising maximal safe surgical resection and chemotherapy +/- radiotherapy. Since it was introduced, radiotherapy has formed the backbone of therapy and was responsible for increasing overall 5-year survival from zero in the 1950's to approximately 53% in the 1970's [63]. Chemotherapy was introduced in the 1980's and resulted in additional improvements, with survival rates reaching 70% [40]. Since then further improvements in outcome have been made through refinements in all three treatment modalities [57, 64]. Notably, for patients with average risk disease, chemotherapy has also permitted reduction in the dose of cranio-spinal radiotherapy required [65]. Infants and young children under the age of three years are generally treated with intensive chemotherapeutic strategies [66], some of which incorporate myeloablative chemotherapy with autologous stem cell rescue [67], with the goal of avoiding or delaying the use of radiotherapy. Overall, patients in this group have a poor prognosis, most likely related to therapeutic restrictions, as well as tumour-specific biological differences. However, recent results have demonstrated significant improvements in survival for distinct groups of patients, namely those with completely resected tumours without macrometastatic disease and those with nodular desmoplastic histology [68].

Whilst significant progress has been made in survival, brain tumours have now surpassed leukaemia as the leading cause of cancer-related death in children, despite being only half as common [11]. This highlights the shortcomings of current non-specific cytotoxic therapies. Additionally, individual responses to treatment remain difficult to predict due the inadequacy of clinical characteristics alone to accurately risk stratify patients. This is underscored by the fact that, despite contemporary treatment approaches, 15 to 20% of patients currently classified as average risk still relapse. Recent advances in genomics have greatly expanded our knowledge of MB biology. High-resolution genome-wide profiling has demonstrated that histologically similar tumours harbour distinct molecular signatures [44-46], which can be both diagnostic and prognostic. Molecular markers have been in clinical use for treatment stratification in other childhood cancers, such as acute lymphoblastic leukaemia [69]. Recently, it has been shown that patients with MB whose tumours demonstrated high nuclear β-catenin immunoreactivity had a more favourable prognosis [70]. This finding was verified by other independent groups [55]. Such molecular features will be incorporated into future clinical trials to allow more accurate risk stratification, with the ultimate aim to increase cure rates whilst preserving the long-term function and quality of life of survivors.

The focus of this chapter is to highlight recent developments in MB and neural stem cell biology and how the integration of this knowledge may be applied to improve the clinical management of children with MB.

Neurogenesis and Cerebellar Development

There has been a major focus on the role of neural stem cells (NSCs) in the development of brain malignancies [71-74]. The parallel study and integration of neural development and MB tumorigenesis may reveal important clues about the mechanisms involved in MB pathogenesis.

Neurogenesis in Humans

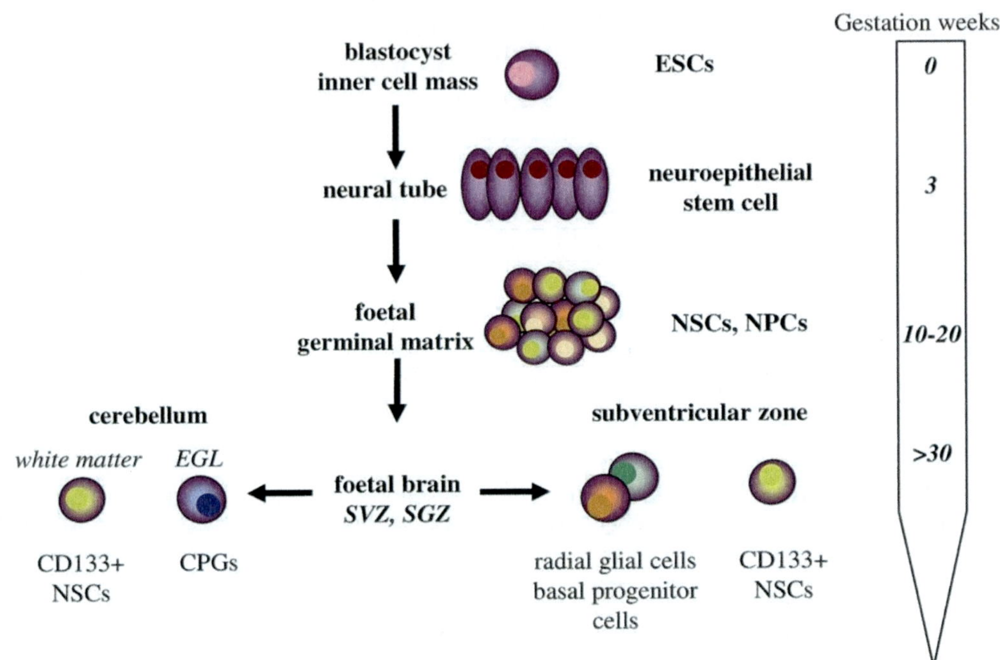

Figure 2. Stem cell populations during human embryonic and foetal brain development. ESCs give rise to neuroepithelial stem cells that form the neural tube. During early gestation a variety of NSC and NPC layers are produced that form the foetal germinal matrix and migrate to form compartments of the brain. During weeks 10-20, the SVZ and SGZ give rise to migrating cells that form the cerebellum and populate the EGL. The EGL contains GPCs and the white matter harbours multipotent CD133+ NSCs. The EGL remains an area of neurogenesis until 12 months after birth, whereas the SVZ and SGZ contain radial glial cells, basal progenitor cells and also CD133+ NSCs that remain active into adulthood. *ESCs = embryonic stem cells, NSC = neural stem cell, NPC = neural progenitor cell, SVZ = subventricular zone, SGZ = subgranular zone, EGL = external granular layer, GPCs = granule precursor cells*

NSCs are one of the earliest tissue-specific stem cells that derive directly from embryonic stem cells (ESCs) present in the inner cell mass of the blastocyst during the first weeks of pregnancy [75] (Fig. 2). The formation of the neural plate from early NSCs in the third week of embryogenesis marks the beginning of neurogenesis. Subsequently, the flat single layer neural plate curls into the neural tube, forming the portion that goes on to form the brain and the spinal cord. The single neuroepithelial layer diversifies into multiple layers of neural cells that expand to form the compartments of the brain by the fifth week of pregnancy (reviewed by Götz and Huttner in 2005 [76]). While neuroepithelial stem cells diversify, they give rise to neural crest cells responsible for the generation of the peripheral nervous system, and several distinct neuroepithelial layers that develop into the fore, mid and hind brain (reviewed by Crane and Trainor in 2006 [77]). At the beginning of week nine the cerebellar plate begins to form and the forebrain region expands to establish the foetal cerebral hemisphere surrounding the lateral ventricles. The ventricular zone, also called the foetal germinal matrix, is the innermost layer of the foetal cerebral hemisphere [78]. The germinal matrix is rich in extracellular matrix (ECM) and densely packed NSCs, which are the source of migrating neuroectodermal cells that generate the neuronal and non-neuronal brain parenchyma [79]. The germinal matrix thickens until the 20th week of gestation, and then begins to regress into the thin subventricular zone (SVZ) lining seen postnatally. Besides residual multipotent neuroepithelial stem cells in this layer, other cell types generated from the neuroepithelium are radial glial and basal progenitor cells. Radial glial cells act as NSCs by generating neurons, astrocytes and oligodendrocytes during brain development [80-82], whereas basal progenitor cells are important contributors to NPC production occurring later in the SVZ [83].

Cerebellar Development

The cerebellum develops around the ninth week of pregnancy from the cerebellar plate in the embryonal hind brain [84]. It is thought that two regions give rise to the cells of the cerebellum. From the ventricular zone, deep nuclear neurons and Purkinje cells migrate to populate the surface of the cerebellum forming the cerebellar cortex [85, 86]. This is followed by migration of granule precursor cells (GPCs), recruited from the rhombic lip during week 11 that form the external granular layer (EGL) until the 27th week of gestation [84]. GPCs undergo significant expansion, initiated by the sonic hedgehog (SHH) secreting Purkinje cells [87]. Following the expansion, GPCs exit the cell cycle and move to the inner zone of the EGL, where they differentiate into immature granule neurons. During the maturation process, granule neurons continue migrating inward and become functional neurons when reaching the internal granular layer of the cerebellum. This process continues up to 12 months postnatally during which the EGL reduces in thickness [85, 87] until it disappears following from the completion of cerebellar development [86]. The length of cerebellar development from early gestation into the postnatal period highlights the vulnerability of the cerebellum to developmental deregulation that could lead to MB.

Histology and Cytogenetics of Medulloblastoma

MB are highly malignant tumours consisting of densely packed cells with large round nuclei. They mostly express neuronal markers, but also show the occasional astrocytic niche positive for glial fibrillary acidic protein (GFAP) [33, 37]. MB have been grouped into five histological sub-types including classic, anaplastic, large cell, desmoplastic/nodular MB and MB with extensive nodularity (MBEN) [88]. Most MB cases fall into the group of classic MB. They are characterised by small round cells with high nuclear-to-cytoplasmic ratio and few signs of morphological differentiation (Fig. 3A). The cells are arranged in sheets that are occasionally interrupted by architectural features such as Homer-Wright rosettes or parallel arrangements resembling a picket fence [89]. The desmoplastic/nodular variant accounts for 15% of cases and is less aggressive in its clinical progression [90]. Desmoplastic means "band forming" and refers to the sheets of tumour cells interrupted by nodular islands [33] (Fig. 3B). The grade of desmoplasia is subjectively defined by the extent of nodular and inter-nodular regions containing areas of cell proliferation, apoptosis and differentiation. Neuronal differentiation is generally greater in the nodular regions, whereas signs of astrocytic differentiation are more commonly detected outside of nodules [41]. At one extreme of the desmoplasia spectrum is the MBEN with extensive features of differentiation. These tumours generally present in children under the age of three and are associated with a more favourable outcome [33]. Evidence for the correlation of the loss of 9q, excessive SHH signalling and desmoplasia suggest a distinct underlying molecular pathogenesis in these tumours, supporting their separate classification [91]. In contrast, large cell and anaplastic MB lack morphological signs of differentiation and present as a monotonous population of cells with large round and/or pleomorphic nuclei and prominent nucleoli [33, 41, 42, 53, 92]. Large cell MB may show overlapping features of anaplasia such as nuclear moulding and cell-cell wrapping but are generally more monomorphic than other variants (Fig. 3C). Both types of MB make up a relatively small proportion of cases but are highly aggressive with early dissemination and poor prognosis despite multi-modality treatment [41, 53, 92].

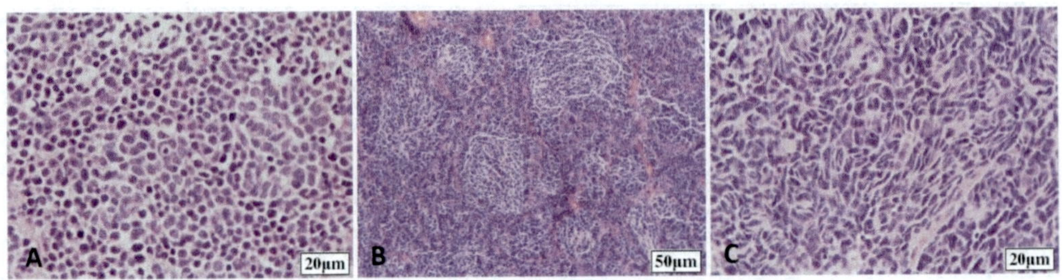

Figure 3. Histology of MB variants. Histopathological features of A: classic MB showing typical arrangement of sheets of undifferentiated tumour cells, B: desmoplastic/ nodular MB with pale nodular areas surrounded by densely packed tumour cells and C: anaplastic MB with prominent cell-cell wrapping of tumour cells. (images kindly provided by Dr. Reimar C Junckerstorff, Section of Neuropathology, PathWest Laboratory Medicine, Royal Perth Hospital, Western Australia) *MB = medulloblastoma.*

MB cells generally exhibit multiple numerical and structural changes affecting different chromosomes randomly and non-randomly [40, 91, 93]. The variety of karyotypic abnormalities spans the entire genome, involving most commonly, chromosomes 17p (23-38%), 1 (19%), 16q (16-17%), 8p (15-22%), 10q (21%), and 11q (11-16%), and others at lower frequency [94-96]. The most frequent aberration is iso-chromosome 17 (i17q), which is defined by the loss of the short arm 17p and the gain of a second long arm 17q [33, 91, 93, 97-99]. The increased occurrence of this mutation may be the result of at least four breakpoint cluster regions on 17p, harbouring TSGs such as *P53* and the *Hyper-methylated in cancer-1* gene [98]. Tumours with chromosome 17 aberrations are mostly classic MB that carry a poorer prognosis than histologically identical variants with an intact chromosome 17 [100]. The second most common chromosomal aberration observed in MB is rearrangement of chromosome 1 [93, 99]. Here, the exact breakpoints remain unknown, and associations with putative TSGs have not been identified so far. Other aberrations include the frequent loss of chromosome 16q and 9q, both harbouring genes encoding elements of the SHH signalling pathway [93, 98]. Deregulation of this pathway is highly consequential for stem cell fate and has been associated with a higher risk of MB development in patients with Gorlin syndrome [101]. In contrast, the loss of chromosome 6 has been associated with the activation of the wingless (WNT) signalling pathway and is mainly observed in classic variants with a more favourable outcome and an intact chromosome 17 [46, 70, 102].

Increased *v-myc myelocytomatosis viral oncogene homolog* (*MYC) C* and *MYCN* expression observed in other cancers has also been described in some MB, CNS-PNETs and pineoblastomas [91, 103-106]. Overexpression derived from amplification of the *MYCC* oncogene is predominantly associated with the large cell phenotype of MB and a poorer outcome [70, 95]. However, overexpression in multiple aggressive cancer types and the association of *MYCC* deregulation with growth acceleration in MB cell lines [107] argues that *MYCC* overexpression may promote tumour progression in general rather than MB initiation in particular. In contrast, in a mouse model *MYCN* overexpression alone initiated MB with classic or large cell characteristics at high penetrance [108], which argues for a specific role of *MYCN* in the initiation of these tumour subtypes.

Histogenesis of Medulloblastoma

MB contain undifferentiated cells resembling stem or progenitor cells in their morphology in addition to cells that show signs of neuronal and glial differentiation. The mixture of undifferentiated and more mature neural cell types in MB and other embryonal tumours led to the assumption that they were derived from NPCs of the CNS. This ignited much speculation about NSCs or NPCs as the cell of origin for different types of embryonal tumours [34, 37]. Subsequently, researchers have attempted to precisely define cell populations in the cerebellum and other regions of the brain to elucidate their potential role in tumorigenesis. With the current understanding of NSC and MB cell differentiation potential, two important questions have arisen. Firstly, are NSCs and NPCs the only potential cell of origin of MB? Secondly, are different MB variants derived from different types of NSCs or NPCs?

The cell of origin of MB was long assumed to be a bi-lineage medulloblast found on the cerebellar surface (reviewed by Fan and Eberhardt 2008 [109]). Besides stem cells that possess multi-lineage potential, the awareness that differentiated cells can regain such potential [110-113] alluded to the fact that committed cells could also be candidates for the cell of origin for MB. However, since no evidence exists to date that relates dedifferentiation to the tumorigenesis of MB, the search for the cell of origin has been increasingly focused on several types of stem and precursor cells residing in the cerebellum. In more recent reports, firm evidence has been presented that NSCs and NPCs in the cerebellum can generate at least one subtype of MB in cell-specific conditional knockout mouse models [114, 115]. More importantly, these studies have shown that the same MB variant can be generated from at least two different precursor cell populations *via* the same molecular mechanism. In addition, the accelerated growth of MB generated in early NSCs in comparison to NPCs suggests that clinical variation of molecularly similar MB may be derived at least in part from a multicellular origin of these tumours. Consequently, the present evidence argues that at least for SHH-driven MB, NSC and NPC populations appear to be the most likely candidates for the cell of origin.

If two different cell types can generate one type of MB, the question arises as to whether different MB variants are derived from different cells of origin and/or through different molecular mechanisms. On the one hand, the different underlying expression signatures of MB subgroups supports the concept that separate molecular variants may arise through different mechanisms despite their similar morphology [34, 44, 46, 116, 117]. On the other hand, different molecular signatures may be an important clue pointing to the potential cell of origin. This is reflected in MB that exhibit signatures resembling cells of the SVZ and the EGL. The SVZ harbours NSCs and gives rise to neuronal and glial cells throughout life [118-121], whereas the EGL contains precursor cells that produce the cerebellar granule neurons mainly in the third trimester of pregnancy until one year postnatally [122]. Some of these precursor cell populations have been linked to MB in mouse models [114, 115], but evidence for such a connection in human MB remains speculative at present. MB located in the EGL region mostly exhibit desmoplastic features and express p75, a marker specifically found in the EGL and not in the SVZ [123, 124]. In contrast, most classic MB arise in the cerebellar vermis and express calbindin, similar to cell populations known to reside in the SVZ and not in the EGL [125-127].

Further support for the latter multi-origin model is derived from molecular studies, in particular through the association of human EGL-derived MB with SHH deregulation, a molecular proliferation program that acts specifically on immature NPCs in the EGL [44, 46]. In contrast, some classic MB exhibit excessive WNT signalling, a pathway crucial for the proliferation of stem and progenitor cells in the foetal and postnatal SVZ, hippocampus [128-131] and regions of the cerebellum outside the EGL [132-134]. This approach divides MB into at least two histogenetic groups with distinct cells of origin. However, the clinical and molecular heterogeneity of MB is more diverse [44] and with novel neural precursor populations being discovered [135-138] it is possible that different cells of origin may be linked to other MB variants.

Molecular Pathogenesis of Medulloblastoma

As described in section 0, several hereditary syndromes predispose to the development of brain tumours including MB [125]. The underlying genetic defects in these syndromes are frequently associated with important regulators of brain development and NSC biology. Critical developmental pathways such as SHH, WNT and NOTCH which regulate the self-renewal and proliferation of NSCs and have been shown to be aberrantly activated in MB (reviewed by de Bont et al. 2008 [125]).

WNT Signalling

The WNT signalling pathway is a regulator of NSC proliferation and differentiation critical during embryonal development [139]. WNT proteins can activate several downstream signalling cascades, of which the canonical cascade through β-catenin activation is currently the best understood [140, 141]. In the canonical pathway, WNT signalling proteins bind frizzled receptors (FZD) on the cell surface and initiate the downstream release of the transcription factor β-catenin, which activates the transcription of target genes such as *cyclins, MYCC* and others. *APC* is a negative regulator of WNT signalling that forms a complex with glycogen synthase kinase (GSK) 1β and AXIN to bind β-catenin and inhibit its translocation to the nucleus [142, 143]. The selectivity of downstream target activation is not well understood, but the evidence suggests a stage and tissue-dependent regulation of different downstream gene sets [144]. In general, WNT signalling activation promotes cell proliferation, tissue expansion and differentiation.

During development, *APC* expression is elevated in the pre-migratory regions of the cerebellum where MB arise [98]. In patients with Turcot syndrome, the *APC* gene is mutated and cannot perform its inhibitory function in developing tissues, which leads to constitutive WNT pathway activation [99]. This manifests itself most commonly in the development of intestinal carcinoma and less frequently glioblastoma multiforme (GBM) and MB. Besides *APC* mutations, the absence of the other inhibitory complex proteins GSK1β, AXIN, as well as the overexpression of β-catenin and other pathway components can lead to similar outcomes. WNT pathway activation converges on the downstream overexpression of β-catenin [46, 70, 116, 145, 146] that can be used to identify the 13-25% of WNT activated sporadic MB [147]. WNT pathway activation appears to be concurrent with the loss of regions of chromosome 6 and mutually exclusive to aberrations on chromosome 17 [102]. This suggests that chromosome 6 may harbour TSGs involved in suppression of WNT signalling and also that the mechanism is distinct from MB with chromosome 17 aberrations. Since patients with losses on chromosome 6 have a more favourable outcome and both groups are indistinguishable by morphology, the WNT signalling status may provide a parameter to identify candidates suitable for therapy reduction in the future [70].

Sonic Hedgehog Signalling

SHH signalling plays an important role in cerebellar development and neurogenesis [148, 149]. *Sonic* (*SHH*), *Desert* (*DHH*) and *Indian hedgehog* (*IHH*) are vertebrate orthologues of the pathway activator *Shh* originally discovered in *Drosophila* that is strictly regulated to allow organ-specific developmental functions. SHH is also one of the key mediators secreted by Purkinje cells in the cerebellum that act on GPCs to initiate proliferation and expansion [89, 137, 150]. SHH signalling molecules bind to PTCH receptors on the cell surface and initiate the dissociation of the transmembrane mediator smoothened (SMOH) [148]. SMOH activates Glioma-associated oncogene (GLI) transcription factors that translocate to the nucleus and activate downstream gene expression. GLI factors can be bound and transported out of the nucleus by human Suppressor of Fused (SUFU), which negatively regulates SHH signalling. SUFU is essential in mammalian brain development and loss of expression has been associated with childhood MB, initiated through excessive SHH signalling [151]. Aberrant activation of the SHH pathway can also be initiated by loss of PTCH or increased expression of SMOH, GLI and SHH [152]. Studies of sporadic MB showed that about 8-10% of cases carry *PTCH* mutations most often associated with a desmoplastic phenotype [33, 98, 99, 150]. Besides *PTCH1* (9q22.3) and *PTCH2* (1p33-1p34), other genes encoding SHH pathway components such as *SHH* (7q36), *IHH* (2q35), *GLI1* (12q13), *GLI2* (2q14), *GLI3*

(7p13), *SUFU* (10q24) and *SMOH* (7q32.3) are spread throughout the genome and coincide with some areas of previously identified chromosomal aberrations. This suggests that besides the most frequent *PTCH* receptor mutations [153] other mutations could potentially lead to a similar SHH-activated MB phenotype [154]. Array based expression studies have generally supported the concept of multiple components impacting on SHH signalling deregulation especially for the desmoplastic subtype of MB in humans [43, 46]. In addition, mouse models demonstrated that targeted mutation of *PTCH* and *SMOH* in cerebellar cells can cause MB [114, 115], which can be treated by SMOH inhibitors GDC-0449 and HhAntag [155, 156]. Following from these successes, the first clinical trials are now in progress to assess efficacy of the SHH-inhibitor GDC-0449 in patients with SHH-dependent MB (http:// clinicaltrials. gov/ct2/show/NCT00822458).

NOTCH Signalling

The NOTCH signalling pathway regulates cell proliferation and differentiation of NSCs and NPCs. The NOTCH transmembrane bound receptors are activated by ligands Jagged (JAG) 1 and 2, and Delta-like (DLL) 1, 3 and 4 through cell-cell contact. The receptors NOTCH 1-4 are heterodimers that when activated release the NOTCH intracellular domain (NICD). NICD translocates to the nucleus where it activates C-repeat binding factor 1 (CBF1), Suppressor of Hairless and Lag-1 (CSL) and other transcription factors (reviewed in detail by Koch and Radtke 2007 [157]). Known NOTCH downstream targets are genes that promote proliferation of GPCs in the cerebellum during CNS development [158]. The role of NOTCH signalling in the expansion of the GPC pool has led to the idea that pathway deregulation may be oncogenic in MB. However, the outcome of NOTCH signalling is variable and depends on the cellular context [157, 159, 160]. Fan and colleagues [161] demonstrated that activation of NOTCH1 reduced, whereas activation of NOTCH2 increased the proliferation of MB cells as well as NPCs during embryonal development. Consistent with these observations, *NOTCH2* was reported to be overexpressed in a group of MB [162], which led to the speculation that overexpression may be induced by SHH activation through shared pathway components [163-165]. However, recent comprehensive studies proved that NOTCH signalling does not impact on initiation, engraftment, or maintenance of SHH-driven MB [166, 167]. Consequently, it may rule out NOTCH signalling in SHH-dependent MB pathogenesis, though whether this also applies to human and other MB variants remains controversial.

From Genes to Genomes - Molecular Signatures

The emergence of genome-wide molecular profiling technology, including microarray expression analysis, has facilitated rapid progress in comprehensive molecular studies. The detection of point mutations, copy number variations, quantitative snapshots of global gene expression, and the detailed investigation of other regulatory elements are now possible, and an enormous amount of genetic data have been generated. Using these databases, any group of samples can be investigated for links with deregulated pathways or novel groups of genes associated with criteria such as histology, relapse or clinical prognosis. Since histological

Table 1. Microarray analyses of MB and related brain tumours

year	sample type (no.)	control type	platform (no. of probe sets)	outcome	Ref.
2010	MB (103)	-	Affymetrix Human 1.0 exon array (1.4 million)	Four molecular subgroups were identified with distinct demographics, clinical presentation, transcriptional profiles, genetic abnormalities, and clinical outcome MB variants: WNT (*DKK1+*), SHH (*SFRP1+*), group C (*NPR3+*), and group D (*KCNA1+*) SHH, C and D variants overexpress *MYCN*; WNT and C variants overexpress *MYCC*	[45]
2009	MB (40)	-	Affymetrix U133plus2.0 (54,675)	MB with activation of the WNT/β-catenin pathway represent a distinct molecular subgroup WNT/β-catenin activated cases carry a more favorable outcome Clinical diagnosis of WNT/β-catenin cases using β-catenin expression or mutation, and loss of heterozygosity of chromosome 6	[116]
2008	MB (62)	-	Affymetrix U133plus2.0 (54,675)	MB were grouped into five molecular sub-types (A-E) Variants were characterized by WNT and TGFβ signalling (A) , SHH activation (B), neuronal differentiation (C+D) and photoreceptor differentiation (D+E)	[44]
2007	MB (27), ependymoma (13)	adult cerebellum	Affymetrix U133plus2.0 (54,675)	Most discriminative genes for MB in comparison to the normal cerebellum included genes involved in cell cycle regulation, replication and response to stimuli Discriminating genes overexpressed in MB but not ependymoma included *SOX4, SOX11*	[147]
2006	MB (46)	-	Affymetrix U133av2 (22,212)	Five distinct molecular subgroups associated with distinct overexpression of gene sets One group contained WNT-activtaed MB , SHH-dependent MB contained mainly desmoplastic variants, and group C contained mainly classic MB with 17q alterations	[46]
2004	MB (35)	-	mitosis, oncogenesis, apoptosis DNA spot arrays (2,600)	54 genes predicted poor outcome, including *MYC, stathmin1* and *cyclinD1*	[169]
2004	MB (10)	mouse cerebellum (P1-P60)	HuGeneFl array (Pomeroy data set) Affymetrix Mu11K (for mouse cerebellum) (5,920)	Human MB resembled early postnatal mouse cerebellum to (P1-P10) Metastatic MB most closely mirrored postnatal week five mouse cerebellum (P5)	[117]
year	sample type (no.)	control type	platform (no. of probe sets)	outcome	Ref.
2003	MB (6)	adult cerebellum	Atlas 1.2 human cancer array (1,200)	Deregulated candidate genes included *cyclinD2, TRK-B, FGFR*	[178]
2003	MB (20)	paediatric and adult cortex/ cerebellum	SAGE (236,718 transcripts)	Eight tags were consistently higher expressed in MB in comparison to normal brain. The tagged regions included genes such *MYCN, thymosin-β,CD24, TOP2A and PRAME*	[179]
2003	ptc-/- driven mouse models	mutant cerebellum, wildtype cerebellum	MG_U47Av2 (3,119)	Expression patterns of MB derived from different mouse models were similar. MB profiles were similar to developing cerebellum (post-natal <3 weeks). Up-regulated genes included *PTCH2, MATH1*	[173]

Table 1. (Continued)

year	sample type (no.)	control type	platform (no. of probe sets)	outcome	Ref.
2002	MB (10), GBM (10), rhabdoid BTs (10), CNS-PNETs (8)	normal cerebellum	HuGeneFl array (5,920)	MB were molecularly distinct from CNS-PNETs and other paediatric tumours. MB may derive from cerebellar granule cells through SHH activation. Clinical outcome can be predicted on basis of gene expression profile.	[43]
2001	MB (23)	-	Affymetrix G110 Cancer Array (1,992)	identified 85 genes predictive for metastatic MB. *PDGFRA* and *RAS* overexpression was associated with metastases	[168]

features of MB have generally proven unreliable for prognosis, microarray technology represents a powerful approach for identifying molecular markers with a more accurate prediction power [43, 168, 169]. Using this approach, early MB microarray studies revealed a large number of novel outcome predictors with unknown functions [168, 170]. Identified expression signatures included genes important in angiogenesis and cell proliferation that were useful for the prediction of patient survival or disease dissemination.

Microarray technology has also provided valuable insights into the molecular signatures of different MB subgroups in comparison to other paediatric brain tumours (Tab.1). Instead of tumour histology, the idea was to define molecular subtypes based solely on gene expression features as was successfully achieved in leukaemia [171]. These strategies involved the search for common expression patterns that may be derived from deregulated pathways distinctive to each tumour type. Utilising this strategy, MB could be distinguished molecularly from other morphologically similar embryonal tumours CNS-PNETs, AT/RTs as well as from GBM [43]. Further, combining of expression profiling, cytogenetics, and mutation analysis identified at least five distinct molecular classes of MB, highlighting the complexity of this disease [44, 46, 116]. Two molecular subtypes were associated with the activation of WNT and SHH signalling respectively, and a third group was marked by aberrations affecting chromosome 17 and mainly classic histology. The remaining two groups showed clear associations with hyperactive early neuronal and retinal differentiation expression patterns [44].

In parallel, less malignant brain tumour types appeared to resemble their normal tissue counterpart more closely, whereas more aggressive types resembled local undifferentiated progenitor cells [170, 172-175]. A similar investigation in GBM revealed that tumours mimic parallel stages of forebrain development, each arrested at a distinct stage [170]. By analogy, other cancers show an overlap with early developmental stages of their corresponding tissues [176, 177]. MB, at the molecular level, resembles the developing cerebellum more than the mature cerebellum, as shown by projection of their expression profiles onto those of mouse cerebellar development [117]. This suggests that human MB expression profiles mirror those of early stages of cerebellar development, strengthening the link between MB tumorigenesis and deregulated neurodevelopmental pathways.

Medulloblastoma and the Cancer Stem Cell Theory

Two fundamentally different models have been proposed to explain the cellular heterogeneity observed in diverse solid tumours, including MB. The first, referred to as the stochastic model, is based on the theory that different cancer cell phenotypes are derived from the original cancer-initiating cell through ongoing mutation and selection pressure (Fig. 4). Thereafter, tumorigenicity is conserved in all cells, which means that in this scenario all cells have to be targeted to eliminate the cancer. In contrast, the cancer stem cell (CSC) model describes the tumour bulk as a hierarchical mix of cell types containing a subset of cancer cells with stem cell characteristics that are responsible for driving tumour growth (Fig. 4B). These CSCs give rise to progeny with limited proliferation and tumorigenic potential that make up the majority of the tumour bulk. As CSCs hold all the tumorigenic potential, their response to therapy determines the overall treatment outcome.

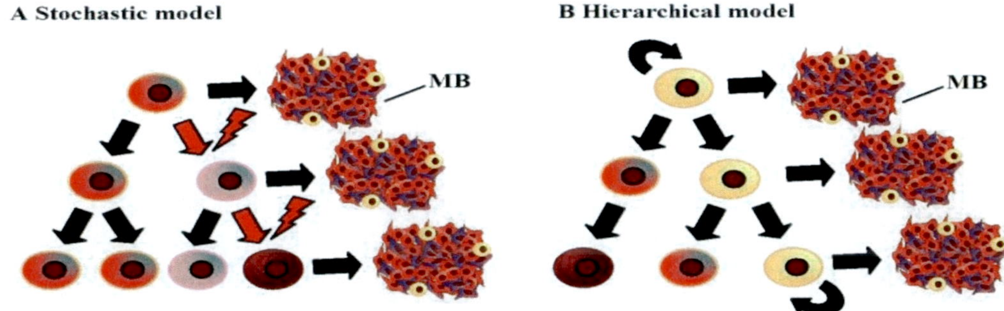

Figure 4. Tumour progression models. A: The stochastic model explains tumour heterogeneity by the emergence of many clones through additional mutations (red arrow), but all clones (red, purple, burgundy) are tumorigenic. B: The hierarchical model assigns tumorigenic properties only to a small subpopulation of CSCs (yellow) that sustain tumour growth. (adapted from Vescovi *et al.* 2006 [180]) *CSCs = cancer stem cells, MB = medulloblastoma.*

Although the CSC theory was initially proposed over 50 years ago [181, 182], it is only relatively recently that reagents and technology have been developed to further investigate this hypothesis. The availability of extensive lineage maps and highly specific antibodies for studying the haematopoetic system facilitated the detailed investigation of the CSC theory in leukaemias in the 1990s [183, 184]. Using hematopoietic stem cell markers, distinct cells were isolated from human acute myeloid leukaemias (AML) that were able to regenerate the cancer hierarchy in immunocompromised mice. Increasing knowledge about cell hierarchies in other tissues and advances in flow cytometry have enabled the isolation of putative CSC populations from solid tumours [180, 185]. Breast cancer was the first solid tumour type from which isolated stem cell-like cells demonstrated tumorigenic capacity and were able to regenerate the cell hierarchy of the original tumour [186]. Since the principles of the CSC theory are consistent with many aspects of brain tumour biology, increasing effort has focused on clarifying the role of CSCs in brain tumour pathogenesis. Firstly, as discussed, embryonal brain tumours such as MB, exhibit an undifferentiated phenotype similar to those of NPCs and NSCs. Secondly, MB establish very early during childhood, throughout the developmental stage of the brain associated with high proliferative activity of NSCs and

NPCs. Lastly, some sporadic MB cells show excessive signalling of important stem cell regulatory pathways such as SHH, WNT and NOTCH and deregulation of some of these pathways has also been linked to hereditary syndromes causing MB. The latter observations suggest that MB cells have characteristics of CSCs and may arise from normal NSCs present during brain development.

Brain Tumour Stem Cells

Putative brain tumour stem cells (BTSCs) were initially isolated using techniques adapted from the field of developmental neuroscience on the basis of phenotypic similarity to NSCs and the ability of these cells to be cultured under similar conditions [187]. One of the crucial approaches was the selective growth of BTSCs in serum-free defined media that had been developed for the *in vitro* maintenance of NSCs [188]. Using this technique, GBM stem cells were isolated for the first time in 2002 [189]. Subsequently, similar cells were isolated from MB, ependymomas and astrocytomas [190-192]. Singh and colleagues were the first to use the CD133 antibody for the direct isolation of putative BTSCs that self-renewed and proliferated in culture [191, 193]. These CD133+ cells isolated from paediatric brain tumours including MB possessed the ability to regrow the tumour when injected intracranially into immunodeficient mice. As few as 100 cells from the CD133+ subpopulation isolated from primary tumours were sufficient to initiate a new tumour identical to the parent tumour, whereas 10^5 cells of the CD133 negative fraction failed to generate tumours [193]. Since then, CD133 expression has been used to isolate putative CSCs from prostate, colorectal, pancreatic and breast cancers [186, 194-197].

BTSC-enriched cultures facilitated the study of BTSC self-renewal, proliferation and differentiation capacities, and improved our understanding of the relationship between BTSCs and normal NSCs [198, 199]. BTSCs and NSCs share the capacity for unlimited self-renewal potential and symmetric as well as asymmetric division to produce identical or differentiated progeny (Fig. 5). However, in contrast to NSCs, for which self-renewal is tightly regulated, in BTSCs it appears to be uncontrolled. The mechanism by which cell division is deregulated in BTSCs is thought to be associated at least in part with the loss of cell polarity, which has been shown to cause neoplastic overgrowth in the *Drosophila* CNS and in mammalian epithelial tumours [200-202]. The dysplasia of neural cells reported in polarity gene lgl1-knock out mice [203] and asymmetrically dividing BTSCs [204] in MB indicate that deregulated polarity pathways may also play a role in MB pathogenesis in humans.

Another hallmark of BTSCs and NSCs is the capacity to give rise to progenitor cells that differentiate along neuronal or glial lineages. However, a considerable proportion of tumour cells co-express neuronal (βIII-tubulin) and astrocytic (GFAP) markers [190]. Since the expression of these markers is mutually exclusive in normal neural cells, lineage co-expression suggests that differentiation programs may be aberrant in MB. Apart from abnormal lineage co-expression, differentiation of GBM cells seems to be reversible as shown *in vitro* [205, 206]. The lineage co-expression and stem cell-like character of MB cells suggests a similar obstruction of differentiation, preventing tumour cells from reaching the terminal differentiation stage. Finally, the hallmark of genuine BTSCs is the exclusive ability to regenerate the tumour *in vivo* as shown by serial transplantation experiments (discussed by Singh and Dirks in 2007 [199]).

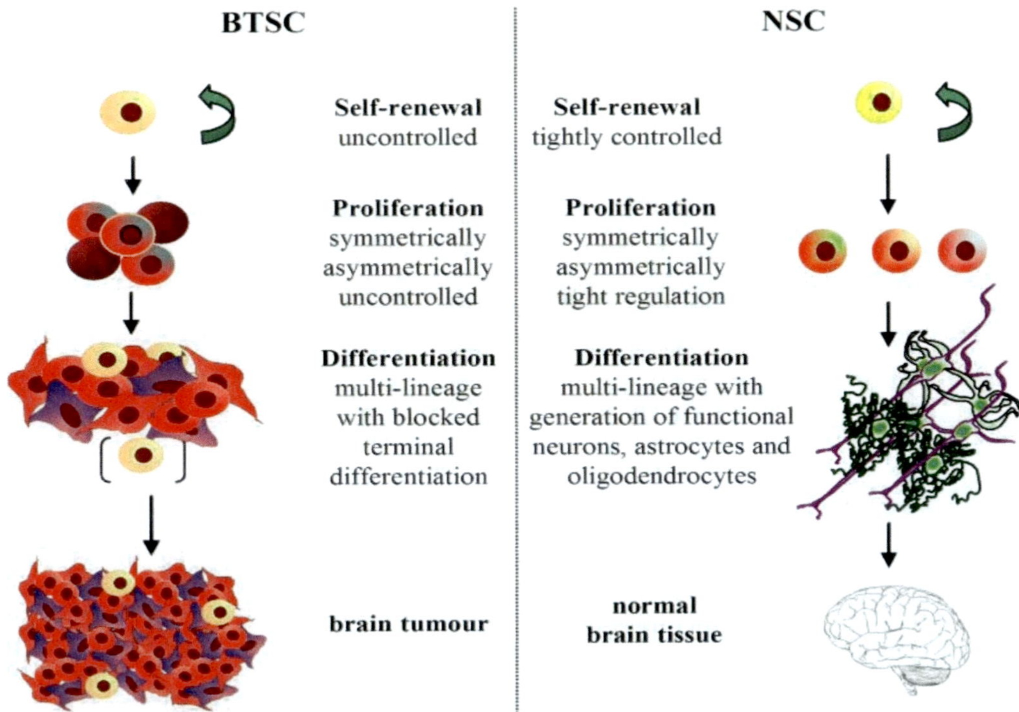

Figure 5. Hallmarks of BTSCs and NSCs. BTSC resemble NSCs in their self-renewal, proliferation and differentiation properties but show aberrant proliferation regulation and incomplete differentiation. *BTSC = brain tumour stem cell, NSC = neural stem cell.*

CD133 Expression is a Marker for Cells with Stem Cell Characteristics

The surface marker CD133+ is encoded by the *Prominin-1* gene and was originally identified in human hematopoietic stem cells and in mouse neuroepithelial stem cells [207, 208]. The gene was named after its "prominent" location in plasma membrane protrusions. *Prominin-1* genes are conserved from zebra fish to mammals [209] with about 60% amino acid-sequence identity between the human and mouse orthologs [210]. *Prominin-1* is widely expressed in epithelial cells as well as in plasma membrane protrusions of non-epithelial cells [207, 211]. It is a single copy gene on chromosome 4 in humans and contains 37 exons, which allow for multiple splicing variants [212, 213]. The protein has a unique structure with an N-terminal extracellular domain and five transmembrane domains containing two large extracellular loops and a cytoplasmic tail (Fig. 6). The extracellular loops contain at least eight known glycosylation sites. Structurally, the protein is a receptor, but the ligand is not known, nor is the interaction with intracellular molecules fully understood [214]. Given that Prominin-1 is expressed on self-renewing epithelial, neuroepithelial, and radial glial cells [210, 215, 216], it has been suggested to play an important role in stem cell division as discussed at a later stage in more detail.

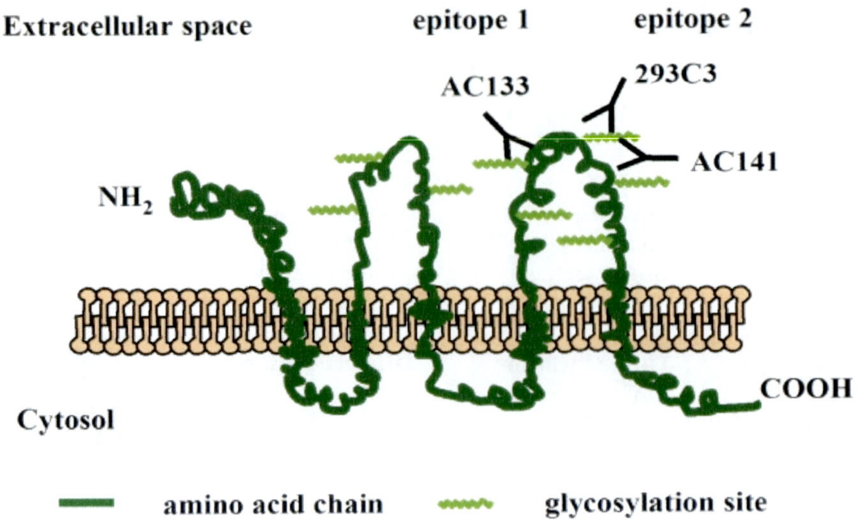

Figure 6. Prominin-1 and CD133 epitopes. The Prominin-1 protein contains eight possible glycosylation sites and can be bound at two epitopes by targeting antibodies. The glycosylated Prominin-1 protein is referred to as CD133.

In humans, Prominin-1 was originally identified using CD34+ selected haematopoietic stem cells to generate the AC133/1 antibody. With this antibody, AC133/1+ retinoblastoma cells were isolated and a second antibody called AC141, was generated using these purified cells [217]. The two antibodies recognise the glycosylated forms of Prominin-1 at two different epitopes within the extracellular loops [208, 218](Fig. 6). Later, cloning of the *Prominin-2* isoform led to the generation of a new antibody called AC133/2 [219]. Epitopes recognised by AC133/1 and AC133/2 are both highly expressed in the foetal brain, and some expression is detectable in adult stem cell niches in the brain. The glycosylated form of Prominin-1 was termed CD133 to distinguish it from the unmodified Prominin-1 [211, 220]. Overall, it is mainly CD133 expression levels that have been associated with stem cell properties as the glycosylation pattern can change upon differentiation resulting in the loss of detection by these antibodies independent of *Prominin-1* down-regulation. This suggests that CD133 expression is a better indicator of stem cell activity than *Prominin-1* expression and that glycosylation may be an important component of the stem cell regulatory activity of the *Prominin-1* gene product [212].

At the cellular level, the initiation of symmetric or asymmetric cell division is mainly determined by cell polarity, spindle orientation and cell cycle regulation [76]. Unequal distribution of cellular constituents, such as NOTCH signalling components during cell division causes unequal signalling in the daughter cells which in turn impacts on cell identity [221]. Membrane associated and transmembrane proteins such as β-catenin and CD133 may mediate this process by interacting with membrane cholesterol and a specific cholesterol-based membrane micro-domain [209, 222]. The cytoskeletal organisation along the basal membrane can direct the cleavage furrow for the NSC to divide symmetrically or asymmetrically. In the case of an asymmetric division, CD133 and other membrane proteins are distributed unequally along the cleavage furrow and can be depleted in the differentiating daughter cells. The depletion is achieved by removing parts of the membrane containing

CD133 from the differentiated daughter cell into the extracellular space, in contrast to the undifferentiated cell, which retains this section including all membrane proteins. Secretion of membrane proteins including the CD133 molecules has been measured in the extracellular space [223] and may explain the observed accumulation of CD133+ particles in the neural tube before onset of neurogenesis and their disappearance thereafter [224]. Furthermore, the over-representation of CD133 in the cerebrospinal fluid in brain tumour patients [225] may reflect a disturbance of such cell cycle control mechanisms. The importance of spatial and temporal organisation of cell components during normal NSC division points to the possibility that deregulation of this mechanism may be a key event in tumorigenesis of MB.

Brain Tumour Stem Cells and Their Relationship to Neural Stem Cells

The parallel investigation of BTSC and NSC biology has identified many similarities that emphasise the relationship between the two. BTSCs and NSCs share surface marker expression, are controlled by similar signalling pathways and behave similarly in culture [180, 198, 204, 226, 227]. The resemblance has led to speculation that BTSCs may indeed be derived from normal NSCs. However, this remains controversial, since it is unclear whether stem cell features are retained or re-acquired by transformed MB cells. Singh and colleagues [228] have proposed that BTSCs may be derived from reverted terminally differentiated neural cells, mutated NPCs, or aberrantly proliferating NSCs.

Terminally Differentiated Cells as Precursors for Brain Tumour Stem Cells

Experimental evidence for terminally differentiated neural cells as the cell of origin for MB has not been presented so far. In GBM mouse models, forced activation of Ras and Akt signalling generated GBM from neural progenitors but failed to transform terminally differentiated astrocytes [229]. However, the reversion of terminally differentiated astrocytes into multipotent cells by the overexpression of *B lymphoma Mo-MLV insertion region 1 (BMI1)* was demonstrated more recently [230]. This highlights the possibility of dedifferentiation events in the brain and suggests that terminally differentiated cells may be required to undergo several mutational events leading to a stem cell phenotype prior to neoplastic transformation. Considering that cancer cells usually bear multiple mutations, this concept seems possible, but remains to be proven in MB.

Committed Neural Progenitor Cells as Precursors for Brain Tumour Stem Cells

SHH activation has been implicated in the pathogenesis of some MB and pathway inhibition has been discussed as a new approach to therapy [109]. Studies in *PTCH* knockout mouse models with constitutive SHH signalling have proven that MB can arise from NSCs and more committed NPCs [114, 115]. These models have been used extensively to develop SHH-inhibitors [155, 156] that are currently under investigation in clinical trials [231]. However, the origins of other MB variants are less well understood, and it remains possible that they arise from a different type of neural cell than SHH-dependant variants.

Neural Stem Cells as Precursors of Brain Tumour Stem Cells

In human and mouse, NSCs functionally resemble BTSCs and reside in regions of neurogenesis, where brain tumours commonly arise [74, 232]. As mentioned above, MB can be generated from transgenic NSCs in mice and exhibit a more aggressive phenotype than those generated from more committed precursor cells [114, 115]. An additional argument supporting a NSC origin is the apparent deregulation in differentiation of BTSCs [180, 206] that may be explained by the developmental arrest of NSCs through genetic alterations. Brain tumour cultures respond normally to mitogenic stem cell conditions, which select for CD133+ BTSC populations but differentiate incompletely even when exposed to strict differentiating conditions *in vitro* [206, 233]. Overall, the evidence derived from mouse models supports the view that there may be more than one cell type of origin for BTSCs. The existence of several multipotent stem and precursor cell types in the developing and postnatal brain of humans suggests a similar scenario may exist for human MB [137, 234].

Identification and Isolation of Neural Stem Cells and Brain Tumour Stem Cells

Insights from the parallel study of neural development and MB pathogenesis have led to an increased interest in using NSCs for the study of MB biology. To learn more about the relationship between BTSCs and NSCs, underlying molecular programs need to be identified that regulate NSC functions, and may be deregulated in BTSCs. The functional assessment of these pathways in NSCs may provide more insights into the underlying molecular mechanisms that promote the neoplastic transformation into BTSCs consistent with the CSC theory in MB. For the investigation of expression signatures, the availability of purified NSCs and BTSCs are crucial. Despite the fact that the isolation of NSCs and BTSCs from primary specimens has become technically feasible, such an undertaking remains a challenge due to the lack of unequivocal surface markers for both cell types.

Hallmarks of NSCs include self-renewal, multi-lineage differentiation potential and the ability to repopulate ablated brain regions [77, 235]. NSCs have been isolated using the neurosphere assay, which reflects the self-renewal capacity of NSCs. This assay was developed by Reynolds and Weiss in 1992 [188] and is currently the most widely accepted isolation technique for NSCs [236, 237]. According to the protocol, brain cells are cultured in defined serum-free media, which is selective for NSCs and does not support differentiated neural cell types. NSCs will proliferate and self-renew to form a spherical aggregate of NSCs and NPCs referred to as neurospheres. Because NPCs also have limited neurosphere-forming potential, it is necessary to perform serial passaging until these become exhausted. The "clonal" cells that emerge from this procedure are then assessed for multi-differentiation potential and integration into the rodent brain [237].

The isolation of neurosphere-forming cells has facilitated the identification of surface markers useful for the direct isolation of NSCs from other sources. NSCs express *SRY (sex determining region Y)-box (SOX) 2, SOX3, BMI1, Paired box (PAX) 6*, and *Musashi-1* (*MSI1*) that can be used as markers for stemness [238]. However, since the definitive marker phenotype of NSCs has yet to be determined, the use of several markers or techniques is recommended. The surface antigens stage-specific embryonic antigen 1 (SSEA1), integrins, and CD133 have been useful in isolating stem cells from neural tissues [234, 238, 239]. Further, the intracellular filament Nestin is expressed in neural stem and progenitor cells and can be used to identify cells of neuroepithelial origin. Currently, Nestin and CD133 co-expression is the most widely used combination of markers for the identification of NSCs.

Similar to NSCs, BTSCs express markers associated with stemness such as *MSI1, SOX2, Maternal embryonic leucine zipper kinase* (*MELK*), Nestin, and CD133 [190, 240-242] that can be used for identification and isolation. CD133 expression has been very useful for the isolation of BTSCs particularly when used in combination with the NPC marker Nestin and stem cell marker SOX2. Several studies have confirmed that CD133+ cells in GBM possess BTSC characteristics, are clinically and molecularly distinct and are more radioresistant than CD133- cells [243-245]. However, xenograft studies have demonstrated that increasing the cell number injected or allowing more time for tumour development suggests that CD133- GBM cells may also regenerate tumours [244, 246, 247]. In MB, the initial results presented by Singh and colleagues supported the exclusive role of CD133+ MB cells in maintaining and regenerating the whole tumour [191, 193]. Recently, SSEA1, also referred to as CD15, was proposed as an alternative marker for BTSCs in GBM and MB specimens [248-250]. SSEA1+ cell populations isolated from human GBM lines and murine SHH-dependent MB fulfilled most BTSC characteristics but were not able to form tumours *in vivo*. Similar to CD133, SSEA1 is a surface antigen, suitable for direct live cell isolation, and is expressed by small subpopulations in most human MB [248]. However, the relative merits of using SSEA1 over CD133 for BTSC or NSC isolation remain to be investigated.

Therapeutic Implications of the Brain Tumour Stem Cell Theory

The importance of BTSCs in driving MB growth, their resemblance to slow cycling NSCs, and the increase of BTSCs in relapsed patients raises the question of whether current treatment regimes are targeting BTSCs efficiently. Most therapies target fast proliferating cells and spare NSCs and may therefore also spare proliferating BTSCs. The failure to eradicate CD133+ BTSCs may be the reason for relapse in some patients (Fig. 7). The apparent increase in CD133+ cancer cells in relapse GBM specimens in comparison to the time of diagnosis supports this theory [251, 252] and may also explain higher treatment failure in tumours with high CD133 expression [253, 254]. *In vitro* studies of CD133+ MB and GBM cells have confirmed their higher resistance to chemotherapy and radiotherapy by means of DNA damage response, and expression of anti-apoptosis or multi-drug resistance proteins [243, 251, 252, 255]. This suggests that clinical outcome can be altered by modulating the percentage of BTSCs in MB and also that treatment more tailored to these cells may be more efficacious.

Work is in progress to find agents that target BTSCs by screening small molecule libraries for molecules targeting NSCs as well as tumour cells [256]. Further, Bone *morphogenic proteins* (*BMPs*) were suggested for treatment therapy due to their differentiation-inducing effects on BTSCs, which may decrease BTSC number or alternatively make them more susceptible to chemotherapy [257]. Nevertheless, in GBMs, BTSCs are desensitised to differentiation cues [258] and the differentiation potential may have to be restored before such therapy can be applied successfully.

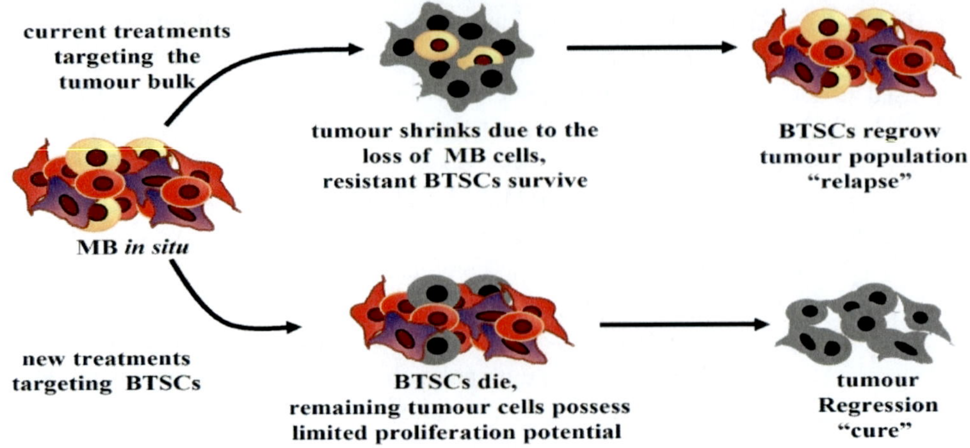

Figure 7. Hypothetical effects of cancer treatment. Current treatment targets the bulk of the tumour but may spare BTSCs. Limited sensitivity of BTSCs to chemo- and radiotherapy may enable their survival and lead to re-growth of the tumour. If treatment is specifically targeted at BTSCs, the tumour is expected to lose the ability to regenerate itself and regress. *BTSC = brain tumour stem cell, MB = medulloblastoma.*

Another strategy for targeting of BTSCs aims to disrupt niche components that are thought to act as a physical shield. Recently, sectioning tumourspheres from mouse and human glioma showed a variety of structural areas that may harbour BTSCs and promote their proliferation similar to the NSC niche [259].The idea that the proliferation of BTSCs in defined tissue niches is influenced by environmental factors has posed questions about the role of niche regulation in tumorigenesis and tumour progression. Initial studies in GBM revealed that BTSCs rely on aberrant vascular niches [260] and CD133+ GBM cells similar to NSCs secrete *Vascular endothelial growth factor* (*VEGF*) in response to hypoxia, which promotes recruitment of mesenchymal cells for the stimulation of brain tumour vascularisation [261-264]. Such recruited mesenchymal stem cells express *von Willebrand factor* (*vWF*), *Platelet-derived growth factor* (*PDGF*) and other factors that contribute significantly to abnormal angiogenesis and vasculogenesis [265]. The interaction of BTSCs and the niche seems also important in MB, since Nestin+/CD133+ BTSCs have been shown to reside next to capillaries, and exhibit increased expansion and higher tumorigenicity when co-cultured with epithelial cells [264]. Initial evidence supports the idea that niche assembly plays a crucial role in brain tumour pathogenesis by promoting tumour growth, vascularisation and particularly protection from external insults [266, 267]. The identification of bioeffectors that specifically target niche signalling may help to disrupt the BTSC niche and may be a valuable strategy for targeting BTSCs in the future.

Modelling Medulloblastoma Pathogenesis

The search for MB cell(s) of origin and the understanding of molecular mechanisms that contribute to brain tumorigenesis has been greatly facilitated by the development of genetically engineered mouse models (GEMMs). GEMMs overexpress oncogenes or carry specific mutations inactivating TSGs that enable the monitoring of tumorigenesis from initiation through to metastasis. They are effective systems for investigating diverse aspects of cancer biology [268]. The development of mouse models for MB research was initiated in 1997 with the generation of the *PTCH-/-* homozygous knockout mouse [269]. Because of severe CNS defects, the homozygous deletion was embryonic lethal and hemizygous *PTCH* mutants were generated. The phenotype of the *PTCH+/-* mouse resembled that seen in humans with Gorlin syndrome, including low penetrance of MB and *PTCH* mutation in all somatic cells. Since these models were less representative of sporadic MB, where patients are thought to only carry mutations in a proportion of their cells, more sophisticated mouse models were generated using the Cre/LoxP conditional knockout system. In these models, the Cre recombinase is placed under the control of cell-specific promoters to create cell-type specific knockouts or overexpression. Such models were used to induce constitutive SHH pathway activation *via PTCH* deletions or *SMOH* overexpression in NPCs and NSCs [114, 115, 270]. The SHH pathway is activated in a subgroup of MB with mostly desmoplastic phenotype [44, 46] and thus the conditional SHH-activation in mouse NSCs and NPCs may represent this human variant well. Indeed, utilising such models, new successful treatment strategies using SHH inhibitors like ShhAntag [155] are specifically targeting the underlying pathogenic mechanisms of MB cells. Currently, several pre-clinical trials using SHH-antagonists GDC-0449 or XL-139 which target the key SHH signalling transducer *SMOH*, are showing promise [271-273]. In a recent case of advanced MB, the SMOH antagonist GDC-0449 achieved only transient regression of the tumour [231]. While the acquired resistance of the patient to the reagent was disheartening, the initial rapid regression of the disseminated tumour was encouraging and suggested that the development of a more tailored approach towards MB subtypes may be effective. However, desmoplastic SHH-driven MB make up only 15-30% of all cases and efforts are currently focused on the development of representative models for the remaining MB subtypes. Mouse models with alternative mutations in *TP53*, *DNA ligase IV* and *XRCC4* exist that induce MB with very high penetrance [155, 274, 275]. However, most of these GEMMs target genes rarely mutated in human MB and their relevance to the human disease is controversial [268, 276-278]. In the future, further insight into the underlying molecular mechanisms of the different subtypes of MB will lead to the identification of new target genes useful to generate GEMMs which develop specific subtypes of MB.

Neural Stem Cell-Based Approaches for *In Vitro* Modelling of Human Medulloblastoma

Generation and Isolation of Neural Stem Cells

Until recently, sources of human NSCs were almost exclusively embryonal and foetal neural resections. The standard methods for isolating NSCs from foetal or postnatal brain tissue were reviewed by Gage in 2000 [279]. In short, brain tissue is dissected from the SVZ and other proliferative brain regions and disaggregated *in vitro*. Cells are then grown in defined medium usually supplemented with fibroblast growth factor – beta (bFGF) and epidermal growth factor (EGF), and the culture is evaluated for neurosphere formation. Multipotent NSCs can also be directly isolated from any part of the embryonal brain and some areas of the postnatal mammalian brain using the surface marker CD133 [137, 213, 234, 280]. This procedure, in combination with functional differentiation assays is a widely accepted approach for the identification of NSCs [224, 281, 282]. However, NSCs isolated from different brain sections exhibit differences in neurosphere forming potential [232], surface markers and gene expression profiles [283-285]. Similarly, differences between embryonic, foetal and postnatal derived NSCs have been detected in gene expression studies [286, 287]. Overall, these differences complicate the interpretation of data gained from analyses and manipulation of NSCs obtained from different sources. Further disadvantages lie in the ethical and technical challenges associated with human foetal resections.

More recently, several protocols have been developed to generate NSCs from ESCs [288-290]. Firstly, ESCs maintained *in vitro* have been shown to spontaneously give rise to a small population of NSCs when cultured in serum [288], although a better yield was achieved using serum-free defined media supplemented with EGF and bFGF [291-293]. In these conditions, ESCs differentiate into a mixture of NSCs, NPCs, and neurons, astrocytes and oligodendrocytes. In addition, the BMP2-inhibitor Noggin has proven more effective for neural induction producing a high percentage of NSCs and NPCs [238 , 294, 295]. Noggin treatment induces an exclusive shift of ESCs to the neural lineage providing an advantage over the spontaneous protocol that yields a more mixed population [296]. BMP2 inhibition paired with defined serum-free culture conditions for NSCs represents a very efficient protocol for the reliable production and maintenance of human NSCs.

NSCs generated using this protocol grow as neurospheres that express the neural and stem cell markers Nestin, *MSI1, SOX1, SOX2, PAX6* and CD133 but not markers of the mesoderm or endoderm, and can differentiate into cells of all three neural lineages [238]. High purity NSC populations can be isolated from neurospheres by selection for CD133 and SSEA1 and by negative selection for ESC markers. Since ESC-derived NSCs behave very similarly to foetal NSCs and express similar gene signatures, these cells have been proposed to be a good representation of foetal NSCs [238, 286, 287]. In addition, ESC lines and neurospheres can be stably maintained and expanded in culture for extended periods [297-299], making this an attractive and reliable system for the study of NSCs.

Genetic Manipulation of Neural Stem Cells

Genetic manipulation of NSCs has been useful for the delineation of crucial components of NSC biology and neuropathology. Overall, a varied array of reagents have been used to manipulate NSCs (reviewed by Jandial et al. 2008 [235] and Rajan and Snyder 2006 [280]). Most non-viral transfection techniques are inefficient and better results have been achieved by using a variety of viral vectors (reviewed by Davidson and Breakefield 2003[300]). The most popular viral vectors are lentiviruses, adeno-associated viruses (AAVs) and adenoviruses (AdVs).

In comparison to AAVs and AdVs that contain DNA-genomes, lentiviral vectors are derived from RNA-based retroviruses such as the human immunodeficiency and feline immunodeficiency virus [301]. After infecting the host cell, the viral RNA genome is reverse transcribed and integrated into the host cell genome, where viral genes become activated by cellular mechanisms. Most viral genes have been removed from the wild type virus to generate the lentivector system [302, 303]. Lentiviruses have been useful to immortalise

NSCs [304] or generate transgenic animal models [305, 306]. Although lentiviral vectors are superior to AdVs in regards to low toxicity [307-309] and stable transgene expression [310], the frequent integration in or near proto-oncogenes in the genome complicates experimental interpretation [311]. Nevertheless, methodologies have been established for the efficient and stable integration of lentiviral vectors in rodent and human NPCs with transgene expression consistently observed in the majority of cells [312].

AAV and AdV vectors have also been useful tools for high level transgene expression studies in rodent cells *in vitro* and *in vivo* [313-315]. Both vectors infect mouse and human ESCs without affecting their stem cell properties. Similarly, NSCs from embryonic mouse brains were successfully manipulated with these vectors [316]. The promising utility of AdV vectors in therapy has driven considerable progress on issues such as target cell specificity, long-term expression and immunotoxicity *in vivo* [310, 317].

Wild type AdVs are a family of double stranded DNA viruses that infect most human cells efficiently. AdVs bind to the Coxsackie-adenovirus receptor (CAR) on the cell surface and are internalised by the subsequent interactions with $\alpha_v\beta_3$ and $\alpha_v\beta_5$ integrins [317]. The CAR is structurally related to other cell adhesion proteins and is highly expressed in the developing CNS, where it was suggested to play a role in cell-cell interactions [318-322]. The expression of CAR on stem cells in the CNS [320, 323, 324] and its implicated importance in NSC biology [320] highlights the potential utility of adenoviral vectors in human NSC research.

Initially, recombinant AdVs were constructed that retained their viral infection ability but were devoid of the replicative competence genes E1/E3 (reviewed by Russell in 2000 [325]). These vectors efficiently shuttled transgenes into target cells and initiated their expression. Following these first generation AdV vectors, the second generation contained truncated E2 elements in order to increase insert capacity. The third and higher generations lacked additional sequences to minimise immunogenic effects and maximise carrying capacity. The efficient entry mechanisms and low pathogenicity for humans made adenoviral vectors a popular tool for the delivery of transgenes with clinical implications [317]. AdV vectors are episomal vectors that don't integrate into the genome and can infect quiescent NSCs, rapidly proliferating precursor cells, as well as post-mitotic mature brain tissue [313, 326, 327]. However, while the episomal status prevents disruption of the genome, it also results in rapid elimination of the vector from proliferating cells. To address this problem, other virus elements have been engineered that allow controlled episome replication synchronized with the cell cycle [328, 329]. Nevertheless, the high-capacity of the virus inserts of up to 30kb and the possibility of producing high titres in a short amount of time make them promising candidates for research and gene therapy. Additional interest has been fuelled by the tropism of NSCs to brain tumours *in vivo*, which highlights their potential as targeted delivery shuttles to the brain [330-333]. This may help to overcome some current limitations of *in vivo* delivery and toxicity of the "naked" virus. Still, controversy about the safety of the virus remains. Even though residual effects of immuno-toxicity have proven manageable by transient removal of specific immune cells before injecting [334, 335], the death of a clinical trial participant due to an adenoviral induced toxic-shock highlighted serious safety concerns [336, 337].

Table 2. The use of adenoviral vectors in neural cells

cell type	species	MOI	infection%	effects on cell properties	study, year
hESC derived NSCs	human	0.1-100	70%	negligible effects on surface marker expression, proliferation and differentiation at MOI 10	[338] 2010
MSC derived NSCs	human	10-100	-	overexpression of GLI1 increased NSC proliferation indicating successful infection.	[339] 2007
foetal SVZ NPCs	human	500	65%	No observed toxicity NPCs engrafted into rat brain exhibiting poor survival	[340] 1995
foetal NSC and NPC	rat	100	90%	GFP was visible for 4 weeks, retained differentiation potential, slowed growth.	[341] 2008
adult NSCs	mouse	500-1000	<1%	adult NSCs do not express CAR and are resistant to infection.	[342] 2005
ESC-derived NSCs	rat	50	80-90%	unchanged tropism to GBM *in vivo* observed after infection	[343] 2003
embryonic NSCs	mouse	20	>50%	maintained GFP expression for one month changes in morphology and astroglial induction	[316] 2002
adult NSCs	mouse	161 PFU	68%	infection occasionally altered differentiation.	[344] 2002
mature neurons, astrocytes and oligodendrocytes	rat	50	>50%	not addressed	[326] 1997

ESC = embryonic stem cells, MOI = multiplicity of infection, MSC = mesenchymal stem cells, NSC = neural stem cells, NPC = neural progenitor cells, PFU = plaque forming units, SVZ = subventricular zone

Despite the array of studies and applications for AdV vectors and NSCs, standardised protocols for the infection of humans NSCs have only been addressed recently [338]. Prior to this report, a large range of viral concentrations, conditions and cell sources have been applied in similar studies (Tab. 2). In addition, most work has been performed in rodent models [316, 341, 344]. Different model systems and viral concentrations may be responsible for variable infection efficiency and viral effects on cell properties. In this context, a recent study using human NSCs demonstrated that the effects of AdV infection on human NSC function are variable and depend on multiple experimental parameters [338]. Overall, the study showed that efficient AdV mediated gene transfer into human ESC-derived neurosphere cells can be achieved without significant effects on CD133 and Nestin expression, cell proliferation, or differentiation potential. With the current interest in CD133+ and Nestin+ human NSCs for modelling neurodegenerative and neoplastic disorders, it is anticipated that this *in vitro* model will provide a practical new approach to study human NSC function in these and other disease contexts.

Future Directions for Modelling Human Medulloblastoma

Apart from promising animal models discussed earlier, pre-clinical testing and drug screening procedures in the human system still relies heavily on human MB cell lines. However, as mostly two-dimensional cultures, they do not recapitulate the genetic and cellular heterogeneity of tumours and tumour stroma that exist *in vivo*. MB cell lines are an easily accessible and an expandable source of tumour cells for the investigation of tumour biology, but contain only a small proportion of BTSCs [247, 345-349], which appear crucial for tumour regrowth and survival following chemo- and radiotherapeutical insults [233, 251, 252, 261]. Fewer insights are available specifically for MB but initial screening of DAOY MB cells was consistent with the latter findings [255]. BTSCs may have major implications for the clinic, since the CSC theory proposes that BTSCs may not be targeted sufficiently by current cytotoxicity-based therapies and that they directly influence the clinical outcome [350]. Potential strategies for CSC-directed therapy have been investigated [351], but the resemblance of BTSCs to NSCs adds an additional challenge to sparing healthy NSCs in the developing brain. If, in fact, MB rely on CD133+ BTSCs for growth, more insights must be gained to distinguish them from healthy NSCs.

Serum-free MB cell lines represent an alternative source of CD133+ BTSCs. BTSCs from several types of brain tumours cultured under serum-free NSC conditions exhibited similar surface marker expression and *in vitro* behaviour [190, 191, 352]. These BTSCs resembled normal NSCs phenotypically and functionally and were capable of generating a hierarchy of tumour cells that reflect different stages of neural differentiation [114, 193]. Similar results have been reported for short-term serum-free MB cultures established from primary specimens [190, 191]. Further, current evidence suggests that MB tumour cells infect efficiently with AdVs [353, 354] and therefore the suitability of BTSC-rich tumoursphere cultures for genetic manipulation and therapeutical screening needs to be explored. This may allow the assessment of putative candidate genes not only in tumorigenesis but also in tumour progression and metastasis, which may reveal possible targets for clinical intervention (Fig. 8).

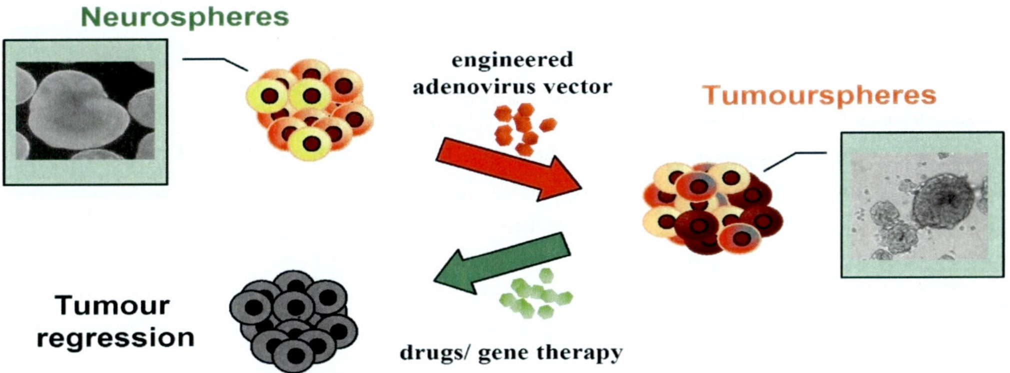

Figure 8. *In vitro* modelling of human MB: The investigation of MB tumorigenesis and intervention using serum-free sphere culture techniques and genetically engineered adenoviral vectors. The manipulation of gene expression in neurosphere cells may initiate neoplastic transformation (red arrow), whereas manipulation of gene expression in tumoursphere cells may abrogate tumoursphere growth (green arrow). *MB = medulloblastoma.*

While molecular analysis techniques are powerful approaches for identifying candidate genes and pathways relevant to disease pathogenesis, the development of appropriate model systems for functional validation is critical. Neurospheres contain a variety of NSCs and NPCs that represent candidates for investigating the cell of origin of MB. Gene expression in neurosphere cell populations can be manipulated using genetically engineered AdV vectors [338]. Assays for NSC proliferation, differentiation and other properties have been established that facilitate the functional investigation of candidate genes identified in previous reports [44, 46, 116]. This makes the proposed system a potentially powerful human model for studying MB tumorigenesis.

Beyond *in vitro* models, it is desirable to also model the *in vivo* behaviour of tumorigenic cells. The engraftment of murine NSCs into brains of NOD/SCID mice has been demonstrated [316], suggesting that this procedure may also be possible for human NSCs [338]. This type of model holds great value for investigating tumour initiation and progression. However, it may be less useful for pre-clinical drug testing, since in some cases drug response was minimally predictive due to transplantation-derived problems [355]. This is the reason why more recent developments have moved towards genetically engineered mouse models with cell-specific promoter-coupled conditional expression cassettes that appear to be more physiologically relevant [277]. The majority of such conditional models target neural stem and progenitor cells and are based on SHH signalling modulation [114, 115, 275, 356, 357]. These are currently the most credible models for SHH-driven MB in pre-clinical studies [155, 156, 358] that have helped the transition of SHH-inhibitors into clinical trials [155, 231] and may improve treatment outcome for 15-20% of MB cases in the near future. Current efforts are aimed at developing *in vivo* mouse models that represent other MB subtypes. Preliminary screening of the functional effects of multiple candidate gene combinations *in vitro* will spare time intensive generation of several GEMMS. With an *in vitro* to *in vivo* combined strategy it will be possible to investigate a larger number of candidate genes and narrow them down to a small selection before moving to GEMMS.

Conclusion

Despite major progress in the understanding of the genetic and molecular pathogenesis of MB, treatment regimes continue to be determined by clinical parameters [87]. In contrast to the five histopathological and at least four molecular subgroups described for MB [46, 89], there are currently only two clinical categories for risk assessment and treatment. This discrepancy suggests that it may be possible to design more tailored therapies to individual subtypes. Recent insights have advanced our understanding of the underlying molecular pathogenesis of individual MB variants. In order to increase our knowledge about these subgroups and their histogenesis, it will be crucial to assess them separately with regard to their cell of origin, genetic aberrations and clinical presentation. The implementation of molecular subgrouping in clinical assessment and classification will be essential. Secondly, better treatment regimes must be developed which are more tailored towards the molecular mechanisms driving progression of the particular molecular MB subgroup in order to eradicate all tumorigenic MB cells and prevent relapse. Murine conditional knockout models for SHH-dependent MB are currently the state of art, which may have direct impact on tumour management for this subgroup of MB in the near future. Comparable murine models for other MB subtypes are being developed, based on an improved understanding of the underlying molecular pathogenesis of the MB subtype. The identification of the cell of origin and the deregulated genes with potential involvment in the neoplastic transformation of individual MB subtypes in combination with the development of appropriate *in vitro* modelling systems will enable rapid feedback on the functional role of candidate genes identified *in silico*. Promising candidate genes with functional implications can then be transferred to whole organisms such as GEMMS. With the promising efficiency and reduced ethical challenges, the combination of a human *in vitro* model and *in vivo* animal model may be advantageous for translating new knowledge from research into clinical practice.

References

[1] Eden T, Eiser C, Hery C, Holland J, Johansen C, Mortara I, et al. Childhood cancer: Rising to the challenge. 2006; Available from: *http:\\www.uicc.org/index.php?option=com_content&task=view&id=14325&Itemid=266*

[2] Gurney JG, Kadan-Lottick N. Brain and other central nervous system tumors: rates, trends, and epidemiology. *Curr Opin Oncol.* 2001 May;13(3):160-6.

[3] AIHW. Cancer incidence projections, Australia 2002 to 2011. 2005; Available from: *http://www.aihw.gov.au/publications/index.cfm/title/10164*

[4] Anderson DM, Rennie KM, Ziegler RS, Neglia JP, Robison LR, Gurney JG. Medical and neurocognitive late effects among survivors of childhood central nervous system tumors. *Cancer.* 2001 Nov 15;92(10):2709-19.

[5] 5. Barnholtz-Sloan JS, Severson RK, Stanton B, Hamre M, Sloan AE. Pediatric brain tumors in non-Hispanics, Hispanics, African Americans and Asians: differences in survival after diagnosis. *Cancer Causes Control.* 2005 Jun;16(5):587-92.

[6] Feltbower RG, Picton S, Bridges LR, Crooks DA, Glaser AW, McKinney PA. Epidemiology of central nervous system tumors in children and young adults (0-29 years), Yorkshire, United Kingdom. *Pediatr Hematol Oncol.* 2004 Oct-Nov;21(7):647-60.

[7] Levy AS. Brain tumors in children: evaluation and management. *Curr Probl Pediatr Adolesc Health Care.* 2005 Jul;35(6):230-45.

[8] Peris-Bonet R, Martinez-Garcia C, Lacour B, Petrovich S, Giner-Ripoll B, Navajas A, et al. Childhood central nervous system tumours - incidence and survival in Europe (1978-1997): Report from Automated Childhood Cancer Information System project. *Eur J Cancer.* 2006 Sep;42(13):2064-80.

[9] Rickert CH, Paulus W. Epidemiology of central nervous system tumors in childhood and adolescence based on the new WHO classification. *Childs Nerv Syst.* 2001 Sep;17(9):503-11.

[10] Nejat F, El Khashab M, Rutka JT. Initial management of childhood brain tumors: neurosurgical considerations. *J Child Neurol.* 2008 Oct;23(10):1136-48.

[11] AIHW. Cancer in Australia 2001. 2004; Available from: http://www.aihw.gov.au/publications/index.cfm/title/10083

[12] Chang D. Statistics on Incidence, Survival rates and Mortality associated with Brain Tumours in Australia. 2003; Available from: *http://www.ozbraintumour.org.htm*

[13] Bunin G. What causes childhood brain tumors? Limited knowledge, many clues. *Pediatr Neurosurg*. 2000 Jun;32(6):321-6.

[14] Shiraishi T, Tabuchi K. Genetic alterations of human brain tumors as molecular prognostic factors. *Neuropathology*. 2003 Mar;23(1):95-108.

[15] Evans DG, Farndon PA, Burnell LD, Gattamaneni HR, Birch JM. The incidence of Gorlin syndrome in 173 consecutive cases of medulloblastoma. *Br J Cancer*. 1991 Nov;64(5):959-61.

[16] Paraf F, Jothy S, Van Meir EG. Brain tumor-polyposis syndrome: two genetic diseases? *J Clin Oncol*. 1997 Jul;15(7):2744-58.

[17] Gurney JG, van Wijngaarden E. Extremely low frequency electromagnetic fields (EMF) and brain cancer in adults and children: review and comment. *Neuro Oncol*. 1999 Jul;1(3):212-20.

[18] Morgan RW, Kelsh MA, Zhao K, Exuzides KA, Heringer S, Negrete W. Radiofrequency exposure and mortality from cancer of the brain and lymphatic/hematopoietic systems. *Epidemiology*. 2000 Mar;11(2):118-27.

[19] Baldwin RT, Preston-Martin S. Epidemiology of brain tumors in childhood - a review. *Toxicol Appl Pharmacol*. 2004 Sep 1;199(2):118-31.

[20] Napalkov NP. Prenatal and childhood exposure to carcinogenic factors. *Cancer Detect Prev*. 1986;9(1-2):1-7.

[21] Anderson LM, Diwan BA, Fear NT, Roman E. Critical windows of exposure for children's health: cancer in human epidemiological studies and neoplasms in experimental animal models. *Environ Health Perspect*. 2000 Jun;108 Suppl 3:573-94.

[22] McLaughlin CC, Baptiste MS, Schymura MJ, Nasca PC, Zdeb MS. Birth weight, maternal weight and childhood leukaemia. *Br J Cancer*. 2006 Jun 5;94(11):1738-44.

[23] Milne E, Laurvick CL, Blair E, de Klerk N, Charles AK, Bower C. Fetal growth and the risk of childhood CNS tumors and lymphomas in Western Australia. *Int J Cancer*. 2008 Jul 15;123(2):436-43.

[24] LeRoith D. Insulin-like growth factors and the brain. *Endocrinology*. 2008 Dec;149(12):5951.

[25] Murphy VE, Smith R, Giles WB, Clifton VL. Endocrine regulation of human fetal growth: the role of the mother, placenta, and fetus. *Endocr Rev*. 2006 Apr;27(2):141-69.

[26] Ross JA, Perentesis JP, Robison LL, Davies SM. Big babies and infant leukemia: a role for insulin-like growth factor-1? *Cancer Causes Control*. 1996 Sep;7(5):553-9.

[27] Hirschfeld S, Helman L. Diverse roles of insulin-like growth factors in pediatric solid tumors. *In Vivo*. 1994 Jan-Feb;8(1):81-90.

[28] Lee J, Chia KS, Cheung KH, Chia SE, Lee HP. Birthweight and the risk of early childhood cancer among Chinese in Singapore. *Int J Cancer*. 2004 Jun 20;110(3):465-7.

[29] Mogren I, Malmer B, Tavelin B, Damber L. Reproductive factors have low impact on the risk of different primary brain tumours in offspring. *Neuroepidemiology*. 2003 Jul-Aug;22(4):249-54.

[30] Von Behren J, Reynolds P. Birth characteristics and brain cancers in young children. *Int J Epidemiol*. 2003 Apr;32(2):248-56.

[31] Louis DN, Ohgaki H, Wiestler OD, Cavenee WK, editors. *WHO Classification of Tumours of the Central Nervous System*. Lyon, France: IARC; 2007.

[32] Hart MN, Earle KM. Primitive neuroectodermal tumors of the brain in children. *Cancer*. 1973 Oct;32(4):890-7.

[33] Ellison D. Classifying the medulloblastoma: insights from morphology and molecular genetics. *Neuropathol Appl Neurobiol*. 2002 Aug;28(4):257-82.

[34] Rorke LB. The cerebellar medulloblastoma and its relationship to primitive neuroectodermal tumors. *J Neuropathol Exp Neurol*. 1983 Jan;42(1):1-15.

[35] Rorke LB. Classification and grading of childhood brain tumors. Overview and statement of the problem. *Cancer*. 1985 Oct 1;56(7 Suppl):1848-9.

[36] Vandenberg SR, Herman MM, Rubinstein LJ. Embryonal central neuroepithelial tumors: current concepts and future challenges. *Cancer Metastasis Rev*. 1987;5(4):343-65.

[37] Rubinstein LJ. Embryonal central neuroepithelial tumors and their differentiating potential. A cytogenetic view of a complex neuro-oncological problem. *J Neurosurg*. 1985 Jun;62(6):795-805.

[38] Kleihues P, Cavenee WK, editors. *WHO Classification of Tumours, Pathology and Genetics, Tumours of the Nervous System*. Lyon: IARC; 2000.

[39] Rorke LB, Trojanowski JQ, Lee VM, Zimmerman RA, Sutton LN, Biegel JA, et al. Primitive neuroectodermal tumors of the central nervous system. *Brain Pathol*. 1997 Apr;7(2):765-84.

[40] Jakacki RI. Pineal and nonpineal supratentorial primitive neuroectodermal tumors. *Childs Nerv Syst*. 1999 Oct;15(10):586-91.

[41] Kleihues P, Louis DN, Scheithauer BW, Rorke LB, Reifenberger G, Burger PC, et al. The WHO classification of tumors of the nervous system. *J Neuropathol Exp Neurol*. 2002 Mar;61(3):215-25; discussion 26-9.

[42] McManamy CS, Pears J, Weston CL, Hanzely Z, Ironside JW, Taylor RE, et al. Nodule formation and desmoplasia in medulloblastomas-defining the nodular/desmoplastic variant and its biological behavior. *Brain Pathol*. 2007 Apr;17(2):151-64.

[43] Pomeroy SL, Tamayo P, Gaasenbeek M, Sturla LM, Angelo M, McLaughlin ME, et al. Prediction of central nervous system embryonal tumour outcome based on gene expression. *Nature*. 2002 Jan 24;415(6870):436-42.

[44] Kool M, Koster J, Bunt J, Hasselt NE, Lakeman A, van Sluis P, et al. Integrated genomics identifies five medulloblastoma subtypes with distinct genetic profiles, pathway signatures and clinicopathological features. *PLoS ONE*. 2008;3(8):e3088.

[45] Northcott PA, Korshunov A, Witt H, Hielscher T, Eberhart CG, Mack S, et al. Medulloblastoma Comprises Four Distinct Molecular Variants. *J Clin Oncol*. 2010 Sep 7.

[46] Thompson MC, Fuller C, Hogg TL, Dalton J, Finkelstein D, Lau CC, et al. Genomics identifies medulloblastoma subgroups that are enriched for specific genetic alterations. *J Clin Oncol*. 2006 Apr 20;24(12):1924-31.

[47] McLendon RE, Judkins AR, Eberhart CG, Fuller GN, Sarkar C, Ng H-K. Central nervous systm primitive neuroectodermal tumours. In: Louis DN, Ohgaki H, Wiestler OD, Cavenee WK, editors. *WHO Classification of Tumours of the Central Nervous System. 3 ed. Lyon: International Agency for Research on Cancer*; 2007. p. 141-46.

[48] Biegel JA, Kalpana G, Knudsen ES, Packer RJ, Roberts CW, Thiele CJ, et al. The role of INI1 and the SWI/SNF complex in the development of rhabdoid tumors: meeting summary from the workshop on childhood atypical teratoid/rhabdoid tumors. *Cancer Res.* 2002 Jan 1;62(1):323-8.

[49] Judkins AR, Eberhart CG, Wesseling P. Atypical teratoid/rhabdoid tumour. In: Louis DN, Ohgaki H, Wiestler OD, Cavenee WK, editors. *WHO Classification of Tumours of the Central Nervous System. 3 ed. Lyon: International Agency for Research on Cancer;* 2007. p. 147-49.

[50] Kraus JA, Oster C, Sorensen N, Berthold F, Schlegel U, Tonn JC, et al. Human medulloblastomas lack point mutations and homozygous deletions of the hSNF5/INI1 tumour suppressor gene. *Neuropathol Appl Neurobiol.* 2002 Apr;28(2):136-41.

[51] Uno K, Takita J, Yokomori K, Tanaka Y, Ohta S, Shimada H, et al. Aberrations of the hSNF5/INI1 gene are restricted to malignant rhabdoid tumors or atypical teratoid/rhabdoid tumors in pediatric solid tumors. Genes Chromosomes *Cancer.* 2002 May;34(1):33-41.

[52] Gottardo NG, Gajjar A. Current therapy for medulloblastoma. *Curr Treat Options Neurol.* 2006 Jul;8(4):319-34.

[53] Giangaspero F, Rigobello L, Badiali M, Loda M, Andreini L, Basso G, et al. Large-cell medulloblastomas. A distinct variant with highly aggressive behavior. *Am J Pathol.* 1992 Jul;16(7):687-93.

[54] Rivera-Luna R, Lopez E, Rivera-Marquez H, Rivera-Ortegon F, Altamirano-Alvarez E, Mercado G, et al. Survival of children under 3 years old with medulloblastoma: a study from the Mexican Cooperative Group for Childhood Malignancies (AMOHP). *Childs Nerv Syst.* 2002 Feb;18(1-2):38-42.

[55] Gajjar A, Chintagumpala M, Ashley D, Kellie S, Kun LE, Merchant TE, et al. Risk-adapted craniospinal radiotherapy followed by high-dose chemotherapy and stem-cell rescue in children with newly diagnosed medulloblastoma (St Jude Medulloblastoma-96): long-term results from a prospective, multicentre trial. *Lancet* Oncol. 2006 Oct;7(10):813-20.

[56] Packer RJ, Gajjar A, Vezina G, Rorke-Adams L, Burger PC, Robertson PL, et al. Phase III study of craniospinal radiation therapy followed by adjuvant chemotherapy for newly diagnosed average-risk medulloblastoma. *J Clin Oncol.* 2006 Sep 1;24(25):4202-8.

[57] Fossati P, Ricardi U, Orecchia R. Pediatric medulloblastoma: toxicity of current treatment and potential role of protontherapy. *Cancer Treat Rev.* 2009 Feb;35(1):79-96.

[58] Packer RJ. Childhood brain tumors: accomplishments and ongoing challenges. *J Child Neurol.* 2008 Oct;23(10):1122-7.

[59] Packer RJ. Progress and challenges in childhood brain tumors. *J Neurooncol.* 2005 Dec;75(3):239-42.

[60] Crawford JR, MacDonald TJ, Packer RJ. Medulloblastoma in childhood: new biological advances. *Lancet Neurol.* 2007 Dec;6(12):1073-85.

[61] Read TA, Hegedus B, Wechsler-Reya R, Gutmann DH. The neurobiology of neurooncology. *Ann Neurol.* 2006 Jul;60(1):3-11.

[62] Duffner PK, Cohen ME. Long-term consequences of CNS treatment for childhood cancer, Part II: Clinical consequences. *Pediatr Neurol.* 1991 Jul-Aug;7(4):237-42.

[63] Helseth E, Due-Tonnessen B, Wesenberg F, Lote K, Lundar T. Posterior fossa medulloblastoma in children and young adults (0-19 years): survival and performance. *Childs Nerv Syst.* 1999 Sep;15(9):451-5; discussion 6.

[64] Karajannis M, Allen JC, Newcomb EW. Treatment of pediatric brain tumors. *J Cell Physiol.* 2008 Dec;217(3):584-9.

[65] Packer RJ, Goldwein J, Nicholson HS, Vezina LG, Allen JC, Ris MD, et al. Treatment of children with medulloblastomas with reduced-dose craniospinal radiation therapy and adjuvant chemotherapy: A Children's Cancer Group Study. *J Clin Oncol.* 1999 Jul;17(7):2127-36.

[66] Bouffet E. Medulloblastoma in infants: the critical issues of the dilemma. *Curr Oncol.* 2010 Jun;17(3):2-3.

[67] Marachelian A, Butturini A, Finlay J. Myeloablative chemotherapy with autologous hematopoietic progenitor cell rescue for childhood central nervous system tumors. *Bone Marrow Transplant.* 2008 Jan;41(2):167-72.

[68] Rutkowski S, Bode U, Deinlein F, Ottensmeier H, Warmuth-Metz M, Soerensen N, et al. Treatment of early childhood medulloblastoma by postoperative chemotherapy alone. *N Engl J Med.* 2005 Mar 10;352(10):978-86.

[69] Schrappe M. Risk-adapted stratification and treatment of childhood acute lymphoblastic leukaemia. *Radiat Prot Dosimetry.* 2008;132(2):130-3.

[70] Ellison DW, Onilude OE, Lindsey JC, Lusher ME, Weston CL, Taylor RE, et al. beta-Catenin status predicts a favorable outcome in childhood medulloblastoma: the United Kingdom Children's Cancer Study Group Brain Tumour Committee. *J Clin Oncol.* 2005 Nov 1;23(31):7951-7.

[71] Basak O, Taylor V. Stem cells of the adult mammalian brain and their niche. *Cell Mol Life Sci.* 2009 Mar;66(6):1057-72.

[72] Farkas LM, Huttner WB. The cell biology of neural stem and progenitor cells and its significance for their proliferation versus differentiation during mammalian brain development. *Curr Opin Cell Biol.* 2008 Dec;20(6):707-15.

[73] Galli R, Gritti A, Bonfanti L, Vescovi AL. Neural stem cells: an overview. *Circ Res.* 2003 Apr 4;92(6):598-608.

[74] Jackson EL, Alvarez-Buylla A. Characterization of adult neural stem cells and their relation to brain tumors. *Cells Tissues Organs.* 2008;188(1-2):212-24.

[75] Murry CE, Keller G. Differentiation of embryonic stem cells to clinically relevant populations: lessons from embryonic development. *Cell.* 2008 Feb 22;132(4):661-80.

[76] Gotz M, Huttner WB. The cell biology of neurogenesis. *Nat Rev Mol Cell Biol.* 2005 Oct;6(10):777-88.

[77] Crane JF, Trainor PA. Neural crest stem and progenitor cells. *Annu Rev Cell Biol.* 2006;22:267-86.

[78] Glenn OA, Barkovich AJ. Magnetic resonance imaging of the fetal brain and spine: an increasingly important tool in prenatal diagnosis, part 1. *AJNR Am J Neuroradiol.* 2006 Sep;27(8):1604-11.

[79] Chan C, Moore BE, Cotman CW, Okano H, Tavares R, Hovanesian V, et al. Musashi1 antigen expression in human fetal germinal matrix development. *Exp Neurol.* 2006 Oct;201(2):515-8.

[80] Anthony TE, Klein C, Fishell G, Heintz N. Radial glia serve as neuronal progenitors in all regions of the central nervous system. *Neuron.* 2004 Mar 25;41(6):881-90.

[81] Merkle FT, Tramontin AD, Garcia-Verdugo JM, Alvarez-Buylla A. *Radial glia give rise to adult neural stem cells in the subventricular zone.* Proc Natl Acad Sci U S A. 2004 Dec 14;101(50):17528-32.

[82] Noctor SC, Flint AC, Weissman TA, Dammerman RS, Kriegstein AR. Neurons derived from radial glial cells establish radial units in neocortex. *Nature*. 2001 Feb 8;409(6821):714-20.

[83] Noctor SC, Martinez-Cerdeno V, Ivic L, Kriegstein AR. Cortical neurons arise in symmetric and asymmetric division zones and migrate through specific phases. *Nat Neurosci.* 2004 Feb;7(2):136-44.

[84] Wang VY, Zoghbi HY. Genetic regulation of cerebellar development. *Nat Rev Neurosci.* 2001 Jul;2(7):484-91.

[85] Lavezzi AM, Ottaviani G, Terni L, Matturri L. Histological and biological developmental characterization of the human cerebellar cortex. *Int J Dev Neurosci.* 2006 Oct;24(6):365-71.

[86] Zhang SC. Defining glial cells during CNS development. *Nat Rev Neurosci.* 2001 Nov;2(11):840-3.

[87] Polkinghorn WR, Tarbell NJ. Medulloblastoma: tumorigenesis, current clinical paradigm, and efforts to improve risk stratification. *Nat Clin Pract Oncol.* 2007 May;4(5):295-304.

[88] Giangaspero F, Eberhart CG, Haapasalo H, Pietsch T, Wiestler OD, Ellison DW. Medulloblastoma. In: Louis DN, Ohgaki H, Wiestler OD, Cavenee WK, editors. WHO *Classification of Tumours of the Central Nervous System*. 3 ed. Lyon: IARC; 2007. p. 132-40.

[89] Gilbertson RJ, Ellison DW. The origins of medulloblastoma subtypes. *Annu Rev Pathol*. 2008;3:341-65.

[90] Brandes AA, Paris MK. Review of the prognostic factors in medulloblastoma of children and adults. *Crit Rev Oncol Hematol*. 2004 May;50(2):121-8.

[91] Bayani J, Zielenska M, Marrano P, Kwan Ng Y, Taylor MD, Jay V, et al. Molecular cytogenetic analysis of medulloblastomas and supratentorial primitive neuroectodermal tumors by using conventional banding, comparative genomic hybridization, and spectral karyotyping. *J Neurosurg*. 2000 Sep;93(3):437-48.

[92] Jay V, Squire J, Bayani J, Alkhani AM, Rutka JT, Zielenska M. Oncogene amplification in medulloblastoma: analysis of a case by comparative genomic hybridization and fluorescence in situ hybridization. *Pathology* (Phila). 1999 Nov;31(4):337-44.

[93] Russo C, Pellarin M, Tingby O, Bollen AW, Lamborn KR, Mohapatra G, et al. Comparative genomic hybridization in patients with supratentorial and infratentorial primitive neuroectodermal tumors. *Cancer*. 1999 Jul 15;86(2):331-9.

[94] Eberhart CG, Kratz JE, Schuster A, Goldthwaite P, Cohen KJ, Perlman EJ, et al. Comparative genomic hybridization detects an increased number of chromosomal alterations in large cell/anaplastic medulloblastomas. *Brain Pathol*. 2002 Jan;12(1):36-44.

[95] Lamont JM, McManamy CS, Pearson AD, Clifford SC, Ellison DW. Combined histopathological and molecular cytogenetic stratification of medulloblastoma patients. *Clin Cancer Res*. 2004 Aug 15;10(16):5482-93.

[96] Rickert CH, Paulus W. Comparative genomic hybridization in central and peripheral nervous system tumors of childhood and adolescence. *J Neuropathol Exp Neurol.* 2004 May;63(5):399-417.

[97] Biegel JA, Wentz E. No preferential parent of origin for the isochromosome 17q in childhood primitive neuroectodermal tumor (medulloblastoma). *Genes Chromosomes Cancer.* 1997 Feb;18(2):143-6.

[98] Gilbertson R. Paediatric embryonic brain tumours. biological and clinical relevance of molecular genetic abnormalities. *Eur J Cancer.* 2002 Mar;38(5):675-85.

[99] Huang H, Mahler-Araujo BM, Sankila A, Chimelli L, Yonekawa Y, Kleihues P, et al. APC mutations in sporadic medulloblastomas. *Am J Pathol.* 2000 Feb;156(2):433-7.

[100] Pan E, Pellarin M, Holmes E, Smirnov I, Misra A, Eberhart CG, et al. Isochromosome 17q is a negative prognostic factor in poor-risk childhood medulloblastoma patients. *Clin Cancer Res.* 2005 Jul 1;11(13):4733-40.

[101] Lacombe D, Chateil JF, Fontan D, Battin J. Medulloblastoma in the nevoid basal-cell carcinoma syndrome: case reports and review of the literature. *Genet Couns.* 1990;1(3-4):273-7.

[102] Clifford SC, Lusher ME, Lindsey JC, Langdon JA, Gilbertson RJ, Straughton D, et al. Wnt/Wingless pathway activation and chromosome 6 loss characterize a distinct molecular sub-group of medulloblastomas associated with a favorable prognosis. *Cell Cycle.* 2006 Nov;5(22):2666-70.

[103] Aldosari N, Bigner SH, Burger PC, Becker L, Kepner JL, Friedman HS, et al. MYCC and MYCN oncogene amplification in medulloblastoma. A fluorescence in situ hybridization study on paraffin sections from the Children's Oncology Group. *Arch Pathol Lab Med.* 2002 May;126(5):540-4.

[104] Eberhart CG, Kratz J, Wang Y, Summers K, Stearns D, Cohen K, et al. Histopathological and molecular prognostic markers in medulloblastoma: c-myc, N-myc, TrkC, and anaplasia. *J Neuropathol Exp Neurol.* 2004 May;63(5):441-9.

[105] Moriuchi S, Shimizu K, Miyao Y, Hayakawa T. An immunohistochemical analysis of medulloblastoma and PNET with emphasis on N-myc protein expression. *Anticancer Res.* 1996 Sep-Oct;16(5A):2687-92.

[106] Rickert CH, Simon R, Bergmann M, Dockhorn-Dworniczak B, Paulus W. Comparative genomic hybridization in pineal parenchymal tumors. *Genes Chromosomes Cancer.* 2001 Jan;30(1):99-104.

[107] Stearns D, Chaudhry A, Abel TW, Burger PC, Dang CV, Eberhart CG. c-myc overexpression causes anaplasia in medulloblastoma. *Cancer Res.* 2006 Jan 15;66(2):673-81.

[108] Swartling FJ, Grimmer MR, Hackett CS, Northcott PA, Fan QW, Goldenberg DD, et al. Pleiotropic role for MYCN in medulloblastoma. *Genes Dev.* 2010 May 15;24(10):1059-72.

[109] Fan X, Eberhart CG. Medulloblastoma stem cells. *J Clin Oncol.* 2008 Jun 10;26(17):2821-7.

[110] Gao WQ, Hatten ME. Immortalizing oncogenes subvert the establishment of granule cell identity in developing cerebellum. *Development.* 1994 May;120(5):1059-70.

[111] Li L, Connelly MC, Wetmore C, Curran T, Morgan JI. Mouse embryos cloned from brain tumors. *Cancer Res.* 2003 Jun 1;63(11):2733-6.

[112] Okano-Uchida T, Himi T, Komiya Y, Ishizaki Y. *Cerebellar granule cell precursors can differentiate into astroglial cells*. Proc Natl Acad Sci U S A. 2004 Feb 3;101(5):1211-6.

[113] Stadtfeld M, Nagaya M, Utikal J, Weir G, Hochedlinger K. Induced Pluripotent Stem Cells Generated Without Viral Integration. *Science*. 2008 Sep 25.

[114] Schuller U, Heine VM, Mao J, Kho AT, Dillon AK, Han YG, et al. Acquisition of granule neuron precursor identity is a critical determinant of progenitor cell competence to form Shh-induced medulloblastoma. *Cancer Cell*. 2008 Aug 12;14(2):123-34.

[115] Yang ZJ, Ellis T, Markant SL, Read TA, Kessler JD, Bourboulas M, et al. Medulloblastoma can be initiated by deletion of Patched in lineage-restricted progenitors or stem cells. *Cancer Cell*. 2008 Aug 12;14(2):135-45.

[116] Fattet S, Haberler C, Legoix P, Varlet P, Lellouch-Tubiana A, Lair S, et al. Beta-catenin status in paediatric medulloblastomas: correlation of immunohistochemical expression with mutational status, genetic profiles, and clinical characteristics. *J Pathol*. 2009 May;218(1):86-94.

[117] Kho AT, Zhao Q, Cai Z, Butte AJ, Kim JY, Pomeroy SL, et al. Conserved mechanisms across development and tumorigenesis revealed by a mouse development perspective of human cancers. *Genes Dev*. 2004 Mar 15;18(6):629-40.

[118] Alvarez-Buylla A, Seri B, Doetsch F. Identification of neural stem cells in the adult vertebrate brain. *Brain Res Bull*. 2002 Apr;57(6):751-8.

[119] Doetsch F, Garcia-Verdugo JM, Alvarez-Buylla A. Cellular composition and three-dimensional organization of the subventricular germinal zone in the adult mammalian brain. *J Neurosci*. 1997 Jul 1;17(13):5046-61.

[120] Doetsch F, Garcia-Verdugo JM, Alvarez-Buylla A. *Regeneration of a germinal layer in the adult mammalian brain*. Proc Natl Acad Sci U S A. 1999 Sep 28;96(20):11619-24.

[121] Nyfeler Y, Kirch RD, Mantei N, Leone DP, Radtke F, Suter U, et al. Jagged1 signals in the postnatal subventricular zone are required for neural stem cell self-renewal. *EMBO J*. 2005 Oct 5;24(19):3504-15.

[122] Abraham H, Tornoczky T, Kosztolanyi G, Seress L. Cell formation in the cortical layers of the developing human cerebellum. *Int J Dev Neurosci*. 2001 Feb;19(1):53-62.

[123] Buhren J, Christoph AH, Buslei R, Albrecht S, Wiestler OD, Pietsch T. Expression of the neurotrophin receptor p75NTR in medulloblastomas is correlated with distinct histological and clinical features: evidence for a medulloblastoma subtype derived from the external granule cell layer. *J Neuropathol Exp Neurol*. 2000 Mar;59(3):229-40.

[124] Yokota N, Aruga J, Takai S, Yamada K, Hamazaki M, Iwase T, et al. Predominant expression of human zic in cerebellar granule cell lineage and medulloblastoma. *Cancer Res*. 1996 Jan 15;56(2):377-83.

[125] de Bont JM, Packer RJ, Michiels EM, den Boer ML, Pieters R. Biological background of pediatric medulloblastoma and ependymoma: a review from a translational research perspective. *Neuro Oncol*. 2008 Dec;10(6):1040-60.

[126] de Haas T, Oussoren E, Grajkowska W, Perek-Polnik M, Popovic M, Zadravec-Zaletel L, et al. OTX1 and OTX2 expression correlates with the clinicopathologic classification of medulloblastomas. *J Neuropathol Exp Neurol*. 2006 Feb;65(2):176-86.

[127] Di C, Liao S, Adamson DC, Parrett TJ, Broderick DK, Shi Q, et al. Identification of OTX2 as a medulloblastoma oncogene whose product can be targeted by all-trans retinoic acid. *Cancer Res*. 2005 Feb 1;65(3):919-24.

[128] Adachi K, Mirzadeh Z, Sakaguchi M, Yamashita T, Nikolcheva T, Gotoh Y, et al. Beta-catenin signaling promotes proliferation of progenitor cells in the adult mouse subventricular zone. *Stem Cells.* 2007 Nov;25(11):2827-36.

[129] Chenn A, Walsh CA. Regulation of cerebral cortical size by control of cell cycle exit in neural precursors. *Science.* 2002 Jul 19;297(5580):365-9.

[130] Li W, Tanaka K, Morioka K, Uesaka T, Yamada N, Takamori A, et al. Thymidine phosphorylase gene transfer inhibits vascular smooth muscle cell proliferation by upregulating heme oxygenase-1 and p27KIP1. *Arterioscler Thromb Vasc Biol.* 2005 Jul;25(7):1370-5.

[131] Zhou CJ, Borello U, Rubenstein JL, Pleasure SJ. Neuronal production and precursor proliferation defects in the neocortex of mice with loss of function in the canonical Wnt signaling pathway. *Neuroscience.* 2006 Nov 3;142(4):1119-31.

[132] McMahon AP, Bradley A. The Wnt-1 (int-1) proto-oncogene is required for development of a large region of the mouse brain. *Cell.* 1990 Sep 21;62(6):1073-85.

[133] Schuller U, Rowitch DH. Beta-catenin function is required for cerebellar morphogenesis. *Brain Res.* 2007 Apr 6;1140:161-9.

[134] Thomas KR, Capecchi MR. Targeted disruption of the murine int-1 proto-oncogene resulting in severe abnormalities in midbrain and cerebellar development. *Nature.* 1990 Aug 30;346(6287):847-50.

[135] Alcock J, Scotting P, Sottile V. Bergmann glia as putative stem cells of the mature cerebellum. *Med Hypotheses.* 2007;69(2):341-5.

[136] Fink AJ, Englund C, Daza RA, Pham D, Lau C, Nivison M, et al. Development of the deep cerebellar nuclei: transcription factors and cell migration from the rhombic lip. *J Neurosci.* 2006 Mar 15;26(11):3066-76.

[137] Lee A, Kessler JD, Read TA, Kaiser C, Corbeil D, Huttner WB, et al. Isolation of neural stem cells from the postnatal cerebellum. *Nat Neurosci.* 2005 Jun;8(6):723-9.

[138] Wang VY, Rose MF, Zoghbi HY. Math1 expression redefines the rhombic lip derivatives and reveals novel lineages within the brainstem and cerebellum. *Neuron.* 2005 Oct 6;48(1):31-43.

[139] Clevers H. Wnt/beta-catenin signaling in development and disease. *Cell.* 2006 Nov 3;127(3):469-80.

[140] Kohn AD, Moon RT. Wnt and calcium signaling: beta-catenin-independent pathways. *Cell Calcium.* 2005 Sep-Oct;38(3-4):439-46.

[141] Moon RT, Kohn AD, De Ferrari GV, Kaykas A. WNT and beta-catenin signalling: diseases and therapies. *Nat Rev Genet.* 2004 Sep;5(9):691-701.

[142] Fogarty MP, Kessler JD, Wechsler-Reya RJ. Morphing into cancer: the role of developmental signaling pathways in brain tumor formation. *J Neurobiol.* 2005 Sep 15;64(4):458-75.

[143] Pinto D, Clevers H. Wnt control of stem cells and differentiation in the intestinal epithelium. *Exp Cell Res.* 2005 Jun 10;306(2):357-63.

[144] Luo J, Chen J, Deng ZL, Luo X, Song WX, Sharff KA, et al. Wnt signaling and human diseases: what are the therapeutic implications? *Lab Invest.* 2007 Feb;87(2):97-103.

[145] Eberhart CG, Tihan T, Burger PC. Nuclear localization and mutation of beta-catenin in medulloblastomas. *J Neuropathol Exp Neurol.* 2000 Apr;59(4):333-7.

[146] Yokota N, Nishizawa S, Ohta S, Date H, Sugimura H, Namba H, et al. Role of Wnt pathway in medulloblastoma oncogenesis. *Int J Cancer.* 2002 Sep 10;101(2):198-201.

[147] de Bont JM, Kros JM, Passier MM, Reddingius RE, Sillevis Smitt PA, Luider TM, et al. Differential expression and prognostic significance of SOX genes in pediatric medulloblastoma and ependymoma identified by microarray analysis. *Neuro Oncol.* 2008 Oct;10(5):648-60.

[148] Cheng SY, Yue S. Role and regulation of human tumor suppressor SUFU in Hedgehog signaling. *Adv Cancer Res.* 2008;101:29-43.

[149] Wechsler-Reya RJ, Scott MP. Control of neuronal precursor proliferation in the cerebellum by Sonic Hedgehog. *Neuron.* 1999 Jan;22(1):103-14.

[150] Raffel C, Jenkins RB, Frederick L, Hebrink D, Alderete B, Fults DW, et al. Sporadic medulloblastomas contain PTCH mutations. *Cancer Res.* 1997 Mar 1;57(5):842-5.

[151] Taylor MD, Liu L, Raffel C, Hui CC, Mainprize TG, Zhang X, et al. Mutations in SUFU predispose to medulloblastoma. *Nat Genet.* 2002 Jul;31(3):306-10.

[152] Marino S. Medulloblastoma: developmental mechanisms out of control. *Trends Mol Med.* 2005 Jan;11(1):17-22.

[153] Zurawel RH, Allen C, Chiappa S, Cato W, Biegel J, Cogen P, et al. Analysis of PTCH/SMO/SHH pathway genes in medulloblastoma. *Genes Chromosomes Cancer.* 2000 Jan;27(1):44-51.

[154] Villavicencio EH, Walterhouse DO, Iannaccone PM. The sonic hedgehog-patched-gli pathway in human development and disease. *Am J Hum Genet.* 2000 Nov;67(5):1047-54.

[155] Romer JT, Kimura H, Magdaleno S, Sasai K, Fuller C, Baines H, et al. Suppression of the Shh pathway using a small molecule inhibitor eliminates medulloblastoma in Ptc1(+/-)p53(-/-) mice. *Cancer Cell.* 2004 Sep;6(3):229-40.

[156] Sasai K, Romer JT, Lee Y, Finkelstein D, Fuller C, McKinnon PJ, et al. Shh pathway activity is down-regulated in cultured medulloblastoma cells: implications for preclinical studies. *Cancer Res.* 2006 Apr 15;66(8):4215-22.

[157] Koch U, Radtke F. Notch and cancer: a double-edged sword. *Cell Mol Life* Sci. 2007 Nov;64(21):2746-62.

[158] Solecki DJ, Liu XL, Tomoda T, Fang Y, Hatten ME. Activated Notch2 signaling inhibits differentiation of cerebellar granule neuron precursors by maintaining proliferation. *Neuron.* 2001 Aug 30;31(4):557-68.

[159] Carlotti CG, Jr., Smith C, Rutka JT. The molecular genetics of medulloblastoma: an assessment of new therapeutic targets. *Neurosurg Rev.* 2008 Oct;31(4):359-68; discussion 68-9.

[160] Lasky JL, Wu H. Notch signaling, brain development, and human disease. P*ediatr Res.* 2005 May;57(5 Pt 2):104R-9R.

[161] Fan X, Mikolaenko I, Elhassan I, Ni X, Wang Y, Ball D, et al. Notch1 and notch2 have opposite effects on embryonal brain tumor growth. *Cancer Res.* 2004 Nov 1;64(21):7787-93.

[162] Xu P, Yu S, Jiang R, Kang C, Wang G, Jiang H, et al. Differential expression of Notch family members in astrocytomas and medulloblastomas. *Pathol Oncol Res.* 2009 Dec;15(4):703-10.

[163] Chapman G, Liu L, Sahlgren C, Dahlqvist C, Lendahl U. High levels of Notch signaling down-regulate Numb and Numblike. *J Cell Biol.* 2006 Nov 20;175(4):535-40.

[164] Di Marcotullio L, Ferretti E, Greco A, De Smaele E, Po A, Sico MA, et al. Numb is a suppressor of Hedgehog signalling and targets Gli1 for Itch-dependent ubiquitination. *Nat Cell Biol.* 2006 Dec;8(12):1415-23.

[165] Hallahan AR, Pritchard JI, Hansen S, Benson M, Stoeck J, Hatton BA, et al. The SmoA1 mouse model reveals that notch signaling is critical for the growth and survival of sonic hedgehog-induced medulloblastomas. *Cancer Res.* 2004 Nov 1;64(21):7794-800.

[166] Julian E, Dave RK, Robson JP, Hallahan AR, Wainwright BJ. Canonical Notch signaling is not required for the growth of Hedgehog pathway-induced medulloblastoma. *Oncogene.* 2010 Jun 17;29(24):3465-76.

[167] Hatton BA, Villavicencio EH, Pritchard J, LeBlanc M, Hansen S, Ulrich M, et al. Notch signaling is not essential in sonic hedgehog-activated medulloblastoma. *Oncogene.* 2010 Jul 1;29(26):3865-72.

[168] MacDonald TJ, Brown KM, LaFleur B, Peterson K, Lawlor C, Chen Y, et al. Expression profiling of medulloblastoma: PDGFRA and the RAS/MAPK pathway as therapeutic targets for metastatic disease. *Nat Genet.* 2001 Oct;29(2):143-52.

[169] Neben K, Korshunov A, Benner A, Wrobel G, Hahn M, Kokocinski F, et al. Microarray-based screening for molecular markers in medulloblastoma revealed STK15 as independent predictor for survival. *Cancer Res.* 2004 May 1;64(9):3103-11.

[170] Phillips HS, Kharbanda S, Chen R, Forrest WF, Soriano RH, Wu TD, et al. Molecular subclasses of high-grade glioma predict prognosis, delineate a pattern of disease progression, and resemble stages in neurogenesis. *Cancer Cell.* 2006 Mar;9(3):157-73.

[171] Golub TR, Slonim DK, Tamayo P, Huard C, Gaasenbeek M, Mesirov JP, et al. Molecular classification of cancer: class discovery and class prediction by gene expression monitoring. *Science.* 1999 Oct 15;286(5439):531-7.

[172] Clement V, Sanchez P, de Tribolet N, Radovanovic I, Ruiz IAA. HEDGEHOG-GLI1 Signaling Regulates Human Glioma Growth, Cancer Stem Cell Self-Renewal, and Tumorigenicity. *Curr Biol.* 2006 Dec 26.

[173] Lee Y, Miller HL, Jensen P, Hernan R, Connelly M, Wetmore C, et al. A molecular fingerprint for medulloblastoma. *Cancer Res.* 2003 Sep 1;63(17):5428-37.

[174] Mischel PS, Cloughesy TF, Nelson SF. DNA-microarray analysis of brain cancer: molecular classification for therapy. *Nat Rev Neurosci.* 2004 Oct;5(10):782-92.

[175] Sharma MK, Mansur DB, Reifenberger G, Perry A, Leonard JR, Aldape KD, et al. Distinct genetic signatures among pilocytic astrocytomas relate to their brain region origin. *Cancer Res.* 2007 Feb 1;67(3):890-900.

[176] Ben-Porath I, Thomson MW, Carey VJ, Ge R, Bell GW, Regev A, et al. An embryonic stem cell-like gene expression signature in poorly differentiated aggressive human tumors. *Nat Genet.* 2008 May;40(5):499-507.

[177] Naxerova K, Bult CJ, Peaston A, Fancher K, Knowles BB, Kasif S, et al. Analysis of gene expression in a developmental context emphasizes distinct biological leitmotifs in human cancers. *Genome Biol.* 2008;9(7):R108.

[178] Park PC, Taylor MD, Mainprize TG, Becker LE, Ho M, Dura WT, et al. Transcriptional profiling of medulloblastoma in children. *J Neurosurg.* 2003 Sep;99(3):534-41.

[179] Boon K, Edwards JB, Siu IM, Olschner D, Eberhart CG, Marra MA, et al. Comparison of medulloblastoma and normal neural transcriptomes identifies a restricted set of activated genes. *Oncogene.* 2003 Oct 23;22(48):7687-94.

[180] Vescovi AL, Galli R, Reynolds BA. Brain tumour stem cells. *Nat Rev Cancer*. 2006 Jun;6(6):425-36.

[181] Bruce WR, Van Der Gaag H. A Quantitative Assay for the Number of Murine Lymphoma Cells Capable of Proliferation *in Vivo*. *Nature*. 1963 Jul 6;199:79-80.

[182] Furth J, Kahn M. The transmission of leukaemia of mice with a single cell. *Am J Cancer*. 1937;31:276–82.

[183] Bonnet D, Dick JE. Human acute myeloid leukemia is organized as a hierarchy that originates from a primitive hematopoietic cell. *Nat Med.* 1997 Jul;3(7):730-7.

[184] Lapidot T, Sirard C, Vormoor J, Murdoch B, Hoang T, Caceres-Cortes J, et al. A cell initiating human acute myeloid leukaemia after transplantation into SCID mice. *Nature*. 1994 Feb 17;367(6464):645-8.

[185] Ward RJ, Dirks PB. Cancer stem cells: at the headwaters of tumor development. *Annu Rev Pathol.* 2007;2:175-89.

[186] Al-Hajj M, Wicha MS, Benito-Hernandez A, Morrison SJ, Clarke MF. *Prospective identification of tumorigenic breast cancer cells*. Proc Natl Acad Sci U S A. 2003 Apr 1;100(7):3983-8.

[187] Pardal R, Clarke MF, Morrison SJ. Applying the principles of stem-cell biology to cancer. *Nat Rev Cancer.* 2003 Dec;3(12):895-902.

[188] Reynolds BA, Weiss S. Generation of neurons and astrocytes from isolated cells of the adult mammalian central nervous system. *Science*. 1992 Mar 27;255(5052):1707-10.

[189] Ignatova TN, Kukekov VG, Laywell ED, Suslov ON, Vrionis FD, Steindler DA. Human cortical glial tumors contain neural stem-like cells expressing astroglial and neuronal markers *in vitro*. *Glia*. 2002 Sep;39(3):193-206.

[190] Hemmati HD, Nakano I, Lazareff JA, Masterman-Smith M, Geschwind DH, Bronner-Fraser M, et al. *Cancerous stem cells can arise from pediatric brain tumors.* Proc Natl Acad Sci U S A. 2003 Dec 9;100(25):15178-83.

[191] Singh SK, Clarke ID, Terasaki M, Bonn VE, Hawkins C, Squire J, et al. Identification of a cancer stem cell in human brain tumors. *Cancer Res*. 2003 Sep 15;63(18):5821-28.

[192] Taylor MD, Poppleton H, Fuller C, Su X, Liu Y, Jensen P, et al. Radial glia cells are candidate stem cells of ependymoma. *Cancer Cell.* 2005 Oct;8(4):323-35.

[193] Singh SK, Hawkins C, Clarke ID, Squire JA, Bayani J, Hide T, et al. Identification of human brain tumour initiating cells. *Nature*. 2004 Nov 18;432(7015):396-401.

[194] Collins AT, Berry PA, Hyde C, Stower MJ, Maitland NJ. Prospective identification of tumorigenic prostate cancer stem cells. *Cancer Res.* 2005 Dec 1;65(23):10946-51.

[195] Hermann PC, Huber SL, Herrler T, Aicher A, Ellwart JW, Guba M, et al. Distinct populations of cancer stem cells determine tumor growth and metastatic activity in human pancreatic cancer. *Cell Stem Cell.* 2007 Sep 13;1(3):313-23.

[196] O'Brien CA, Pollett A, Gallinger S, Dick JE. A human colon cancer cell capable of initiating tumour growth in immunodeficient mice. *Nature*. 2007 Jan 4;445(7123):106-10.

[197] Ricci-Vitiani L, Lombardi DG, Pilozzi E, Biffoni M, Todaro M, Peschle C, et al. Identification and expansion of human colon-cancer-initiating cells. *Nature*. 2007 Jan 4;445(7123):111-5.

[198] Dirks PB. Brain tumour stem cells: the undercurrents of human brain cancer and their relationship to neural stem cells. *Philos Trans R Soc Lond B Biol Sci.* 2008 Jan 12;363(1489):139-52.

[199] Singh S, Dirks PB. Brain tumor stem cells: identification and concepts. *Neurosurg Clin N Am.* 2007 Jan;18(1):31-8, viii.

[200] Clarke MF, Fuller M. Stem cells and cancer: two faces of eve. *Cell.* 2006 Mar 24;124(6):1111-5.

[201] Clevers H. Stem cells, asymmetric division and cancer. *Nat Genet.* 2005 Oct;37(10):1027-8.

[202] Lee CY, Robinson KJ, Doe CQ. Lgl, Pins and aPKC regulate neuroblast self-renewal versus differentiation. *Nature.* 2006 Feb 2;439(7076):594-8.

[203] Klezovitch O, Fernandez TE, Tapscott SJ, Vasioukhin V. Loss of cell polarity causes severe brain dysplasia in Lgl1 knockout mice. *Genes Dev.* 2004 Mar 1;18(5):559-71.

[204] Xie Z, Chin LS. Molecular and cell biology of brain tumor stem cells: lessons from neural progenitor/stem cells. *Neurosurg Focus.* 2008;24(3-4):E25.

[205] Yuan X, Curtin J, Xiong Y, Liu G, Waschsmann-Hogiu S, Farkas DL, et al. Isolation of cancer stem cells from adult glioblastoma multiforme. *Oncogene.* 2004 Dec 16;23(58):9392-400.

[206] Zhang QB, Ji XY, Huang Q, Dong J, Zhu YD, Lan Q. Differentiation profile of brain tumor stem cells: a comparative study with neural stem cells. *Cell Res.* 2006 Dec;16(12):909-15.

[207] Weigmann A, Corbeil D, Hellwig A, Huttner WB. *Prominin, a novel microvilli-specific polytopic membrane protein of the apical surface of epithelial cells, is targeted to plasmalemmal protrusions of non-epithelial cells.* Proc Natl Acad Sci U S A. 1997 Nov 11;94(23):12425-30.

[208] Yin AH, Miraglia S, Zanjani ED, Almeida-Porada G, Ogawa M, Leary AG, et al. AC133, a novel marker for human hematopoietic stem and progenitor cells. *Blood.* 1997 Dec 15;90(12):5002-12.

[209] Corbeil D, Roper K, Fargeas CA, Joester A, Huttner WB. Prominin: a story of cholesterol, plasma membrane protrusions and human pathology. *Traffic.* 2001 Feb;2(2):82-91.

[210] Fargeas CA, Corbeil D, Huttner WB. AC133 antigen, CD133, prominin-1, prominin-2, etc.: prominin family gene products in need of a rational nomenclature. *Stem Cells.* 2003;21(4):506-8.

[211] Corbeil D, Roper K, Hellwig A, Tavian M, Miraglia S, Watt SM, et al. The human AC133 hematopoietic stem cell antigen is also expressed in epithelial cells and targeted to plasma membrane protrusions. *J Biol Chem.* 2000 Feb 25;275(8):5512-20.

[212] Shmelkov SV, St Clair R, Lyden D, Rafii S. AC133/CD133/Prominin-1. Int *J Biochem Cell Biol.* 2005 Apr;37(4):715-9.

[213] Yu S, Zhang JZ, Zhao CL, Zhang HY, Xu Q. Isolation and characterization of the CD133+ precursors from the ventricular zone of human fetal brain by magnetic affinity cell sorting. *Biotechnol Lett.* 2004 Jul;26(14):1131-6.

[214] Maw MA, Corbeil D, Koch J, Hellwig A, Wilson-Wheeler JC, Bridges RJ, et al. A frameshift mutation in prominin (mouse)-like 1 causes human retinal degeneration. *Hum Mol Genet.* 2000 Jan 1;9(1):27-34.

[215] Richardson GD, Robson CN, Lang SH, Neal DE, Maitland NJ, Collins AT. CD133, a novel marker for human prostatic epithelial stem cells. *J Cell Sci.* 2004 Jul 15;117(Pt 16):3539-45.

[216] Kosodo Y, Roper K, Haubensak W, Marzesco AM, Corbeil D, Huttner WB. Asymmetric distribution of the apical plasma membrane during neurogenic divisions of mammalian neuroepithelial cells. *EMBO J.* 2004 Jun 2;23(11):2314-24.

[217] Miraglia S, Godfrey W, Yin AH, Atkins K, Warnke R, Holden JT, et al. A novel five-transmembrane hematopoietic stem cell antigen: isolation, characterization, and molecular cloning. *Blood.* 1997 Dec 15;90(12):5013-21.

[218] Bidlingmaier S, Zhu X, Liu B. The utility and limitations of glycosylated human CD133 epitopes in defining cancer stem cells. *J Mol Med.* 2008 Sep;86(9):1025-32.

[219] Yu Y, Flint A, Dvorin EL, Bischoff J. AC133-2, a novel isoform of human AC133 stem cell antigen. *J Biol Chem.* 2002 Jun 7;277(23):20711-6.

[220] Peichev M, Naiyer AJ, Pereira D, Zhu Z, Lane WJ, Williams M, et al. Expression of VEGFR-2 and AC133 by circulating human CD34(+) cells identifies a population of functional endothelial precursors. *Blood.* 2000 Feb 1;95(3):952-8.

[221] Chenn A, McConnell SK. Cleavage orientation and the asymmetric inheritance of Notch1 immunoreactivity in mammalian neurogenesis. *Cell.* 1995 Aug 25;82(4):631-41.

[222] Roper K, Corbeil D, Huttner WB. Retention of prominin in microvilli reveals distinct cholesterol-based lipid micro-domains in the apical plasma membrane. *Nat Cell Biol.* 2000 Sep;2(9):582-92.

[223] Marzesco AM, Janich P, Wilsch-Brauninger M, Dubreuil V, Langenfeld K, Corbeil D, et al. Release of extracellular membrane particles carrying the stem cell marker prominin-1 (CD133) from neural progenitors and other epithelial cells. *J Cell Sci.* 2005 Jul 1;118(Pt 13):2849-58.

[224] Dubreuil V, Marzesco AM, Corbeil D, Huttner WB, Wilsch-Brauninger M. Midbody and primary cilium of neural progenitors release extracellular membrane particles enriched in the stem cell marker prominin-1. *J Cell Biol.* 2007 Feb 12;176(4):483-95.

[225] Huttner HB, Janich P, Kohrmann M, Jaszai J, Siebzehnrubl F, Blumcke I, et al. The stem cell marker prominin-1/CD133 on membrane particles in human cerebrospinal fluid offers novel approaches for studying central nervous system disease. *Stem Cells.* 2008 Mar;26(3):698-705.

[226] Reya T, Morrison SJ, Clarke MF, Weissman IL. Stem cells, cancer, and cancer stem cells. *Nature.* 2001 Nov 1;414(6859):105-11.

[227] Sutter R, Yadirgi G, Marino S. Neural stem cells, tumour stem cells and brain tumours: dangerous relationships? *Biochim Biophys Acta.* 2007 Dec;1776(2):125-37.

[228] Singh SK, Clarke ID, Hide T, Dirks PB. Cancer stem cells in nervous system tumors. *Oncogene.* 2004 Sep 20;23(43):7267-73.

[229] Holland EC, Celestino J, Dai C, Schaefer L, Sawaya RE, Fuller GN. Combined activation of Ras and Akt in neural progenitors induces glioblastoma formation in mice. *Nat Genet.* 2000 May;25(1):55-7.

[230] Moon JH, Yoon BS, Kim B, Park G, Jung HY, Maeng I, et al. Induction of neural stem cell-like cells (NSCLCs) from mouse astrocytes by Bmi1. *Biochem Biophys Res Commun.* 2008 Jun 27;371(2):267-72.

[231] Rudin CM, Hann CL, Laterra J, Yauch RL, Callahan CA, Fu L, et al. Treatment of medulloblastoma with hedgehog pathway inhibitor GDC-0449. *N Engl J Med.* 2009 Sep 17;361(12):1173-8.

[232] Golmohammadi MG, Blackmore DG, Large B, Azari H, Esfandiary E, Paxinos G, et al. Comparative analysis of the frequency and distribution of stem and progenitor cells in the adult mouse brain. *Stem Cells*. 2008 Apr;26(4):979-87.

[233] Shervington A, Lu C. Expression of multidrug resistance genes in normal and cancer stem cells. *Cancer Invest.* 2008 Jun;26(5):535-42.

[234] Uchida N, Buck DW, He D, Reitsma MJ, Masek M, Phan TV, et al. *Direct isolation of human central nervous system stem cells.* Proc Natl Acad Sci U S A. 2000 Dec 19;97(26):14720-5.

[235] Jandial R, Singec I, Ames CP, Snyder EY. Genetic modification of neural stem cells. *Mol Ther.* 2008 Mar;16(3):450-7.

[236] Chaichana K, Zamora-Berridi G, Camara-Quintana J, Quinones-Hinojosa A. Neurosphere Assays: Growth Factors and Hormone Differences in Tumor and Non-tumor Studies. *Stem Cells.* 2006 Aug 31.

[237] Reynolds BA, Rietze RL. Neural stem cells and neurospheres--re-evaluating the relationship. *Nat Methods.* 2005 May;2(5):333-6.

[238] Peh GS, Lang R, Pera M, Hawes S. CD133 expression by neural progenitors derived from human embryonic stem cells and its use for their prospective isolation. *Stem Cells Dev.* 2009 Mar;18(2):269-82.

[239] Hall PE, Lathia JD, Miller NG, Caldwell MA, Ffrench-Constant C. Integrins are markers of human neural stem cells. *Stem Cells.* 2006 May 11.

[240] Alvarez-Buylla A, Temple S. Stem cells in the developing and adult nervous system. *J Neurobiol.* 1998 Aug;36(2):105-10.

[241] Galli R, Binda E, Orfanelli U, Cipelletti B, Gritti A, De Vitis S, et al. Isolation and characterization of tumorigenic, stem-like neural precursors from human glioblastoma. *Cancer Res.* 2004 Oct 1;64(19):7011-21.

[242] Jackson EL, Garcia-Verdugo JM, Gil-Perotin S, Roy M, Quinones-Hinojosa A, VandenBerg S, et al. PDGFR alpha-positive B cells are neural stem cells in the adult SVZ that form glioma-like growths in response to increased PDGF signaling. *Neuron.* 2006 Jul 20;51(2):187-99.

[243] Bao S, Wu Q, McLendon RE, Hao Y, Shi Q, Hjelmeland AB, et al. Glioma stem cells promote radioresistance by preferential activation of the DNA damage response. *Nature*. 2006 Oct 18.

[244] Beier D, Hau P, Proescholdt M, Lohmeier A, Wischhusen J, Oefner PJ, et al. CD133(+) and CD133(-) glioblastoma-derived cancer stem cells show differential growth characteristics and molecular profiles. *Cancer Res.* 2007 May 1;67(9):4010-5.

[245] Gunther HS, Schmidt NO, Phillips HS, Kemming D, Kharbanda S, Soriano R, et al. Glioblastoma-derived stem cell-enriched cultures form distinct subgroups according to molecular and phenotypic criteria. *Oncogene*. 2008 May 1;27(20):2897-909.

[246] Joo KM, Kim SY, Jin X, Song SY, Kong DS, Lee JI, et al. Clinical and biological implications of CD133-positive and CD133-negative cells in glioblastomas. *Lab Invest.* 2008 Aug;88(8):808-15.

[247] Wu A, Oh S, Wiesner SM, Ericson K, Chen L, Hall WA, et al. Persistence of CD133+ cells in human and mouse glioma cell lines: detailed characterization of GL261 glioma cells with cancer stem cell-like properties. *Stem Cells Dev.* 2008 Feb;17(1):173-84.

[248] Read TA, Fogarty MP, Markant SL, McLendon RE, Wei Z, Ellison DW, et al. Identification of CD15 as a marker for tumor-propagating cells in a mouse model of medulloblastoma. *Cancer Cell.* 2009 Feb 3;15(2):135-47.

[249] Son MJ, Woolard K, Nam DH, Lee J, Fine HA. SSEA-1 is an enrichment marker for tumor-initiating cells in human glioblastoma. *Cell Stem Cell.* 2009 May 8;4(5):440-52.

[250] Ward RJ, Lee L, Graham K, Satkunendran T, Yoshikawa K, Ling E, et al. Multipotent CD15+ cancer stem cells in patched-1-deficient mouse medulloblastoma. *Cancer Res.* 2009 Jun 1;69(11):4682-90.

[251] Liu G, Yuan X, Zeng Z, Tunici P, Ng H, Abdulkadir IR, et al. Analysis of gene expression and chemoresistance of CD133+ cancer stem cells in glioblastoma. *Mol Cancer.* 2006;5:67.

[252] Salmaggi A, Boiardi A, Gelati M, Russo A, Calatozzolo C, Ciusani E, et al. Glioblastoma-derived tumorospheres identify a population of tumor stem-like cells with angiogenic potential and enhanced multidrug resistance phenotype. *Glia.* 2006 Dec;54(8):850-60.

[253] Cheng JX, Liu BL, Zhang X. How powerful is CD133 as a cancer stem cell marker in brain tumors? *Cancer Treat Rev.* 2009 Aug;35(5):403-8.

[254] Zhang M, Song T, Yang L, Chen R, Wu L, Yang Z, et al. Nestin and CD133: valuable stem cell-specific markers for determining clinical outcome of glioma patients. *J Exp Clin Cancer Res.* 2008;27:85.

[255] Blazek ER, Foutch JL, Maki G. Daoy medulloblastoma cells that express CD133 are radioresistant relative to CD133- cells, and the CD133+ sector is enlarged by hypoxia. *Int J Radiat Oncol Biol Phys.* 2007 Jan 1;67(1):1-5.

[256] Diamandis P, Wildenhain J, Clarke ID, Sacher AG, Graham J, Bellows DS, et al. Chemical genetics reveals a complex functional ground state of neural stem cells. *Nat Chem Biol.* 2007 May;3(5):268-73.

[257] Piccirillo SG, Reynolds BA, Zanetti N, Lamorte G, Binda E, Broggi G, et al. Bone morphogenetic proteins inhibit the tumorigenic potential of human brain tumour-initiating cells. *Nature.* 2006 Dec 7;444(7120):761-5.

[258] Lee J, Son MJ, Woolard K, Donin NM, Li A, Cheng CH, et al. Epigenetic-mediated dysfunction of the bone morphogenetic protein pathway inhibits differentiation of glioblastoma-initiating cells. *Cancer Cell.* 2008 Jan;13(1):69-80.

[259] Bleau AM, Howard BM, Taylor LA, Gursel D, Greenfield JP, Lim Tung HY, et al. New strategy for the analysis of phenotypic marker antigens in brain tumor-derived neurospheres in mice and humans. *Neurosurg Focus.* 2008;24(3-4):E28.

[260] Gilbertson RJ, Rich JN. Making a tumour's bed: glioblastoma stem cells and the vascular niche. *Nat Rev Cancer.* 2007 Oct;7(10):733-6.

[261] Bao S, Wu Q, Sathornsumetee S, Hao Y, Li Z, Hjelmeland AB, et al. Stem Cell-like Glioma Cells Promote Tumor Angiogenesis through Vascular Endothelial Growth Factor. *Cancer Res.* 2006 Aug 15;66(16):7843-8.

[262] De Palma M, Naldini L. Role of haematopoietic cells and endothelial progenitors in tumour angiogenesis. *Biochim Biophys Acta.* 2006 Aug;1766(1):159-66.

[263] Jin K, Zhu Y, Sun Y, Mao XO, Xie L, Greenberg DA. *Vascular endothelial growth factor (VEGF) stimulates neurogenesis in vitro and in vivo.* Proc Natl Acad Sci U S A. 2002 Sep 3;99(18):11946-50.

[264] Calabrese C, Poppleton H, Kocak M, Hogg TL, Fuller C, Hamner B, et al. A perivascular niche for brain tumor stem cells. *Cancer Cell.* 2007 Jan;11(1):69-82.

[265] Santarelli JG, Sarkissian V, Hou LC, Veeravagu A, Tse V. Molecular events of brain metastasis. *Neurosurg Focus.* 2007;22(3):E1.

[266] Garcia-Barros M, Paris F, Cordon-Cardo C, Lyden D, Rafii S, Haimovitz-Friedman A, et al. Tumor response to radiotherapy regulated by endothelial cell apoptosis. *Science.* 2003 May 16;300(5622):1155-9.

[267] Vredenburgh JJ, Desjardins A, Herndon JE, 2nd, Marcello J, Reardon DA, Quinn JA, et al. Bevacizumab plus irinotecan in recurrent glioblastoma multiforme. *J Clin Oncol.* 2007 Oct 20;25(30):4722-9.

[268] Abate-Shen C. A new generation of mouse models of cancer for translational research. *Clin Cancer Res.* 2006 Sep 15;12(18):5274-6.

[269] Goodrich LV, Milenkovic L, Higgins KM, Scott MP. Altered neural cell fates and medulloblastoma in mouse patched mutants. *Science.* 1997 Aug 22;277(5329):1109-13.

[270] Hatton BA, Villavicencio EH, Tsuchiya KD, Pritchard JI, Ditzler S, Pullar B, et al. The Smo/Smo model: hedgehog-induced medulloblastoma with 90% incidence and leptomeningeal spread. *Cancer Res.* 2008 Mar 15;68(6):1768-76.

[271] Tremblay MR, Nesler M, Weatherhead R, Castro AC. Recent patents for Hedgehog pathway inhibitors for the treatment of malignancy. *Expert Opin Ther Pat.* 2009 Aug;19(8):1039-56.

[272] Yauch RL, Dijkgraaf GJ, Alicke B, Januario T, Ahn CP, Holcomb T, et al. Smoothened mutation confers resistance to a Hedgehog pathway inhibitor in medulloblastoma. *Science.* 2009 Oct 23;326(5952):572-4.

[273] Von Hoff DD, LoRusso PM, Rudin CM, Reddy JC, Yauch RL, Tibes R, et al. Inhibition of the hedgehog pathway in advanced basal-cell carcinoma. *N Engl J Med.* 2009 Sep 17;361(12):1164-72.

[274] Lee Y, McKinnon PJ. DNA ligase IV suppresses medulloblastoma formation. *Cancer Res.* 2002 Nov 15;62(22):6395-9.

[275] Yan CT, Kaushal D, Murphy M, Zhang Y, Datta A, Chen C, et al. *XRCC4 suppresses medulloblastomas with recurrent translocations in p53-deficient mice.* Proc Natl Acad Sci U S A. 2006 May 9;103(19):7378-83.

[276] Fomchenko EI, Holland EC. Mouse models of brain tumors and their applications in preclinical trials. *Clin Cancer Res.* 2006 Sep 15;12(18):5288-97.

[277] Huse JT, Holland EC. Genetically engineered mouse models of brain cancer and the promise of preclinical testing. *Brain Pathol.* 2009 Jan;19(1):132-43.

[278] Raffel C. Medulloblastoma: molecular genetics and animal models. *Neoplasia.* 2004 Jul-Aug;6(4):310-22.

[279] Gage FH. Mammalian neural stem cells. *Science.* 2000 Feb 25;287(5457):1433-8.

[280] Rajan P, Snyder E. Neural stem cells and their manipulation. *Methods Enzymol.* 2006;419:23-52.

[281] Coskun V, Wu H, Blanchi B, Tsao S, Kim K, Zhao J, et al. *CD133+ neural stem cells in the ependyma of mammalian postnatal forebrain.* Proc Natl Acad Sci U S A. 2008 Jan 22;105(3):1026-31.

[282] Pfenninger CV, Roschupkina T, Hertwig F, Kottwitz D, Englund E, Bengzon J, et al. CD133 is not present on neurogenic astrocytes in the adult subventricular zone, but on

embryonic neural stem cells, ependymal cells, and glioblastoma cells. *Cancer Res.* 2007 Jun 15;67(12):5727-36.

[283] Doetsch F, Caille I, Lim DA, Garcia-Verdugo JM, Alvarez-Buylla A. Subventricular zone astrocytes are neural stem cells in the adult mammalian brain. *Cell.* 1999 Jun 11;97(6):703-16.

[284] Easterday MC, Dougherty JD, Jackson RL, Ou J, Nakano I, Paucar AA, et al. Neural progenitor genes. Germinal zone expression and analysis of genetic overlap in stem cell populations. *Dev Biol.* 2003 Dec 15;264(2):309-22.

[285] Encinas JM, Vaahtokari A, Enikolopov G. *Fluoxetine targets early progenitor cells in the adult brain.* Proc Natl Acad Sci U S A. 2006 May 23;103(21):8233-8.

[286] Cai J, Shin S, Wright L, Liu Y, Zhou D, Xue H, et al. Massively parallel signature sequencing profiling of fetal human neural precursor cells. *Stem Cells Dev.* 2006 Apr;15(2):232-44.

[287] Shin S, Sun Y, Liu Y, Khaner H, Svant S, Cai J, et al. Whole genome analysis of human neural stem cells derived from embryonic stem cells and stem and progenitor cells isolated from fetal tissue. *Stem Cells.* 2007 May;25(5):1298-306.

[288] Reubinoff BE, Itsykson P, Turetsky T, Pera MF, Reinhartz E, Itzik A, et al. Neural progenitors from human embryonic stem cells. *Nat Biotechnol.* 2001 Dec;19(12):1134-40.

[289] Tabar V, Panagiotakos G, Greenberg ED, Chan BK, Sadelain M, Gutin PH, et al. Migration and differentiation of neural precursors derived from human embryonic stem cells in the rat brain. *Nat Biotechnol.* 2005 May;23(5):601-6.

[290] Wernig M, Benninger F, Schmandt T, Rade M, Tucker KL, Bussow H, et al. Functional integration of embryonic stem cell-derived neurons *in vivo*. *J Neurosci.* 2004 Jun 2;24(22):5258-68.

[291] Joannides AJ, Fiore-Heriche C, Battersby AA, Athauda-Arachchi P, Bouhon IA, Williams L, et al. A scaleable and defined system for generating neural stem cells from human embryonic stem cells. *Stem Cells.* 2006 Nov 9.

[292] Moliner A, Enfors P, Ibanez CF, Andang M. Mouse embryonic stem cell-derived spheres with distinct neurogenic potentials. *Stem Cells Dev.* 2008 Apr;17(2):233-43.

[293] Zhang SC, Wernig M, Duncan ID, Brustle O, Thomson JA. *In vitro* differentiation of transplantable neural precursors from human embryonic stem cells. *Nat Biotechnol.* 2001 Dec;19(12):1129-33.

[294] Colombo E, Giannelli SG, Galli R, Tagliafico E, Foroni C, Tenedini E, et al. Embryonic stem-derived versus somatic neural stem cells: a comparative analysis of their developmental potential and molecular phenotype. *Stem Cells.* 2006 Apr;24(4):825-34.

[295] Pera MF, Andrade J, Houssami S, Reubinoff B, Trounson A, Stanley EG, et al. Regulation of human embryonic stem cell differentiation by BMP-2 and its antagonist noggin. *J Cell Sci.* 2004 Mar 1;117(Pt 7):1269-80.

[296] Cohen MA, Itsykson P, Reubinoff BE. Neural differentiation of human ES cells. *Curr Protoc Cell Biol.* 2007 Sep;Chapter 23:Unit 23 7.

[297] Adewumi O, Aflatoonian B, Ahrlund-Richter L, Amit M, Andrews PW, Beighton G, et al. Characterization of human embryonic stem cell lines by the International Stem Cell Initiative. *Nat Biotechnol.* 2007 Jul;25(7):803-16.

[298] Svendsen CN, ter Borg MG, Armstrong RJ, Rosser AE, Chandran S, Ostenfeld T, et al. A new method for the rapid and long term growth of human neural precursor cells. *J Neurosci Methods.* 1998 Dec 1;85(2):141-52.

[299] Tamaki S, Eckert K, He D, Sutton R, Doshe M, Jain G, et al. Engraftment of sorted/expanded human central nervous system stem cells from fetal brain. *J Neurosci* Res. 2002 Sep 15;69(6):976-86.

[300] Davidson BL, Breakefield XO. Viral vectors for gene delivery to the nervous system. *Nat Rev Neurosci.* 2003 May;4(5):353-64.

[301] Azzouz M, Kingsman SM, Mazarakis ND. Lentiviral vectors for treating and modeling human CNS disorders. *J Gene Med.* 2004 Sep;6(9):951-62.

[302] Dull T, Zufferey R, Kelly M, Mandel RJ, Nguyen M, Trono D, et al. A third-generation lentivirus vector with a conditional packaging system. *J Virol.* 1998 Nov;72(11):8463-71.

[303] Kim VN, Mitrophanous K, Kingsman SM, Kingsman AJ. Minimal requirement for a lentivirus vector based on human immunodeficiency virus type 1. *J Virol.* 1998 Jan;72(1):811-6.

[304] Hoffrogge R, Mikkat S, Scharf C, Beyer S, Christoph H, Pahnke J, et al. 2-DE proteome analysis of a proliferating and differentiating human neuronal stem cell line (ReNcell VM). *Proteomics.* 2006 Mar;6(6):1833-47.

[305] Lois C, Hong EJ, Pease S, Brown EJ, Baltimore D. Germline transmission and tissue-specific expression of transgenes delivered by lentiviral vectors. *Science.* 2002 Feb 1;295(5556):868-72.

[306] McGrew MJ, Sherman A, Ellard FM, Lillico SG, Gilhooley HJ, Kingsman AJ, et al. Efficient production of germline transgenic chickens using lentiviral vectors. *EMBO Rep.* 2004 Jul;5(7):728-33.

[307] Abordo-Adesida E, Follenzi A, Barcia C, Sciascia S, Castro MG, Naldini L, et al. Stability of lentiviral vector-mediated transgene expression in the brain in the presence of systemic antivector immune responses. *Hum Gene Ther.* 2005 Jun;16(6):741-51.

[308] Kordower JH, Emborg ME, Bloch J, Ma SY, Chu Y, Leventhal L, et al. Neurodegeneration prevented by lentiviral vector delivery of GDNF in primate models of Parkinson's disease. *Science.* 2000 Oct 27;290(5492):767-73.

[309] Mazarakis ND, Azzouz M, Rohll JB, Ellard FM, Wilkes FJ, Olsen AL, et al. Rabies virus glycoprotein pseudotyping of lentiviral vectors enables retrograde axonal transport and access to the nervous system after peripheral delivery. *Hum Mol Genet.* 2001 Sep 15;10(19):2109-21.

[310] Wong LF, Goodhead L, Prat C, Mitrophanous KA, Kingsman SM, Mazarakis ND. Lentivirus-mediated gene transfer to the central nervous system: therapeutic and research applications. *Hum Gene Ther.* 2006 Jan;17(1):1-9.

[311] Beard BC, Dickerson D, Beebe K, Gooch C, Fletcher J, Okbinoglu T, et al. Comparison of HIV-derived lentiviral and MLV-based gammaretroviral vector integration sites in primate repopulating cells. *Mol Ther.* 2007 Jul;15(7):1356-65.

[312] Capowski EE, Schneider BL, Ebert AD, Seehus CR, Szulc J, Zufferey R, et al. Lentiviral vector-mediated genetic modification of human neural progenitor cells for ex vivo gene therapy. *J Neurosci Methods.* 2007 Jul 30;163(2):338-49.

[313] Glatzel M, Flechsig E, Navarro B, Klein MA, Paterna JC, Bueler H, et al. *Adenoviral and adeno-associated viral transfer of genes to the peripheral nervous system.* Proc Natl Acad Sci U S A. 2000 Jan 4;97(1):442-7.

[314] Ma HI, Lin SZ, Chiang YH, Li J, Chen SL, Tsao YP, et al. Intratumoral gene therapy of malignant brain tumor in a rat model with angiostatin delivered by adeno-associated viral (AAV) vector. *Gene Ther.* 2002 Jan;9(1):2-11.

[315] Smith-Arica JR, Thomson AJ, Ansell R, Chiorini J, Davidson B, McWhir J. Infection efficiency of human and mouse embryonic stem cells using adenoviral and adeno-associated viral vectors. *Cloning Stem Cells.* 2003;5(1):51-62.

[316] Hughes SM, Moussavi-Harami F, Sauter SL, Davidson BL. Viral-mediated gene transfer to mouse primary neural progenitor cells. *Mol Ther.* 2002 Jan;5(1):16-24.

[317] Volpers C, Kochanek S. Adenoviral vectors for gene transfer and therapy. *J Gene Med.* 2004 Feb;6 Suppl 1:S164-71.

[318] Carson SD. Receptor for the group B coxsackieviruses and adenoviruses: CAR. *Rev Med Viro*l. 2001 Jul-Aug;11(4):219-26.

[319] Coyne CB, Bergelson JM. CAR: a virus receptor within the tight junction. *Adv Drug Del Rev.* 2005 Apr 25;57(6):869-82.

[320] Hauwel M, Furon E, Gasque P. Molecular and cellular insights into the coxsackie-adenovirus receptor: role in cellular interactions in the stem cell niche. *Brain Res Brain Res Rev.* 2005 Apr;48(2):265-72.

[321] Honda T, Saitoh H, Masuko M, Katagiri-Abe T, Tominaga K, Kozakai I, et al. The coxsackievirus-adenovirus receptor protein as a cell adhesion molecule in the developing mouse brain. *Brain Res Mol Brain Res.* 2000 Apr 14;77(1):19-28.

[322] Hotta Y, Honda T, Naito M, Kuwano R. Developmental distribution of coxsackie virus and adenovirus receptor localized in the nervous system. *Brain Res Dev Brain Res.* 2003 Jun 12;143(1):1-13.

[323] Persson A, Fan X, Widegren B, Englund E. Cell type- and region-dependent coxsackie adenovirus receptor expression in the central nervous system. *J Neurooncol.* 2006 May;78(1):1-6.

[324] Raschperger E, Thyberg J, Pettersson S, Philipson L, Fuxe J, Pettersson RF. The coxsackie- and adenovirus receptor (CAR) is an *in vivo* marker for epithelial tight junctions, with a potential role in regulating permeability and tissue homeostasis. *Exp Cell Res.* 2006 May 15;312(9):1566-80.

[325] Russell WC. Update on adenovirus and its vectors. *J Gen Virol.* 2000 Nov;81(Pt 11):2573-604.

[326] Hermens WT, Giger RJ, Holtmaat AJ, Dijkhuizen PA, Houweling DA, Verhaagen J. Transient gene transfer to neurons and glia: analysis of adenoviral vector performance in the CNS and PNS. *J Neurosci Methods.* 1997 Jan;71(1):85-98.

[327] Watanabe TS, Ohtori S, Koda M, Aoki Y, Doya H, Shirasawa H, et al. Adenoviral gene transfer in the peripheral nervous system. *J Orthop Sci.* 2006 Jan;11(1):64-9.

[328] Feng M, Jackson WH, Jr., Goldman CK, Rancourt C, Wang M, Dusing SK, et al. Stable *in vivo* gene transduction via a novel adenoviral/retroviral chimeric vector. *Nat Biotechnol.* 1997 Sep;15(9):866-70.

[329] Reynolds PN, Feng M, Curiel DT. Chimeric viral vectors--the best of both worlds? *Mol Med Today.* 1999 Jan;5(1):25-31.

[330] Aboody KS, Brown A, Rainov NG, Bower KA, Liu S, Yang W, et al. *Neural stem cells display extensive tropism for pathology in adult brain: evidence from intracranial gliomas.* Proc Natl Acad Sci U S A. 2000 Nov 7;97(23):12846-51.

[331] Aboody KS, Najbauer J, Danks MK. Stem and progenitor cell-mediated tumor selective gene therapy. *Gene Ther.* 2008 May;15(10):739-52.

[332] Dickson PV, Hamner JB, Burger RA, Garcia E, Ouma AA, Kim SU, et al. Intravascular administration of tumor tropic neural progenitor cells permits targeted delivery of interferon-beta and restricts tumor growth in a murine model of disseminated neuroblastoma. *J Pediatr Surg.* 2007 Jan;42(1):48-53.

[333] Yip S, Sabetrasekh R, Sidman RL, Snyder EY. Neural stem cells as novel cancer therapeutic vehicles. *Eur J Cancer.* 2006 Jun;42(9):1298-308.

[334] Alemany R, Suzuki K, Curiel DT. Blood clearance rates of adenovirus type 5 in mice. *J Gen Virol.* 2000 Nov;81(Pt 11):2605-9.

[335] Wolff G, Worgall S, van Rooijen N, Song WR, Harvey BG, Crystal RG. Enhancement of *in vivo* adenovirus-mediated gene transfer and expression by prior depletion of tissue macrophages in the target organ. *J Virol.* 1997 Jan;71(1):624-9.

[336] Assessment of adenoviral vector safety and toxicity: report of the National Institutes of Health Recombinant DNA Advisory Committee. *Hum Gene Ther.* 2002 Jan 1;13(1):3-13.

[337] Marshall E. Gene therapy death prompts review of adenovirus vector. *Science.* 1999 Dec 17;286(5448):2244-5.

[338] Bertram CM, Hawes S, Egli S, Peh GS, Dottori M, Kees UR, et al. Effective adenovirus mediated gene transfer into neural stem cells derived from human embryonic stem cells. *Stem Cells Dev.* 2010 Apr;19(4):569-78.

[339] Zeng Z, Yuan X, Liu G, Zeng X, Zeng X, Ng H, et al. Manipulation of proliferation and differentiation of human bone marrow-derived neural stem cells *in vitro* and *in vivo*. *J Neurosci Res.* 2007 Feb 1;85(2):310-20.

[340] Sabate O, Horellou P, Vigne E, Colin P, Perricaudet M, Buc-Caron MH, et al. Transplantation to the rat brain of human neural progenitors that were genetically modified using adenoviruses. *Nat Genet.* 1995 Mar;9(3):256-60.

[341] Fu Y, Wang SQ, Liu YP, Wang GP, Wang JT, Gong SS. Gene transfer into primary cultures of fetal neural stem cells by a recombinant adenovirus carrying the gene for green fluorescent protein. *J Zheijang Univ Sci B.* 2008 Apr;9(4):299-305.

[342] Schmidt A, Bockmann M, Stoll A, Racek T, Putzer BM. Analysis of adenovirus gene transfer into adult neural stem cells. *Virus Res.* 2005 Dec;114(1-2):45-53.

[343] Arnhold S, Hilgers M, Lenartz D, Semkova I, Kochanek S, Voges J, et al. Neural precursor cells as carriers for a gene therapeutical approach in tumor therapy. *Cell Transplant.* 2003;12(8):827-37.

[344] Falk A, Holmstrom N, Carlen M, Cassidy R, Lundberg C, Frisen J. Gene delivery to adult neural stem cells. *Exp Cell Res.* 2002 Sep 10;279(1):34-9.

[345] Komuro H, Saihara R, Shinya M, Takita J, Kaneko S, Kaneko M, et al. Identification of side population cells (stem-like cell population) in pediatric solid tumor cell lines. *J Pediatr Surg.* 2007 Dec;42(12):2040-5.

[346] Kondo T, Setoguchi T, Taga T. *Persistence of a small subpopulation of cancer stem-like cells in the C6 glioma cell line.* Proc Natl Acad Sci U S A. 2004 Jan 20;101(3):781-6.

[347] Patrawala L, Calhoun T, Schneider-Broussard R, Zhou J, Claypool K, Tang DG. Side population is enriched in tumorigenic, stem-like cancer cells, whereas ABCG2+ and ABCG2- cancer cells are similarly tumorigenic. *Cancer Res*. 2005 Jul 15;65(14):6207-19.

[348] Holthouse DJ, Dallas PB, Ford J, Fabian V, Murch AR, Watson M, et al. Classic and desmoplastic medulloblastoma: complete case reports and characterizations of two new cell lines. *Neuropathology*. 2009 Aug;29(4):398-409.

[349] Dallas PB, Terry PA, Kees UR. Genomic deletions in cell lines derived from primitive neuroectodermal tumors of the central nervous system. *Cancer Genet Cytogenet.* 2005 Jun;159(2):105-13.

[350] Donnenberg VS, Donnenberg AD. Multiple drug resistance in cancer revisited: the cancer stem cell hypothesis. *J Clin Pharmacol.* 2005 Aug;45(8):872-7.

[351] Park CY, Tseng D, Weissman IL. Cancer stem cell-directed therapies: recent data from the laboratory and clinic. *Mol Ther.* 2009 Feb;17(2):219-30.

[352] Lee J, Kotliarova S, Kotliarov Y, Li A, Su Q, Donin NM, et al. Tumor stem cells derived from glioblastomas cultured in bFGF and EGF more closely mirror the phenotype and genotype of primary tumors than do serum-cultured cell lines. *Cancer Cell.* 2006 May;9(5):391-403.

[353] Bhoopathi P, Chetty C, Kunigal S, Vanamala SK, Rao JS, Lakka SS. Blockade of tumor growth due to matrix metalloproteinase-9 inhibition is mediated by sequential activation of beta1-integrin, ERK, and NF-kappaB. *J Biol Chem.* 2008 Jan 18;283(3):1545-52.

[354] Stolarek R, Gomez-Manzano C, Jiang H, Suttle G, Lemoine MG, Fueyo J. Robust infectivity and replication of Delta-24 adenovirus induce cell death in human medulloblastoma. *Cancer Gene Ther.* 2004 Nov;11(11):713-20.

[355] Sharpless NE, Depinho RA. The mighty mouse: genetically engineered mouse models in cancer drug development. *Nat Rev Drug Discov*. 2006 Sep;5(9):741-54.

[356] Lee Y, Kawagoe R, Sasai K, Li Y, Russell HR, Curran T, et al. Loss of suppressor-of-fused function promotes tumorigenesis. *Oncogene*. 2007 Sep 27;26(44):6442-7.

[357] Frappart PO, Lee Y, Lamont J, McKinnon PJ. BRCA2 is required for neurogenesis and suppression of medulloblastoma. *EMBO J.* 2007 Jun 6;26(11):2732-42.

[358] Kimura H, Ng JM, Curran T. Transient inhibition of the Hedgehog pathway in young mice causes permanent defects in bone structure. *Cancer Cell.* 2008 Mar;13(3):249-60.

In: Brain Cancer, Tumor Targeting and Cervical Cancer
Editor: Elena K. Salvatti
ISBN: 978-1-61122-738-3

Chapter IV

Laparoscopy and "Robotics" in Cervical Cancer Treatment: Current Insights, Novel Approaches and Future Perspectives in the New Era of Gynecological Surgery

Andrea Tinelli,[*1] ***Antonio Malvasi***[2]***, Maurizio Buscarini***[3]***, Emilio Ruiz Morales***[4] ***and Stark Michael***[5]

[1]Department of Obstetrics and Gynecology, Division of Experimental Endoscopic Surgery, Imaging, Minimally Invasive Therapy and Technology (D.R.e.A.M.), Vito Fazzi Hospital, Lecce, Italy

[2]Department of Obstetrics and Gynecology, Santa Maria Hospital, Bari, Italy

[3]Department of Urology, Campus Biomedico University, Roma, Italy

[4]ALF-X Surgical Robotics Department, Sofar S.p.A., Milan, Italy

[5]The New European Surgical Academy (NESA), Berlin, Germany

Abstract

Cervical cancer is the second most common cancer in women worldwide and is a leading cause of cancer-related death in women in underdeveloped countries. More than 99% of cervical uterine cancer cases show HPV presence. Minimal invasive surgery is increasingly its popularity in surgical oncology. Robotic surgery with its numerous advantages over conventional laparoscopy has assumed an ever-expanding role in minimally invasive surgery. This chapter focuses on the recent adoption, experience, and applications of laparoscopy and robot-assisted laparoscopy in cervical cancer surgery,

[*] Corresponding author. Tel-Fax +39/0832/661511; Cell. +39/339/207408; E-mail: andreatinelli@gmail.com

focusing on radical endoscopic treatments, such as hysterectomy, bilateral salpingo-oophorectomy, pelvic and para-aortic lymphadenectomy.

A computer aided and manual search for clinical and systematic reviews, randomized controlled trials, prospective observational studies, retrospective studies and case reports published between 1980 and April of 2010, assessing robotic surgery was carried out defined by search strings including: laparoscopic surgery; robotics or robot or robot-assisted or da Vinci; radical hysterectomy; cervical cancer, minimally invasive surgery.

Literature has shown the feasibility of a radical resection by laparoscopic surgery and documented an equivalent number of pelvic nodes harvested by laparoscopy and open surgery. With technical advances and emerging devices, as well as accumulating experiences in laparoscopic surgery, radical laparoscopic hysterectomy was achieved as a feasible and safe surgical procedure, with long-term data on the morbidity and survival. Robotic assisted gynecologic surgery is, recently, worldwide increased, as the number of dedicated scientific articles. Few retrospective and prospective studies have demonstrated the feasibility of robotic assisted surgery in radical hysterectomy. In general, robot-assisted gynecologic surgery is often associated with longer operating room time but generally similar clinical outcomes, decreased blood loss, and shorter hospital stay. Robotic-assisted procedures are not, however, without their limitations: the equipment is still very large, bulky, and expensive, the staff must be trained, specifically on draping and docking the apparatus to maintain efficient operative times. Functional limitations include lack of haptic feedback, limited vaginal access, limited instrumentation, and larger port incisions. Exchanging instruments becomes more cumbersome and requires a surgical assistant to change the instruments. Additionally, the current robotic instruments do not include endoscopic staplers or vessel sealing devices. Finally, laparoscopic radical hysterectomy is a feasible and safe procedure that is associated with fewer intraoperative and postoperative complications than abdominal radical hysterectomy. The role of robotic-assisted surgery is continuing to expand, but well-designed, prospective studies with well-defined clinical, long-term outcomes, including complications, cost, pain, return to normal activity, and quality of life, are needed to fully assess the value of this new technology in radical hysterectomy.

Keywords: Robotics, gynecological surgery, Da Vinci, ALF-X, Telap, complication, gynecological cancer, oncology

Introduction

Cervical cancer is the second most common cancer in women worldwide and is a leading cause of cancer-related death in women in underdeveloped countries. Worldwide, approximately 500,000 cases of cervical cancer are diagnosed each year: approximately 13,000 cases of invasive cervical cancer and 50,000 cases of cervical carcinoma in situ (i.e., localized cancer) are diagnosed yearly in USA; over the last 40 years, the death rate from cervical cancer decreased by more than 70% because pre-invasive lesions and cervical cancers were detected at an earlier stage [1].

Cervical cancer starts in the cells on the surface of the cervix: it develops in the lining of the cervix, and is always associated to HPV infection, since carcinogenic human papillomavirus (HPV) infection is necessary for the development of cervical cancer [2].

More than 99% of cervical uterine cancer cases show HPV presence, with more than 100 HPV types identified, and types 16 and 18 being found in approximately 70 percent of cases. But infection alone is not sufficient to cause cervical cancer, since infections become undetectable within 1–2 years [3].

Cervical cancer risk seems to be influenced by other variables too, like smoking, while other factors, like alcohol consumption and diet, don't seem to have any influence. Infection with other sexually transmitted viruses seems to act as a cofactor in the development of cervical cancer [1].

In this chapter, the authors will focus on the area of minimally invasive treatment of cervical cancer, since this tumor can be safely and feasibly managed from minimally invasive endoscopic radical operations, such as hysterectomy, bilateral salpingo-oophorectomy, pelvic and para-aortic lymphadenectomy for surgical treatment.

Cervical Cancer Staging and Radical Hysterectomy Classification

Radical hysterectomy is one of the defining surgeries in the treatment of gynecologic malignancies, starting from cervical cancer staging.

Correct staging of advanced cervical cancer is essential to optimize its oncological treatment. Although, the new FIGO classification is limited to clinical findings and does not include complex imaging such as computed tomography, magnetic resonance imaging, or positron emission tomography.

The rationale for the FIGO approach is to provide a template so that both resource-rich and resource-poor countries can compare data by stage to standardize management of the disease. Women in resource-poor countries, which have a high prevalence of cervical cancer, generally do not have access to advanced imaging techniques and specialized surgical care.

Nevertheless, the limitations of FIGO clinical staging are well appreciated: it can be difficult to accurately assess parametrial and sidewall invasion, as well as metastases to lymph nodes, using clinical staging alone.

The International Federation of Gynecology and Obstetrics (FIGO) staging systems for cervix and other genital malignancies have been revised for the first time in over a decade [4].

The purpose of the staging system is to provide uniform terminology for better communication among health professionals and to provide appropriate prognosis to the patients which results in treatment improvement. This is a constantly evolving process as new therapeutic modalities are developed, new imaging and surgical approaches are applied, and more prognostic information becomes available. The previous system did not reflect the prognosis in some patient subsets where medical research and practice have shown explosive growth of new knowledge in recent years.

The 41st Annual Meeting of the Society of Gynecologic Oncologists was held in March 2010. Several abstracts reported retrospective studies that evaluated the prognostic significance of new 2009 FIGO staging guidelines compared to the old 1988 FIGO system.

The new FIGO staging of carcinoma of the cervix uteri is reported in the table 1.

The term "radical" or "extended" hysterectomy encompasses various types of surgery. Since the first publications of large series of surgeries for cervical cancer by Wertheim in Austria [5] and later by Okabayashi in Japan [6] and Meigs in the USA [7], many radical procedures that accord with different degrees of radicality have been described and done.

These procedures give different names for the same anatomical structures and define these structures according to different interpretations of anatomy.

The Piver–Rutledge–Smith classification published in 1974 has achieved substantial popularity, whereas the 1975 Symmonds classification was not adopted [8]. The former describes five classes of radical hysterectomy, but has several major drawbacks. It is misused by many researchers and surgeons because the tradition is transmitted orally without careful reading of the original paper [9].

The original paper does not refer to clear anatomical landmarks or international anatomical definitions. The vaginal extent of resection is systematically attached to pericervical extent; vaginal resection is excessive—from a third to three-quarters of the vagina.

It includes a class I category, which is not radical hysterectomy, and a class V category, which is no longer used. The rationale and anatomy for differentiation between class III and IV are unclear. Surgeons frequently need to define intermediate classes between that of II and III (eg, II-III or II-and-a-half). The classification by Piver and colleagues [8] does not take into account the idea of nerve preservation that was introduced in the 1950s [10] subsequently refined by Japanese surgeons [11,12] and adopted by European surgeons [13,14].

Finally, the Piver–Rutledge–Smith classification applies only to open surgery, and does not take into account the development of laparoscopic techniques.

Querleu and Morrow recently published a new classification of radical hysterectomy, based, for simplification, on only the lateral extent of uterine resection [15].

Only four types of radical hysterectomy are described, adding when necessary a few subtypes. Relatively stable anatomical landmarks are used to define the limits of resection, such as the crossing of the ureter with the uterine artery and paracervix, and the vascular plane of the internal iliac system. To make a clear distinction with the Piver–Rutledge–Smith [8], in the Querleu and Morrow classification letters are used rather than numbers to define classes (table 2). Simple hysterectomy is not included in the classification. Lymph-node dissection, an essential part of surgical management of cervical cancer, is considered separately.

For lymph-node dissections, the limit between level 1 and level 2 is the bifurcation of the common iliac artery; the limit between level 2 and level 3 is the bifurcation of the aorta; and the limit between level 3 and level 4 is the inferior mesenteric artery.

Evolution of Surgery in Cervical Cancer

From the original publications of this surgery for radical hysterectomy, the surgical technique has had several efforts at re-classification including a variety of techniques via laparotomy [16,17], then via laparoscopy, resulting in reasonable morbidity with similar surgical outcomes [18,19].

However, longer operating times, steep learning curves, and lack of training have prevented laparoscopic treatment from becoming a widely adopted surgical approach to radical hysterectomy [20,21]

Although, laparoscopic surgery for gynecological oncology does not substantially reduce tissue trauma, which makes it possible for extensive and complex operative procedures to be performed through small incisions, and reduces intra-operative blood loss and impact on the body as well as common surgical complications. In addition, laparoscopic surgery is superior to conventional surgery with regard to postoperative mental rehabilitation in gynecological oncology patients.

Hence, laparoscopic surgery is consistent with the concept of minimally invasive surgery, i.e., smaller trauma, milder pain or analgesia, better homeostasis, more accurate operative outcomes, shorter hospital stay and better psychological effects than current standard open surgery. As a technical innovation, laparoscopic techniques for gynecological oncology do not change gynecological oncological surgery fundamentally, but they have improved surgical techniques for gynecological oncology in many aspects, and enhanced the efficacy of surgical treatment of cervical cancer.

Laparoscopic surgery must comply with the principles of traditional open surgery for radical treatment of gynecological tumor [22].

The principles are: to resect tumor and its surrounding tissues en bloc, to use tumor-free techniques when manipulating tumors, to preserve sufficient incised margins, and to perform complete pelvic lymphadenectomy. Lymph node status is the most important prognostic factor in gynecologic tumor and surgical removal of the pelvic and/or para-aortic lymph nodes for histological assessment is a part of staging of gynecologic malignancies.

Additionally, removal of bulky lymph nodes may have therapeutic benefit. According to the requirement established FIGO classification systems, different range lymphadenectomy will be performed depending on the different type of tumor present. The range of lymphadenectomy, consistent with that of open abdominal surgery, depends on disease in cervical cancer. Lymph nodes para-aortic and iliac vessels are resected within the vessel sheath. Obturator and deep inguinal lymph nodes, including lymph nodes below the obturator nerve, must be resected radically.

Laparoscopic and Robotic-assisted Approaches in Gynaecological Malignancies

The goal of laparoscopic surgery is to duplicate traditional open procedures via small incisions in the skin with surgical outcomes equivalent or superior to a traditional surgical approach.

Laparoscopy offers multiple advantages in the management of malignancy, including smaller incisions, shorter hospital stay, quicker recovery, improved visualization, less need for postoperative analgesics, and a lower risk of complications, such as blood loss, wound infection, herniation, and ileus.

These characteristics may prove particularly important in the setting of oncology where a shorter recovery period may facilitate a shorter interval to postoperative treatments such as chemotherapy or radiation. Laparoscopy also has its limitations.

Disadvantages include a long learning curve, counterintuitive motions, and limited depth perception as imaging is limited to 2-dimensional views.

In an effort to overcome these limitations, multiple innovations have evolved over the last decade. Laparoscopic instrumentation has expanded to include several different vessel sealing devices with integrated cutting capabilities, endoscopic staplers, articulating instrument tips, 3-dimensional capabilities, and computer-enhanced technology in the form of robotics.

Therefore, robotic surgical systems seem to be the future in gynecologic surgery, since robotic technology can improve accuracy, enhance dexterity, and provide for faster suturing and a lower training curve than laparoscopy.

All of this makes gynecologic surgeons tend to perform a greater number of gynecologic procedures by robotic approach, if that is available. In the last five years, a large number of robotic surgery related papers have been published, both for benign and oncological cases, which is proof of the rapid spread and high acceptance of this new technology.

Gynecological oncology probably presents the optimal forum for application of robotics, given the complexity of surgical steps involved in performing radical hysterectomies for cervical cancer and lymph node sampling for endometrial cancer.

The History of Laparoscopy in Cervical Cancer

In 1989, Querleu et al. [23] performed the first laparoscopic pelvic lymphadenectomy for cervical cancer, than, in 1990, Canis et al described a radical hysterectomy totally by laparoscopy [24]. In 1991, Querleu et al. reported laparoscopic pelvic lymphadenectomy and vaginal assistant radical hysterectomy for early cervical cancer [25].

In 1992 and in 1993, Nezhat et al. reported the first case of cervical cancer treated with laparoscopic radical hysterectomy and pelvic lymphadenectomy [26,27].

Since then, the techniques have been applied clinically and achieved satisfactory clinical outcomes [28].

In 2000, Dargent et al. reported laparoscopic performed successfully vaginal radical trachelectomy and pelvic lymphadenectomy for young women with cervical cancer, who wanted to preserve their fertility [29].

In 2000, Possover et al. reported modified laparoscopic nerve sparing type III radical hysterectomy for the treatment of cervical cancer, and found that this procedure decreased the incidence of postoperative bladder dysfunction [30].

In 2003, Pomel et al evaluated, in a series of 50 consecutive patients, the feasibility, morbidity, and survival, as outcome of the laparoscopic radical hysterectomy for carcinoma of the uterine cervix, operated between 1993 and 2001 at two cancer centers. Thirty-one patients had prior brachytherapy.

The median overall operative time was 258 min. The mean number of harvested pelvic external iliac nodes was 13.22 per patient. The median postoperative hospital stay was 7.5 days. Two patients had major urinary complications; one had a bladder fistula and one a ureteral stenosis. The median follow-up was 44 months. The overall 5-year survival rate of FIGO stage Ia2 and Ib1 patients was 96%. Their results demonstrated that radical hysterectomy can be performed by laparoscopy in stage IB1 or less advanced node negative cervical cancer patients without compromising survival; moreover, prior brachytherapy did not affect the feasibility of this radical procedure [31].

With technical advances and emerging devices, as well as accumulating experiences in laparoscopic surgery, some surgical procedures that are difficult to carry out even by traditional open procedures can be performed successfully by laparoscopy.

During the past decade some reports, on a limited number of patients, have shown the feasibility of a radical resection by laparoscopic surgery and have documented an equivalent number of pelvic nodes harvested by laparoscopy and open surgery [32-38].

Nevertheless, few long-term data on the morbidity and survival after laparoscopic radical hysterectomy are available.

Although open radical hysterectomy is still considered the gold standard for the treatment of early cervical cancer [39], laparoscopic radical (LRH) and laparoscopy assisted radical vaginal hysterectomy (LARH) are evolving as potential surgical alternatives.

LRH or LARH have been established as standard procedures routinely performed as first line therapy for the treatment of early cervical cancer at specialized centers [28,40-42].

The major advantages associated with minimally invasive laparoscopy are, amongst others, lower intraoperative bleeding rates, less post-operative pain, a shorter period of recovery time and a shorter hospital stay.

In addition the optical devices used for laparoscopic surgery feature a 10 to 15 times magnification and, therefore, provide an excellent view of the anatomy of the pelvis.

Since Liang Z et al reported their initial experience with 57 LRH; they have continuously improved and standardized the technique [43].

The indication for laparoscopic surgery was similar to open surgery in cervical cancer patients.

According to the literature and our experience, the indication for laparoscopic radical hysterectomy (type III) and pelvic lymphadenectomy was early, than FIGO stage IIa in cervical cancer [44].

However, initially, the laparoscopic technique treatment of cervical cancer is associated with a high complication rate, the indication must be carefully assessed, and patients have to be counseled extensively prior to surgery. Apart from the size and stage of the tumor, the indication for this surgical procedure is mainly influenced by the experience of the surgeon [45].

More recently, it was reported that stage IIb or more advanced cervical cancer may be treated with type IV radical hysterectomy under laparoscopy [15,46,47], as the aim of surgery is to stage and radically resect the tumor (including metastases). For certain patients, prolonging the lifespan are the objective of tumor treatment. Through laparoscopic exploration, the feasibility and thoroughness of surgery can be evaluated, and the relative benefits of surgery for the patient can be weighed. However, it is still controversial about what types or what stages of cervical cancer are adopted for laparoscopic surgical treatment.

Total Laparoscopic Radical Hysterectomy for Cervical Cancer

Laparoscopy can be safely and adequately used in the treatment of endometrial, ovarian and cervical cancer [48].

In the setting of gynecologic oncology, laparoscopic approaches have been implemented in the use of radical hysterectomy. Laparoscopic surgery for gynecological oncology is an important progress in the combination of scientific, technological and surgical techniques, which not only enhanced the efficacy of surgical treatment of gynecological oncology, but also is superior to conventional open surgery with regard to postoperative mental rehabilitation in gynecological oncology patients.

The first total laparoscopic radical hysterectomy with lymphadenectomy was reported in June 1989 by Nezhat et al [26,48]. Since then, more than 600 cases of total laparoscopic radical hysterectomy have been reported. A recent prospective case-control series compared total laparoscopic radical hysterectomy with abdominal radical hysterectomy and found shorter hospital stays and lower blood loss as well as a statistically significantly higher nodal yield among total laparoscopic radical hysterectomy cases [49].

Spirtos et al. described 78 patients with early-stage cervical cancer undergoing laparoscopic radical hysterectomy. In that series, 94% of the procedures were completed laparoscopically.

The mean follow-up time was 67 months. The mean operative time was 205 min, and the mean blood loss was 225 mL. One patient required a blood transfusion, three patients had unintended cystotomies, two patients required laparotomy to control bleeding, and one patient suffered an ureterovaginal fistula. Three patients had microscopically positive or close margins. The authors reported a cervical cancer recurrence rate of 5% [50].

Abu-Rustum et al. reported 19 patients with stage IA1 or IB1 cervical cancer who underwent total laparoscopic radical hysterectomy with pelvic lymphadenectomy. The mean estimated blood loss (301 mL) and mean hospital stay (4.5 days) were significantly less than in historical controls. Mean operating time was longer with the laparoscopic approach than with laparotomy (371 min vs. 295 min). Two patients required conversion to laparotomy. Five patients had unintended cystotomies, five patients had iliac vein injury, and one patient had a ureteral transection [18].

Pomel et al. reported 50 patients with stage IA1–IB1 cervical cancer who underwent total laparoscopic radical hysterectomy. The median operating time was 258 min, and the mean number of lymph nodes harvested was 13. No conversions to laparotomy were reported. The median length of hospital stay was 7.5 days. The authors reported that 10 patients had early complications (within 2 months of surgery) and that three of those patients required reoperation. They also reported three patients had late complications (more than 2 months after surgery) and two of those patients required reoperation. Three patients experienced recurrence with a median follow-up time of 44 months [41].

Gil-Moreno et al. reported a series of 12 patients with cervical cancer who underwent total laparoscopic radical hysterectomy with sentinel lymph node identification. The mean operating time was 271 min, and the mean blood loss was 445 mL. The mean length of the hospital stay was 5.25 days. No intraoperative complications were reported, and there were no conversions to laparotomy. No recurrences were detected with a median follow-up of 20 months [51].

Ramirez et al. evaluated 20 patients with stage IA2 or IB1 cervical cancer. The median blood loss was 200 mL (range, 25–700). The median operative time was 332.5 min (range, 275–442). The surgical margins were free of disease in all cases, and no patients required conversion to laparotomy. This was the first study to report a median length of hospital stay of 1 day after surgery [20].

A retrospective study by Li et al. compared the morbidity, recurrence rates, and mortality of patients undergoing laparoscopic radical hysterectomy versus abdominal radical hysterectomy. The authors found that recurrence rates (14% and 12%, respectively; $p \geq 0.05$) and mortality rates (10% and 8%, respectively; $p \geq 0.05$) were similar for the two groups [52].

Frumovitz et al. compared 35 patients undergoing total laparoscopic radical hysterectomy with 54 undergoing total abdominal radical hysterectomy. The mean estimated blood loss was significantly lower in the laparoscopic-surgery group than in the open-surgery group (319 vs. 548 mL; p=0.009). The mean operative time was significantly longer with laparoscopic surgery (344 vs. 307 min; p=0.03), but the median hospital stay was significantly shorter (2.0 vs. 5.0 days, pb0.001). Postoperative infections were much less common after laparoscopic surgery (18% vs. 53%; p=0.001) [53].

Despite these advantages of conventional laparoscopy, however, recent surveys of practicing gynecologic oncologist revealed that most respondents believed minimally invasive surgery by conventional laparoscopy had only a minimal role in the management of cervical cancer [54].

Puntambekar et al reported a retrospective review of 248 patients with FIGO stage Ia2 (n=32) and Ib1 (n=216) cervical cancer who underwent a TLRH (type III) with bilateral pelvic lymphadenectomy, which is the largest single-institution study. The operation was performed entirely by laparoscopy in all of the patients and by the same surgical team. The median operative time was 92 min. The median number of resected pelvic nodes was 18. The median blood loss was 165 ml. The median length of the hospital stay was 3 days. None of the patients required conversion to laparotomy. Seventeen patients had complications within 2 months of surgery. Seven patients had recurrences after a median follow-up of 36 months. They concluded that TLRH can be performed safely, reduces morbidity associated with ARH and is an easily replicable technique [55].

Colombo et al evaluated the surgical outcome and oncologic results of total laparoscopic radical hysterectomy (TLRH) after neoadjuvant chemoradiation therapy (CRT) for locally advanced cervical carcinoma. All patients who underwent TLRH after CRT for stages IIB–IIA and bulky IB diseases were reviewed. The control group for this analysis was a cohort of patients treated with abdominal radical hysterectomy (ARH) after CRT for the same stage cancers. They reviewed 102 patients operated on between 2000 and 2008 (46 TLRH and 56 ARH). Mean age at diagnosis was 44 years, and mean B.M.I was 22.1. There was no difference in tumor characteristics between the two groups. Seven patients in the laparoscopic group required conversion to laparotomy (15%). Mean estimated blood loss (200 vs. 400 mL, pb0.01) and the median duration of hospital stay (5 vs. 8 days, pb0.01) were significantly lower in the laparoscopic group. Morbidity rates and urinary complications were reduced in the laparoscopic group (p=0.04). Local recurrence rates, disease-free and overall survival were comparable in the two groups. Best survival was observed for patients with pathological complete response or microscopic residual disease compared to patients with macroscopic residues (pb0.01). Authors concluded that radical hysterectomy after CRT is known to be difficult with significant morbidity rates and remains controversial in comparison to exclusive CRT. TLRH after preoperative CRT is feasible for patients with locally advanced cervical cancer in 85% of the cases. For these patients, TLRH compared with ARH was associated with a favorable surgical outcome with comparable oncological results [56].

Although laparoscopic surgery has many advantages, it is also associated with a number of potential drawbacks, including limited range of motion intra-abdominally (only 4° of freedom), counterintuitive movements, amplification of tremors in prolonged cases because of the length and rigidity of the instrumentation, and reduced depth perception secondary to a two-dimensional view.

Nevertheless, the total laparoscopic radical hysterectomy is not without its risks and complications.

There are still several new challenges on both theory and surgical skill improvement to be overcome.

Firstly, to form a laparoscopic surgical team, there are a lot of special requirements of both the team members and the equipment which restrict the popularization of the laparoscopic radical hysterectomy.

Secondly, every detail should be taken into careful consideration and the risks should be balanced and benefits ascertained before performing the laparoscopic procedures.

Moreover, it is likely that well-known barriers to the utilization of advanced minimally invasive procedures such as association with a long learning curve, lack of training, complexity of operations, limitation of technology and instrumentation, and the necessity of an expert assistant were responsible for this sentiment.

For these reasons, only a few surgeons have adopted a laparoscopic approach to type III radical hysterectomies for cervical cancer. There are issues with the length of the procedure, as type III laparoscopic hysterectomies take significantly longer than open cases.

Furthermore, intraoperative complications of the urinary tract (for example as a result of thermal injuries) tend to be much more common in laparoscopic than in open cases [57].

A recent experience of Malzoni et al on 71 patients treated by total laparoscopic radical hysterectomy (type II, III) with lymphadenectomy between January 2000 and March 2008. FIGO stage included five patients Ia1 with LVSI (lymph-vascular involvement), 24 patients Ia2, and 48 patients Ib1. 60 patients underwent a class III procedure and 17 patients a class II procedure according to the Piver classification. Histological types included: squamous cell carcinoma in 65 patients, adenocarcinomas in 10 patients, and adenosquamous carcinoma in two. Para-aortic lymphadenectomy was performed up to the level of the inferior mesenteric artery in eight cases with positive pelvic lymph nodes at frozen section evaluation. In the results, surgical margins were free of disease in all cases and no patient required an intraoperative or postoperative blood transfusion. The catheter was removed on day 2 after the surgical procedure; postoperatively, 18 patients with voiding difficulty required intermittent catheterization. The mean time to resumption of normal bladder function was 9 days. One patient had a postoperative complication as ureterovaginal fistula. The patient underwent placement of a ureteral stent 1 week later that was removed after 3 months and there was no sequel. In 23 patients it was observed lymphorrhea at pelvic examination as profuse discharge of lymphatic fluid that was leaking from the cuff. In all the cases, this condition resolved spontaneously. The mean number of resected pelvic lymph nodes was 23 (range 16-39). The mean number of resected aortic lymph nodes was 8 (range 5-11). Nine patients had positive pelvic lymph nodes discovered at the final histological examination. Four patients had unilateral involvement of one lymph node, three patients had unilateral involvement of two lymph nodes, and two patients had bilateral involvement of two lymph nodes. One patient had positive para-aortic lymph nodes discovered at the final histological examination. All of these patients received radio-chemotherapy with pelvic external beam radiation. The authors concluded that total laparoscopic radical hysterectomy can be considered a safe and effective therapeutic procedure for the management of early stage cervical cancer with a low morbidity, offering an alternative option for patients undergoing radical hysterectomy, although multicenter studies and long-term follow-up are required to evaluate the oncologic outcomes of this procedure [58].

A review of the literature on laparoscopic radical hysterectomy demonstrates the procedure is also safe and feasible, but is associated with an operative time range of 205 min–371 min, an estimated blood loss of 200 cc–445 cc, a nodal yield ranging from 13–25, a hospital stay ranging from 1–7.5 days, and an overall complication rate of 11%–20%, as sorted by the referenced papers who provide a good cross section of the data [18, 20, 38, 41, 52, 53].

Contra-indications to Laparoscopic Surgery for Gynecological Oncology

There are very few absolute contra-indications as previously noted. With increased anesthesia ability, even some of these may not be considered as absolute

The severe cardiac diseases could be a problem, since some patients may not tolerate the operation and the deep trendelenburg positions necessary during most operative laparoscopy in order to maintain an adequate pneumoperitoneum.

So the poor condition of the patient, such as severe liver/renal/respiratory dysfunction which cannot be corrected preoperatively are also considered to be absolute contraindications.

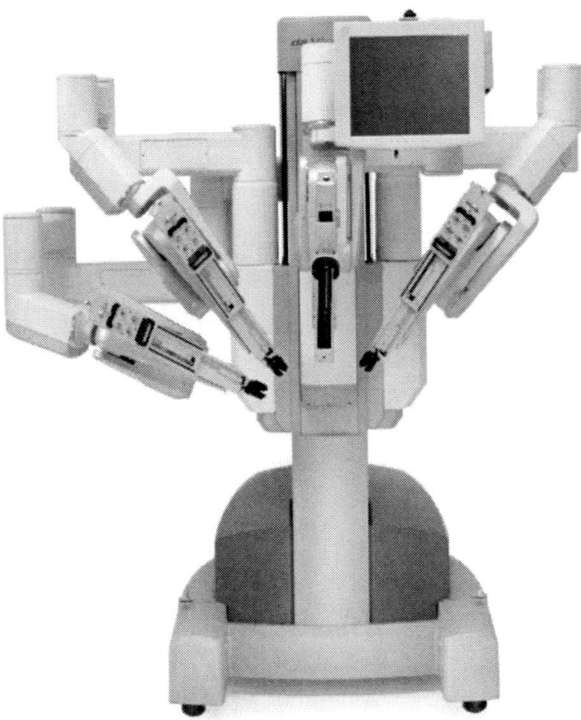

Figure 1. The da Vinci surgical system, developed by Intuitive Surgical, Inc. (Intuitive Surgical, Mountain View, CA, USA).

Figure 2. The da Vinci surgical system has a 3-dimensional vision system, in which double endoscopes generate two images resulting in the perception of a 3D image.

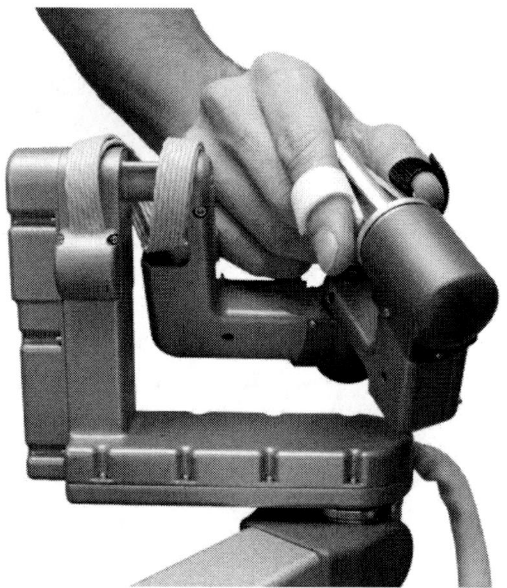

Figure 3. Da Vinci is equipped with the endowrist system: it reproduces the range of motion and dexterity of the surgeon hand, providing high precision, flexibility and the ability to rotate instruments 360 degrees.

Advanced or late stage gynecological tumor could be refused in laparoscopy by surgeons, in patients with the stage III or above cervical carcinoma, or lymph node metastases of cervical cancer which inter-fusion and encapsulate vital vessels, and/or extensive and infiltration of adjacent tissues.

On the other hand, there are few relative contra-indications to laparoscopy for cervical cancer patients; for example, severe abdominal adhesions or morbid obesity and so on.

Total Robotic-assisted Laparoscopic Radical Hysterectomies

Today there is only one U.S. Food and Drug Administration approved device for surgical robotics.

This current robotic platform is known as the "da Vinci" surgical system (Figure 1), developed by Intuitive Surgical, Inc. (Intuitive Surgical, Mountain View, CA, USA).

The da Vinci surgical system is equipped with a 3-dimensional vision system in which double endoscopes generate two images resulting in perception of a 3D image (Figure 2).

In addition, with the development of endowrist, it reproduces the range of motion and dexterity of the surgeon hand, providing high precision, flexibility and the ability to rotate instruments 360 degrees (Figure 3).

Thus, the learning curve of achievement for the surgeons using the da Vinci surgical system was shortened. In 2001, a more advanced da Vinci surgical system with four robotic arms gained US FDA approval and it is now being used in many surgical procedures throughout the world.

Recently, robotic surgery that is FDA approved has become an option in the definitive surgical management of early stage cervical cancers.

Several case series have now been published about robotic-assisted radical hysterectomies for cervical cancer [21,59-61].

The initial experience and the first publications of robot assistance in gynecological oncology date from recent years.

In February, 2006, Boggess performed the first live telecast demonstrating a technique for performing robotic assisted radical hysterectomy and subsequently presented data for a series of 13 radical hysterectomies at the Society of Gynecologic Oncologists annual meeting in March of the same year [62].

Since this initial demonstration of feasibility and technique, the interest in robotic assisted gynecologic oncology procedures has spread rapidly.

Recently, Nezhat et al prospectively compared the outcomes of 13 total robotic-assisted laparoscopic radical hysterectomies to 30 cases of traditional total laparoscopic radical hysterectomy with pelvic lymphadenectomy in patients with stage IB to IIA cervical cancer. There was no difference in operative time, hospital stay, blood loss, complications, or number of nodes retrieved. This study suggests that robotic radical hysterectomy may be a feasible alternative to total laparoscopic radical hysterectomy (unpublished data). However, there were no advantages of robotic-assisted procedures compared to traditional total laparoscopic radical hysterectomy when performed by an experienced laparoscopic gynecologic oncologist [63].

Fanning et al reported the first series of robotic radical hysterectomy on 20 women with stage IB–IIA cervical carcinoma. Median operative time was 6.5 hours, and median blood loss was 300 mL.

No complications were encountered, and all patients were discharged home on the first postoperative day. A retrospective cohort study of robotic vs. open radical hysterectomy found that the mean blood loss was significantly lower for the robotic group (81.9 vs. 665 mL, p<0.0001), but operative time was longer (4.5 vs. 3.39 hours, p=0.0002) [64].

The mean number of lymph nodes resected did not differ, and no complications were reported in the robotic assisted group [65].

In a comparative study of laparotomy, standard laparoscopy, and robotic technique for hysterectomy and staging for endometrial cancer, Boggess et al reported a higher lymph node yield, shorter hospital stay, and lower estimated blood loss in the robotic cohort.

Additionally, postoperative complications were significantly lower in the robotic group compared to the open group (5.9% vs. 29.7%, p<0.001) [66].

Furthermore, laparoscopic radical hysterectomy studies, although variable, report an operating time from 92 to 350 minutes, an estimated blood loss of 165 ml, and a length of stay of three days, whereas on robotic surgery comparable figures are 190-370 minutes, 140 ml, and two days [67-70].

Veljovich et al. published their first year experience, performing 118 robot assisted oncology procedures for different gynecologic oncologic conditions. They compared their robot procedures with open and conventional laparoscopic procedures. They found less blood loss, shorter hospital stay and longer operating time in the robot procedures (but operating time significantly decreasing with experience) [71].

The first radical hysterectomy in cervical cancer with robot assistance was described by Sert and Abeler. They concluded that radical dissection could be performed much more precise than with conventional laparoscopy. In 2007, they described 15 women with early-stage cervical cancer as a pilot case–control study and compared robotic-assisted laparoscopic radical hysterectomy with conventional total laparoscopic radical hysterectomy. There was a significant difference in mean operating time (241 minutes in the robot group and 300 minutes in the conventional group). No difference in the number of lymph nodes and size of parametrial tissue was found. In the robot group, there was significant less bleeding and shorter hospital stay [72].

A prospective analysis of 27 women with early-stage cervical cancer undergoing a robotic radical hysterectomy was performed by Margina et al., and a comparison was made with matched patients operated by conventional laparoscopy and laparotomy. They concluded that operating times for robot surgery and laparotomy were similar (189 and 166 minutes) and significantly shorter compared with laparoscopy (220 minutes). Blood loss and length of hospital stay were similar in robotics and laparoscopy and significantly longer in laparotomy. At a mean follow up of 31 months, there were no recurrences at all [60].

Kim et al. performed robotic radical hysterectomy and pelvic lymphadenectomy in ten cases and found a mean operating time of 207 minutes. The mean docking time was 26 minutes, but this was reduced significantly with experience (from 35 to 10 minutes) [61].

Kowalski et al. compared 14 robotic versus 17 open radical hysterectomies. Their mean operating time was significantly longer in the robot group (204 versus 121 minutes) [73].

These small series, however, do not report the outcome of surgery in terms of lymph node yield and radicality and also lack sufficient oncological follow up.

Boggess found no difference in the operating time (242 versus 240 minutes) [66]. He performed 13 robot-assisted radical hysterectomies and compared them with 48 open radical hysterectomies. Significantly more lymph nodes were collected in the robot group (33 versus

22). All the robotic patients were discharged within 24 hours. He also describes how to set up a robotic program in gynaecological oncology.

A robotic radical hysterectomy in 20 women was described by Fanning et al. Operating console time reduced from 8.5 to 3.5 hours in 20 cases. Less blood loss and shorter hospital stay was found. There were no recurrences at a median follow up of 2 years [64].

Recently, a study of Lowe MP et al reported a multi-institutional experience with robotic-assisted radical hysterectomy in patients with early stage cervical cancer with respect to peri-operative outcomes. In their investigation, a multi-institutional robotic surgical consortium consisting of five board-certified gynecologist oncologists in distinct geographical regions of the United States was created, in order to evaluate the utility of robotics for gynecologic surgery (benign and malignant). Between April 2003 and August 2008, a total of 835 patients underwent robotic surgery for benign gynecologic disorders and/or gynecologic malignancies by a surgeon in the consortium. For the purposes of the study, a multi-institutional HIPPA compliant database was then created for all of the patients. This database was queried for all patients who underwent a robotic-assisted type II or III radical hysterectomy for Stage IA1 (+vsi)-IB2 cervical carcinoma. Forty-two patients were identified. Records were then reviewed for demographic data, medical conditions, prior abdominal or pelvic surgeries, and follow-up. The perioperative outcomes analyzed included: operative time (skin–skin), estimated blood loss (EBL), length of hospital stay, total lymph node count, conversion to laparotomy, and operative complications. In the results, they identified from a database of 835 patients who underwent robotic surgery by a gynecologic oncologist, a total of 42 patients who underwent a robotic-assisted type II (n=10) or type III (n=32) radical hysterectomy for early stage cervical cancer. Demographic data demonstrated a median age of 41 and a median BMI of 25.1. With regard to stage, seven patients (17%) were Stage IA2, twenty-eight patients (67%) were Stage IB1 and six patients (14%) were Stage IB2. There was a single patient with Stage IA1 cervical cancer with vascular space invasion who underwent a type II radical hysterectomy. The overall median operative time was 215 min. The overall median estimated blood loss was 50 cc. No patient received a blood transfusion. The median lymph node count was 25. The median hospital stay was 1 day. Positive lymph nodes were detected in 12% of the patients. Pelvic radiotherapy or chemo-radiation was given to 14% of the patients based on final surgical pathology. Intraoperative complications occurred in 4.8% of the patients and included one conversion to laparotomy (2.4%) and one ureteral injury (2.4%). Postoperative complications were reported in 12% of the patients and included a DVT (2.4%), infection (7.2%), and bladder/urinary tract complication (2.4%) The conversion rate to laparotomy was 2.4%.

In their conclusions, Lowe et al reported that robotic-assisted radical hysterectomy is associated with minimal blood loss, a shortened hospital stay, and few operative complications. Operative time and lymph node yields are acceptable. This data suggests that robotic-assisted radical hysterectomy may offer an alternative to traditional radical hysterectomy [74].

Persson et al reported their experience in 80 robot assisted laparoscopic radical hysterectomy to evaluate its feasibility and morbidity, from December 2005 to September 2008, by using a prospective protocol, and an active investigation policy for defined adverse events, perioperative, and short and long term data. In the results, the time for surgery (skin to skin) reached 176 and 132 min after 9 and 34 procedures respectively, all tumors were radically removed, and median number of retrieved lymph nodes was 26 (range 15–55). All women had an early follow up (1–3 months) and 43 of eligible 46 women (93%) had a long term follow up (≥12 months). In 33 of 80 women (41%) the peri/postoperative period was uneventful. The remainder had one or more mainly mild adverse events, most commonly from the vaginal cuff (n=17, 21%) or the lymphatic system (n=16, 20%). The proportion of uneventful cases increased significantly over time. Five women were resutured for dehiscence of the vaginal cuff, two women were reoperated for trocar site hernias and one woman had a ureter stricture that resolved following stent treatment. Eight women (14%) needed 60 days or more to resume spontaneous voiding. One 72-year old woman with disseminated endometrial cancer on autopsy died of pulmonary embolism 31 days after surgery. In their conclusions, authors showed that robot assisted laparoscopic radical hysterectomy could be a feasible alternative to conventional laparoscopy and open surgery, even if efforts should be made to ensure proper closure of the vaginal cuff, trocar sites and to develop nerve sparing techniques [75].

Cantrell et al assessed progression-free (PFS) and overall survival (OS) for women with cervical cancer who underwent type III robotic radical hysterectomy (RRH), in a retrospective analysis of women who underwent RRH from 2005 to 2008 was performed. Comparison was made to a group of historical open radical hysterectomies. They analyzed seventy-one women underwent attempted RRH during the study period. Eight were excluded from analysis, 4 for non-cervical primary and 4 cases aborted due to extent of disease. Squamous was the most common histology (62%) followed by adenocarcinoma (32%). Median patient age was 43 years. There was one intraoperative complication (asystole after induction) and two postoperative complications (ICU admission to rule out myocardial infarction and reoperation for cuff dehiscence). Of the patients who underwent RRH, 32% received whole-pelvis radiation with chemo sensitization. The median follow-up was 12.2 months (range 0.2–36.3 months). Kaplan–Meier survival analysis demonstrated 94% PFS and OS at 36 months due to the recurrence and the death of one patient. As compared with a historical cohort at our institution, there was no statistically significant difference in PFS (P=0.27) or OS (P=0.47).

In the conclusions, Cantrell et al reported that the RRH is safe and feasible and has been shown to be associated with improved operative measures. This study showed that at 3 years, RRH appears to have PFS and OS equivalent to that of traditional laparotomy. While the 5-year data are not yet available for the cohort of patients treated with robotic surgery, the survival of 94% after 3 years of performing RRH is comparable, to other surgical methods and radiation [76].

The Problem of Lymph-node in Endoscopic Lymphadenectomy

The limitations of FIGO clinical staging involve the possibility to detect metastases to lymph nodes, using clinical staging alone. This leads to understaging of some patients [77,78].

One option consists of evaluating the lymph node invasion using imaging techniques. However, as reported recently, false negative rates as high as 11% have been reported when comparing PET to lymphadenectomy in advanced cervical cancer [79].

Therefore, another option is to perform surgical staging [80-82].

Laparoscopic para-aortic node sampling has been shown to be feasible in gynecological malignancies. In addition, it is associated with a lower morbidity than staging using laparotomy [83-88].

Its only technical limitation, that which occurs in obese patients. However, using the classical laparoscopic approach, the surgeon is limited in the degrees of his movements. By assisted-robotic surgery, this problem could be passed. The feasibility of a robotically assisted retroperitoneal approach, to dissect lower lumbo-aortic lymph nodes, was reported recently.

Simultaneously to Vergote et al. [89], Fastrez et al evaluated the feasibility and safety of a robot assisted laparoscopic transperitoneal approach of the para-aortic lymph node dissection on 8 patients with advanced cervical carcinoma who were eligible for primary pelvic radiotherapy combined with concurrent cisplatin chemotherapy or pelvic exenteration underwent a pre-treatment robot assisted transperitoneal laparoscopic para-aortic lymphadenectomy.

Authors isolated from 1 to 38 para-aortic nodes per patient and had one para-aortic node positive patient who was treated with extended doses of pelvic radiotherapy. We did not encounter any major complications and post-operative morbidity was low.

In the conclusions, Fastrez reported that robot assisted transperitoneal laparoscopic para-aortic lymphadenectomy is feasible and provides the surgeon with greater precision than classical laparoscopy, even if larger prospective multicentric trials are needed to validate the generalized usefulness of this technique [90].

The technique of the robotic retroperitoneal para-aortic lymphadenectomy has been described by Vergote et al., who reported the feasibility of robot assisted laparoscopic retroperitoneal para-aortic lymphadenectomy in five patients [89].

Ramirez et al described in 2008 a series of patients diagnosed with invasive cervical cancer after undergoing simple hysterectomy who subsequently underwent robotic radical parametrectomy and bilateral pelvic lymphadenectomy, by a retrospective review was performed of all patients who underwent robotic radical parametrectomy and bilateral pelvic lymphadenectomy at our institution during the period December 2006 to February 2008. The authors analyzed our data to evaluate the safety and feasibility of performing robotic radical parametrectomy. In their results, authors included 5 patients in analysis, with invasive squamous cell carcinoma of the cervix. The median body mass index was 23.8 kg/m2 (range, 17.7 to 26.5). The median operative time was 365 min (range, 331 to 430). The median estimated blood loss was 100 mL (range, 50 to 175). There were no conversions to laparotomy. There was 1 intraoperative complication—cystotomy. No patient required blood transfusion. The median length of hospital stay was 1 day (range, 1 to 2). One patient experienced two postoperative complications, a vesicovaginal fistula and a lymphocyst. No patient had residual tumor in the parametrectomy specimen, and no patient underwent adjuvant therapy. The median number of pelvic lymph nodes removed was 14 (range, 6 to 16). The median follow-up for all patients was 7.5 months (range, 1.3 to 13.8), without recurrences. In conclusion, Ramirez assessed that robotic radical parametrectomy and bilateral pelvic lymphadenectomy is feasible and safe and can be performed with an acceptable complication rate [91].

Robotic assistance with the Da Vinci system provides the surgeon with more precise dissection conditions, thanks to the three dimensional visualization, instrumentation with articulating tips that allows the surgeon's hands more mobility and the decreased tremor movements.

This increased precision in procedure as compared with a classical laparoscopy is particularly important in the para-aortic region and may enhance safety and decrease intraoperative morbidity.

Robotic-Assisted Trachelectomy

Radical trachelectomy is performed in select patients diagnosed with early-stage cervical cancer who wish to preserve their fertility.

Since the procedure was first described by Dargent et al. in 1994 [92], numerous reports have documented the safety and feasibility of the vaginal approach [93-96].

Alternatively, the procedure may also be performed successfully via the abdominal approach. Several groups have published works on the success rate and feasibility of the abdominal approach [97,98].

The first to report on robotic radical trachelectomy were Geisler et al., who reported on a patient with stage IB1 adenosarcoma of the cervix. The total operative time in that case was 172 min, and the estimated blood loss was 100 ml. There was no residual tumor found in the surgical specimen, and all lymph nodes were negative for any evidence of disease [99].

Persson et al published on 2 patients who underwent robotic radical trachelectomy. One patient was diagnosed with a stage IB1 adenocarcinoma of the cervix and the other with a stage IA2 squamous carcinoma of the cervix. This group of investigators was the first to publish on robotic radical trachelectomy in conjunction with lymphatic mapping and sentinel node identification. In that study, the console time was 387 min for the first patient and 359 min for the second patient. The estimated blood loss was 150 ml for the first patient and 100 ml for the second patient. The authors reported no intraoperative complications. Neither patient had residual cancer or evidence of disease in the lymph nodes [100].

Chuang et al described a robotic radical trachelectomy in a young woman with cervical cancer who desired preservation of fertility; the patient had previously undergone a cervical conization procedure, with negative margins.

Findings at positron emission tomography and computed tomography were normal, and there was no evidence of metastasis before surgery. The operation lasted 345 minutes, with blood loss of 200 mL. Final pathologic analysis showed no evidence of residual cancer. All reported cases were completed successfully and highlight robotic-assisted laparoscopy as a useful approach to trachelectomy in appropriate patients [101].

Ramirez et al described their surgical technique in a retrospective review on 4 patients who underwent robotic radical trachelectomy and bilateral pelvic lymphadenectomy from October 2008 to May 2009. Their analysis included 4 patients with early-stage squamous cell carcinoma of the cervix. The median body mass index was 27.1 kg/m2 (range, 22.7 to 39.1). Three patients had stage IA2 adenocarcinoma; 1 patient had stage IA1 adenocarcinoma with lymph-vascular space invasion. The median operative time was 339.5 min (range, 245 to 416). The median console time was 282.5 min (range, 217 to 338). The median estimated blood loss was 62.5 ml (range, 50 to 75). There were no conversions to laparotomy. There were no intraoperative complications. No patient required blood transfusion. The median length of hospital stay was 1.5 days (range, 1 to 2). One patient experienced a postoperative complication, transient left lower extremity sensory neuropathy. No patient had residual tumor in the trachelectomy specimen, and no patient underwent adjuvant therapy. The median number of pelvic lymph nodes removed was 20 (range, 18 to 27). The median time to a successful voiding trial was 8 days (range, 7 to 9). The median follow-up was 105 days (range, 82 to 217). There were no recurrences.

Ramirez et al concluded that robotic radical trachelectomy and bilateral pelvic lymphadenectomy is feasible and safe and should be considered for patients desiring fertility-sparing surgery [102].

There is increasing evidence in the literature that not only is radical trachelectomy feasible and safe but the oncologic outcomes are similar to those of equivalent patients undergoing radical hysterectomy.

In a recent article, Diaz et al. [103] compared the outcomes of 40 patients with stage IB1 cervical cancer who underwent radical trachelectomy and 110 patients with stage IB1 cervical cancer who underwent radical hysterectomy. The median follow-up time was 44 months. The 5-year recurrence-free survival rate was 96% for the radical trachelectomy group compared to 86% for the radical hysterectomy group (p=NS). The authors concluded that for select patients with stage IB1 cervical cancer, fertility-sparing radical trachelectomy appears to produce oncologic outcomes similar to those after radical hysterectomy.

Comparison between Laparoscopy and Robotic in Radical Hysterectomy

Magrina and colleagues recently presented the first prospective study comparing the perioperative results of patients undergoing radical hysterectomy and lymph node dissection by robotics, laparoscopy and laparotomy. The mean operating time for a robotic, laparoscopic and radical hysterectomy per laparotomy was 190, 220 and 167 min, respectively; the mean blood loss was 133, 208 and 444 mL respectively; the mean number of removed lymph nodes was 26, 26 and 28, respectively, and the mean hospital stay was 1.7, 2.4 and 3.6 days, respectively.

There were no significant differences in intra- or postoperative complications among the three groups and no conversions in the robotics or laparoscopic groups. At a mean follow up of 31 months, none of the patients with cervical cancer had experienced a recurrence. The authors concluded that laparoscopy and robotics are preferable to laparotomy for patients requiring radical hysterectomy, with advantages like significantly shorter operating times noted for robotics over laparoscopy [60].

Their results are in accordance with other groups, although some reported the feasibility of a more radical lymph node dissection with the robotic system when compared to conventional laparoscopy [21,61].

Several case reports suggest that debulking surgery is a further potential application for robotics surgery in gynaecological oncology [104,105].

In a recent paper, Cho and Nezhat compared robotic and laparoscopy in gynecological oncology. The objectives of their article were to review the published scientific literature about robotics and its application to gynecologic oncology to date and to summarize findings of this advanced computer enhanced laparoscopic technique. Relevant sources were identified by a search of PUBMED from January 1950 to January 2009 using the key words Robot or Robotics and Cervical cancer, Endometrial cancer, Gynecologic oncology, and Ovarian cancer. Appropriate case reports, case series, retrospective studies, prospective trials, and review articles were selected. A total of 38 articles were identified on the subject, and 27 were included in the study. The data for gynecologic cancer show comparable results between robotic and laparoscopic surgery for estimated blood loss, operative time, length of hospital stay, and complications. Overall, there were more wound complications with the laparotomy approach compared with laparoscopy and robotic-assisted laparoscopy. There were more lymphocysts, lymphoceles, and lymphedema in the robotic-assisted laparoscopic group compared with the laparoscopy and laparotomy groups in patients with cervical cancer. Overall, 126 robotic-assisted laparoscopies, 68 laparoscopies, and 136 laparotomies were performed in the cohort studies of cervical cancer, each with varying intraoperative and postoperative issues.

In the robotic-assisted laparoscopy group, there were 24 complications, with lymphocysts or lymphoceles, infections, and vaginal cuff complications most commonly reported. In the traditional laparoscopy group, there were 16 complications, with lymphocysts or lymphoceles and infections most commonly reported. In the laparotomy group, there were 24 complications, with adverse wound and gastrointestinal sequelae most commonly reported.

Computer-enhanced technology may enable more surgeons to convert laparotomies to laparoscopic surgery with its associated benefits. It seems that in the hands of experienced laparoscopic surgeons, final outcomes are the same with or without use of the robot [106].

Lambaudie et al compared the feasibility and efficacy of 22 robot-assisted laparoscopies with 20 traditional laparotomy and 16 conventional laparoscopy in a series of patients with locally advanced cervical cancer managed in two institutions.

In the results, no significant difference was between the three groups in terms of body mass index, FIGO stage, or tumor histology. The complication rate was similar in the three groups of patients, although there was a trend towards more lymphatic complications in the robot assisted subgroup managed medically. There was no significant difference in the recurrence rate between the robot-assisted laparoscopy, conventional laparoscopy and laparotomy groups (27.3%, 29.4% and 30%, respectively).

Authors concluded that robot-assisted laparoscopy is feasible after concurrent chemoradiation and brachytherapy in cases of locally advanced cervical cancer, reduces hospital stay, and seems to result in less severe complications than conventional laparotomy without modifying the oncological outcome [107].

Complications of Endoscopy in Cervical Cancer

Robotics, like any novel technology, has its advantages, disadvantages and reported complications.

The initial experience with case series and cohort studies seems promising.

In the modern era of minimally invasive techniques, current developments in surgical robotics represent only the initial attempts to simplify complex laparoscopic procedures, providing precision in dexterity and perfection of repetitive tasks such as suturing.

The complication rates of robotic radical hysterectomy are lower compared to our historical cohort of radical hysterectomy by standard laparotomy.

Many authors and surgeons affirm that some complications may be associated with robotic or laparoscopic surgery *per se*.

The frequent reported complications are on vaginal cuff (dehiscence, lymphatic leaking, infection , hematoma, vault prolapse, short vagina), on lymphatic system (proximal lymphedema, mild distal lymphedema, severe distal lymphedema, lymphocyst), on neural peripheral system (genitofemoral nerve injury, partial obturator nerve palsy), on abdominal wall (port site hernia, port site muscle rupture, hematoma, port site metastases) and on vascular system (postop hemoglobin and/or transfusion, ovarian vein thrombosis and pulmonary embolism).

Moreover there are also risks of infection by Pneumonia, with pyelonephritis and/or fever of unknown origin, a risk of ureter stenosis and of uncorrected positioning, with arm/shoulder/leg pain.

In the Piver study, there were 2 deaths (one from pneumonia and one from pelvic abscess), 2 pulmonary emboli, one ureterovaginal fistula and one ureteral stricture and a total of 15 complications in 55 patients who underwent type III open radical hysterectomy (27%).

Moreover, Piver and colleagues showed a 5-year disease-free interval of 87.5–100% in women with cervical cancers less than 3 cm treated by type III radical hysterectomy [8].

Debate on Robotic Approach in Cervical Cancer Surgery

The most important question for patients faced with cervical cancer and for the physicians caring for them is whether the oncologic outcomes and survival are at least equivalent to other surgical modalities.

Multiple studies have shown comparable long term outcomes between radical hysterectomy performed either via laparoscopy or laparotomy. It seems logical that robotic surgery should be equivalent to laparoscopy with regard to outcome, but this need to be best realized in literature.

Operative laparoscopy is an accepted approach to performing extirpative procedures, which in the past were traditionally performed via laparotomy. Laparoscopy has several advantages including shorter hospitalizations, less blood loss, faster return of bowel function, faster overall recovery time, fewer complications, and better results. These benefits maybe particularly meaningful in a patient with a malignancy who requires additional adjuvant therapy after recuperation from surgery.

In gynecologic oncology, the robot has been used increasingly in a variety of applications, notably for radical hysterectomy in patients with cervical cancer but also for pelvic and para aortic lymphadenectomy, trachelectomy, simple hysterectomy, and oophorectomy.

It has still to be evaluated if the robotic assisted approach is oncological reasonable and a safe alternative to open surgery.

The first, by Marchal et al evaluated 12 malignant cases (five endometrial adenocarcinomas and seven cervical carcinomas). The mean number of pelvic lymph nodes removed was 11 (range 4–21). No port-site metastasis or recurrences were found with a mean follow-up of 10 months (range 2–23) [108].

Early experiences clearly demonstrated the feasibility of applying robotic assistance to laparoscopic cancer staging without an increase in complication rates or compromise to surgical technique.

Since these early reports, several other case series confirming feasibility have been published [71,109].

Magrina et al published the first comparative study of robotic radical hysterectomy.

This study evaluated all three approaches: laparotomy, conventional laparoscopic, and robotic. There were no statistically significant differences between the three groups with respect to mean age, body mass index, or lymph node count.

However, the authors did find significantly less estimated blood loss and shorter length of stay associated with the robotic approach. In particular, operative times were comparable to open surgery and better than conventional laparoscopy. There were also no conversions or intraoperative complications in the robotic group [60].

Similarly, Boggess et al published a study comparing robot-assisted, conventional laparoscopic and open hysterectomy with staging for endometrial cancer. The robotic cohort was associated with higher lymph node retrieval, shorter operative time, lower estimated blood loss, and shorter length of stay. Ultimately, 5-year survival rates will need to be evaluated to truly assess the effect of robotics on gynecologic cancer staging. Conversion rates for robotic and laparoscopic groups were similar [66].

Drawbacks of Literature Concerning Robotic Surgery in Gynaecological Oncology

There are currently no randomized trials concerning robotic surgery in gynaecological oncology, and none of the published studies provides information on long-term effects, such as survival, lymph edema, continence and sexual function.

There is not enough evidence to conclude that robotic surgery improves outcomes in all gynecologic diseases [110].

Estape R et al compared robotic radical hysterectomy to laparoscopic and radical abdominal hysterectomy in the treatment of cervical cancer, by prospective analyses of thirty-two consecutive patients undergoing robotic radical hysterectomy, compared to 17 patients undergoing laparoscopic radical hysterectomy and 14 patients undergoing radical abdominal hysterectomy.

In their results, the operative time for the robotic group was 2.4 h±0.8 and not significantly different from the laparoscopic group at 2.2 h±0.7, nor the laparotomy group (1.9 h±0.6, $p=0.05$). The estimated blood loss for patients undergoing robotic hysterectomy was 130 ml ±119.4. This was significantly less than the laparotomy group (621.4 mL±294.0, $p<0.0001$), but not the laparoscopic group (209.4 mL±169.9, $p=0.09$). The robotic group had an average of 32.4 total nodes retrieved, as compared to 18.6 and 25.7 nodes retrieved in the laparoscopy and laparotomy cohorts, respectively. All differences were significant ($p<0.0001$ and $p<0.05$). Mean length of hospital stay was 2.6, 2.3 and 4.0 days in the robotic, laparoscopic, and laparotomy cohorts respectively. The incidence of postoperative complications was less in the robotic cohort (18.8%) as compared to the laparoscopic (23.5%), and laparotomy cohorts (28.6%).

In the conclusions, authors affirmed that robotic total laparoscopic radical hysterectomy with pelvic and para-aortic lymphadenectomy is feasible and may be preferable over laparoscopic or radical abdominal hysterectomy.

Authors reported advantages to both minimally invasive approaches, with decreased average estimated blood loss, length of stay, number of catheter days, days on pain medication, and a faster return to work in both the robotic and laparoscopic groups when compared to laparotomy. More specifically, the robotic approach resulted in a more complete lymphadenectomy as measured by number of nodes removed, with an equivalent rate of surgical margins when compared to the traditional laparoscopic approach.

Patients in the robotic group were on average significantly older than patients in the laparotomy group; however, since patients were matched for weight, stage of cancer and histology, the differences seen could be due to surgical approach rather than severity of disease or surgical complexity [111].

Tables 3 summarize the current literature detailing experiences of robotic procedures versus laparoscopic hysterectomy for the treatment of cervical cancer.

While these series are relatively small and non-randomized, they consistently demonstrate safety and efficacy with respect to complications, blood loss, operative time and patient convalescence comparing robotic assisted surgery with laparoscopy [60,63,72,111].

The problem of this comparison is the lack of reference on prospective comparison between robotics and laparoscopy. Although robotic-assisted technology is supposedly an enhancement of conventional laparoscopy, several studies have reported conversion of the robotic approach to laparotomy, not laparoscopy. There were 6 conversions from robotic-assisted laparoscopy to laparotomy in 2 studies [66,112].

Pro and Contra of Robotic Assisted Surgery in Cervical Cancer Surgery

A natural progression of robotic-assisted technology in gynecology has been to the area of oncology.

In 2005, the first feasibility studies in both Europe and the United States were published, since robotic-assisted procedures have several advantages.

Binocular vision and 3-dimensional views permit improved depth perception, which may facilitate advanced laparoscopic procedures, such as intracorporeal suturing. The console is located away from the patient and permits the surgeon to operate in a comfortable, seated position, thus making operator positioning more ergonomic.

Tremor filtration is another benefit of robotic-assisted surgery, as the video laparoscope is no longer in a human hand, which may tire or move, but rather is fixed in position by the robotic arm.

This feature permits finer surgical movements with more precise dissections.

The articulating instrument tips that are utilized in traditional laparoscopy are taken to a new level with not only rotational capabilities but an independent 90-degree articulation of the tip.

These features make robotic surgery more intuitive, with a shorter learning curve. Additional robotic arms have been introduced to further minimize the need for surgical assistants in institutions that may have limited staff.

Robotic-assisted procedures are not, however, without their limitations.

The equipment is still very large, bulky, and expensive. The staff must be trained specifically on draping and docking the apparatus to maintain efficient operative times.

Functional limitations include lack of haptic feedback, limited vaginal access, limited instrumentation, and larger port incisions requiring fascial closure. In terms of haptic feedback, visual cues become imperative to ensure that tissue manipulation is not performed with undue force. Intracorporeal knots must be tied carefully such that the suture is not avulsed by the strength imposed by the robotic arm.

Limited vaginal access can be problematic in gynecologic surgery as frequent uterine or vaginal manipulation is necessary, particularly in extirpative procedures.

Once the robot has ascended into place, access to the vagina becomes markedly limited. Robotic accessory trocars are 8 mm in size with a 12 mm laparoscope.

These incision sizes are larger than the 5 mm accessories that are frequently used in traditional laparoscopy and also require fascial closure, with higher risks of herniation.

The robot is also limited in its instrumentation. Exchanging instruments becomes more cumbersome and requires a surgical assistant to change the instruments.

Additionally, the current robotic instruments do not include endoscopic staplers or vessel sealing devices. Moreover, the trocars required for robotic procedures are larger than those used for traditional laparoscopy. In radical hysterectomy, the dissection of the uterine arteries, ureteral tunneling, and vaginal cuff closure are among the most useful indications for robotic-assisted procedures. The greater range of motion afforded by the robotic instruments permits easier maneuverability for these dissections. With these multiple advantages of robotic-assisted surgery, the clinical equivalence must be evaluated.

Future Developments and New Frontiers: the Alf-x Telelap – The renaissance of abdominal surgery

The role of so-called "robotically" assisted surgery is continuously expanding and may reach a higher degree of applicability and cost-efficiency thanks to novel concepts as the one of the Alf-x Telelap system, developed by Sofar S.p.A. (Milan, Italy) under the academic assistance of the New European Surgical Academy (NESA). However, well-designed, prospective studies with well-defined clinical, long-term outcomes, including complications, cost, pain, return to normal activity, and quality of life, are needed to fully assess the value of this new technology in radical hysterectomy. The main advantages of the Alf-x Telelap over existing so-called "robotic" systems are the existence of haptic sense, the surgeon's possibility to see his/her own hands and, simultaneously, the operative site. This, together with the 3D vision, simulates laparotomy while operating endoscopically. Hence the term "renaissance of abdominal surgery".

The system is called Telelap rather than a robot in order to avoid any misunderstanding, as it is manipulated by the surgeon rather than through artificial intelligence.

All the previously mentioned considerations on the use of "robotically" assisted surgery demonstrate that there is a clear understanding of advantages and drawbacks of the currently available systems, but also that there are no randomized statistical scientific evidences of the benefits in terms of clinical outcome. In addition, the current approach of transforming a surgeon specialized in opened surgery into an endoscopic surgeon for just a few tele-surgical interventions is not efficient and creates a strong dependence on a specific technology. Indeed, the surgeon becomes proficient for a limited number of procedures after a long learning curve, but the technique that the surgeon learned on the tele-surgical systems does not enable him to carry out these procedures 'by hand' as a laparoscopist. Is it acceptable that tele-surgical technologies propose a third discipline in-between laparotomy and laparoscopy with so many operative limitations and constrains? Why not using a tele-surgical system for training junior surgeons in the art of laparoscopy?

Other questions are centered on the limited focus of tele-surgical procedures that may be justified only in a context of highly specialized health care organizations that can afford a dedicated OR team. Therefore, what would be a balanced approach for tele-surgery? How shall tele-surgery be designed to improve intervention safety, efficiency and quality with respect to laparoscopy? Taking apart operation costs, a significant step towards versatility and ease of use appear mandatory for the diffusion of such a technology.

Following this mainstream, a new development is emerging in the field of tele-surgery presenting a radically different concept oriented towards users' practical needs.

This new tele-surgical system, the Alf-x Telelap (Figure 4), is the result of a fruitful research and development collaboration between SOFAR S.p.A. and the EC-Joint Research Centre under academic assistance of the NESA. An International Advisory Board consisting of leading endoscopic surgeons from the United States and Europe, is developing optimal standardized evidence-based operative methods to use the Alf-x Telelap.

The Alf-x Telelap transmits an accurate estimation of forces applied at the instrument tip to the fingers of the surgeon who remotely operates the system from a haptic device. The Alf-x Telelap's tactile feedback contributes to the safety and delicacy of the surgical gesture, which is essential for finely palpating tissues when searching hidden structures, and for tuning applied forces through surgical instruments when stitching and tying suture knots or when pushing or pulling out organs or tissues. An advantage with respect to manual laparoscopy is the improved tactile perception and its constant quality throughout the surgical intervention. Thanks to its real-time monitoring of forces applied to the surgical instrument, the Alf-x Telelap introduces an additional layer of safety by limiting the applied forces at the instrument tip but also at the trocar level.

The Alf-x Telelap is not limited to the haptic sensation: it also proposes a new concept of multi-disciplinary tele-surgical system dramatically enhancing safety, versatility, ergonomics, ease of use, operation time and cost-effectiveness. All these aspects aim to improve the quality and efficiency of tele-surgery but also to justify its every-day use in easy or challenging surgical procedures.

The Alf-x Telelap surgeon console has an ergonomic and fully immersive design incorporating a 3D stereo vision and a haptic console with two handles replicating a surgical instrument inserted through an access port. From the Alf-x Telelap console (Figure 5), the surgeon remotely controls the surgical instruments held by the system's arms with augmented perception, improved dexterity and accuracy. The surgeon can seamlessly assign the haptic console's handles to the surgical instruments. Moreover, the Alf-x Telelap implements an innovative and efficient method for the control of the endoscopic view through the surgeon's gaze.

The Alf-x Telelap is a modular configurable tele-surgical system consisting of three to up to five separated arms according to the specific needs of the surgical intervention (Figure 6). Each manipulator arm holds either a surgical instrument or the endoscope. The surgeon assistant interacts with the arms to exchange instruments in an intuitive and fast way. He/she can also easily drive by hand the manipulator arms and insert the attached surgical instruments through a standard 5mm or 10mm trocar. Afterwards, in a few seconds, the system's arm identifies the optimum fulcrum with respect to which the inserted instruments move with an accuracy lower than 2mm. The optimum fulcrum minimizes the interaction forces with the abdominal fascia and thereby reduces wear on the incision. The Alf-x Telelap has currently a basic set of reusable monopolar and bipolar instruments including needle holders with articulated tips. Skilled laparoscopic surgeons and some of the authors experienced a very fast learning process of a few minutes for operating the Alf-x Telelap and managed to execute suturing exercises very easily. In fact, the Alf-x Telelap's current haptic console is based on a training device for laparoscopy maintaining the fulcrum effect but simplifying the hand-instrument coordination. A resulting potential advantage is to train surgeons in laparoscopy while using the Alf-x Telelap.

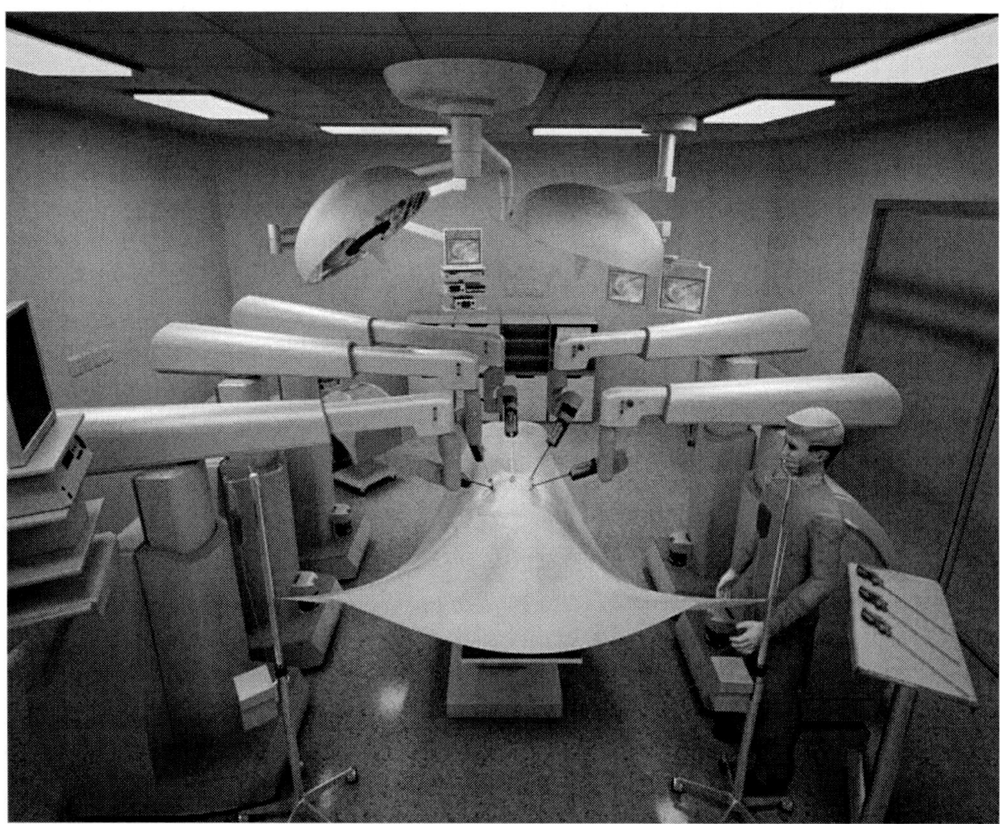

Figure 4. A general view of an Operation Room with a Alf-x Telelap system consisting of 5 arms.

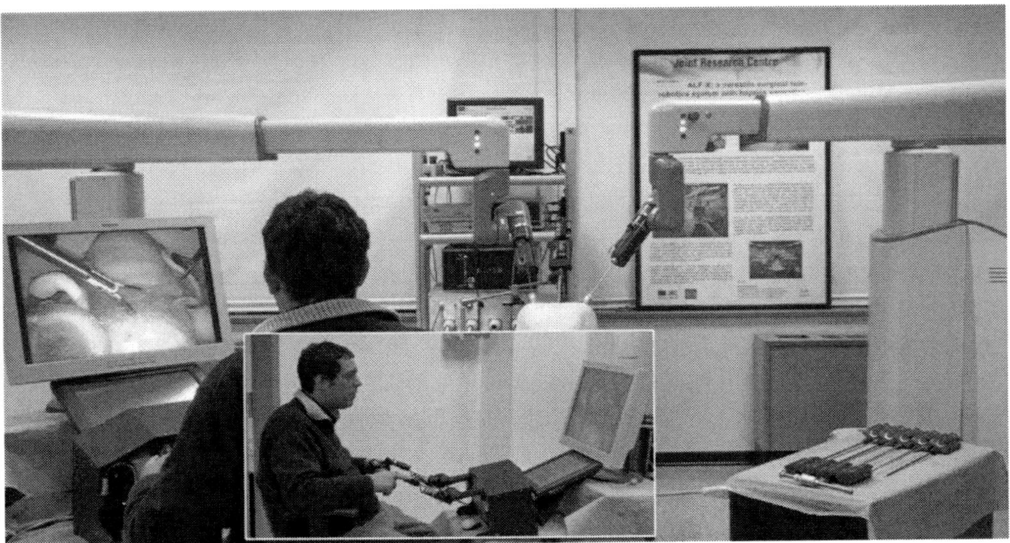

Figure 5. The surgeon operating two arms from a console.

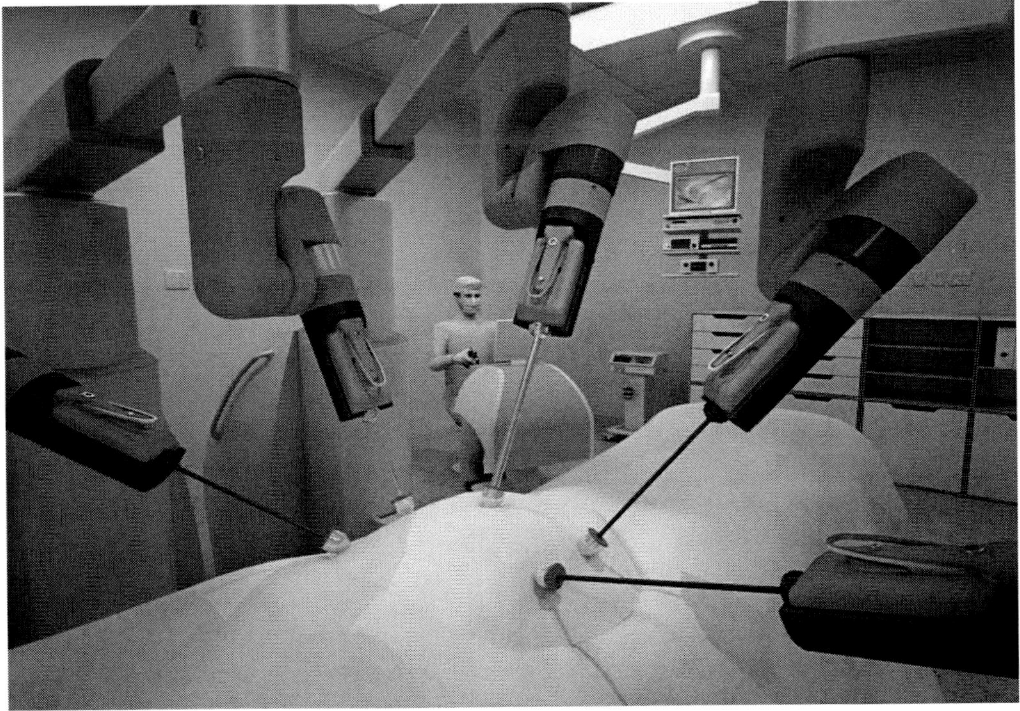

Figure 6. View of the patient's area with 5 inserted instruments held by the Alf-x Telelap.

In terms of versatility, the surgeon can place trocars in a configuration similar to the one he/she would normally use in manual laparoscopy thanks to the mechanical design and the extended range of motion of the Alf-x Telelap's arms. Therefore, the system docking on the patient (at the beginning of the surgical procedure) and the change of orientation of the patient (during the surgical procedure) can be carried out each in less than one minute and have a negligible impact on the operation time.

Safety is a fundamental criterion considered throughout the design of the new system. For example, in case of urgent access to the patient the arms can be straightforwardly withdrawn in a few seconds thanks to their mechanical reversibility.

The authors had the opportunity to execute several trials on live porcine models and appreciated the above-mentioned features and in particular its ease of use and versatility. As a conclusion, although the Alf-x Telelap is currently in an advanced development stage and undergoing a preclinical validation, the authors believe that this tele-surgical system has a high potential to become a valuable tool for minimally invasive surgery. As a conclusion, although the Alf-x Telelap is currently in an advanced development stage and undergoing a preclinical validation, the authors believe that this tele-surgical system has a high potential to become a valuable and efficient tool for laparoscopic gynecological surgery.

Conclusion

There is good evidence that tele-surgery facilitates laparoscopic surgery, with equivalent if not better operative time and comparable surgical outcomes, shorter hospital stay, and fewer major complications than with surgeries using the laparotomy approach.

The role of tele-surgery is continuing to expand.

In addition to radical hysterectomy, gynecologic oncologists are applying tele-surgical technology to ovarian transposition, lymphadenectomy, and even tumor debulking.

Some future directions that will expand the scope of tele-surgery include incorporation of the system in the operating room facility to permit better accessibility to the patient during the procedure as well as expansion in instrumentation.

Total laparoscopic radical hysterectomy is a feasible and safe procedure that is associated with fewer intraoperative and postoperative complications than abdominal radical hysterectomy.

Longer follow-up is needed, but early data are supportive of at least equivalent oncologic outcomes compared with other surgical modalities. The role of tele-surgery is continuing to expand and new promising approaches with added benefits are emerging, as the one of the the Alf-x Telelap. Surgeons await results from additional series of radical hysterectomy performed by tele-surgery and from International prospective randomized trial evaluating outcomes in patients randomly assigned to either open or laparoscopic/tele-surgical radical hysterectomy.

References

[1] Tinelli A, Malvasi A, Lorusso V, Martignago R, Vergara D, Vergari U, Guido M, Zizza A, Pisanò M, Leo G. An outlook on uterine neoplasms: from dna damaging to endometrial and cervical cancer development and minimally invasive management. In: Matsumoto, A, Nakano, M. The human genome: features, variations and genetic disorders. Nova Science Publishers Editor, Hauppauge (NY), USA 2009; Chapter 12: 227-255.

[2] Tinelli A, Vergara D, Leo G, Malvasi A, Casciaro S, Leo E, Montanari MR, Maffia M, Marsigliante S, Lorusso V. Human papillomavirus genital infection in modern gynaecology: genetic and genomic aspects. *Eur Clinics Obstet Gynaecol* 2007; 3:1-6.

[3] Tinelli A, Leo G, Pisanò M, Leo S, Storelli F, Vergara D, Malvasi A. HPV viral activity by mRNA-HPV molecular analysis to screen the transforming infections in precancer cervical lesions. *Curr Pharm Biotechnol* 2009; 10 (8), 767-71.

[4] Pecorelli S, Zigliani L, Odicino F. Revised FIGO staging for carcinoma of the cervix. *Int J Gynaecol Obstet* 2009;105:107-8.

[5] Wertheim E. The extended abdominal operation for carcinoma uteri (based on 500 operative cases). *Am J Obstet Dis Women Child* 1912;66: 169–232.

[6] Okabayashi H. Radical abdominal hysterectomy for cancer of the cervix uteri. *Surg Gynecol Obstet* 1921; 33: 335–41.

[7] Meigs JV. Carcinoma of the cervix—the Wertheim operation. *Surg Gynecol Obstet* 1944; 78: 195–98.

[8] Piver MS, Rutledge F, Smith JP. Five classes of extended hysterectomy for women with cervical cancer. *Obstet Gynecol* 1974; 44: 265–72.

[9] Symmonds RE. Some surgical aspects of gynecologic cancer. *Cancer* 1975; 36: 649–60.

[10] Kobayashi T. *Abdominal radical hysterectomy with pelvic lymphadenectomy for cancer of the cervix.* Tokyo: Nanzando, 1961: 178.

[11] Sakuragi N, Todo Y, Kudo M, Yamamoto R, Sato T. A systematic nerve-sparing radical hysterectomy technique in invasive cervical cancer for preserving postsurgical bladder function. *Int J Gynecol Cancer* 2005;15: 389–97.

[12] Fujii S, Tanakura K, Matsumura N, Higuchi T, Yura S, Mandai M, Baba T. Precise anatomy of the vesico-uterine ligament for radical hysterectomy. *Gynecol Oncol* 2007; 104: 186–91.

[13] Trimbos JB, Maas CP, Deruiter MC, Peters AA, Kenter GG. A nerve-sparing radical hysterectomy: guidelines and feasibility in Western patients. *Int J Gynecol Cancer* 2001; 11: 180–86.

[14] Raspagliesi F, Ditto A, Fontanelli R, Solima E, Hanozet F, Zanaboni F, Kusamura S. Nerve-sparing radical hysterectomy: a surgical technique for preserving the autonomic hypogastric nerve. *Gynecol Oncol* 2004; 93: 307–14.

[15] Querleu D, Morrow CP. Classification of radical hysterectomy. *Lancet Oncol* 2008; 9: 297–303.

[16] Averette HE, Nguyen HN, Donato DM, Penalver MA, Sevin BU, Estape R, Little WA. Radical hysterectomy for invasive cervical cancer. A 25-year prospective experience with the Miami technique. *Cancer* 1993;71(4 Suppl):1422-37.

[17] Magrina JF, Goodrich MA, Lidner TK, Weaver AL, Cornella JL, Podratz KC. Modified radical hysterectomy in the treatment of early squamous cervical cancer. *Gynecol Oncol* 1999;72(2):183–6.

[18] Abu-Rustum NR, Gemignani ML, Moore K, Sonoda Y, Venkatraman E, Brown C, et al. Total laparoscopic radical hysterectomy with pelvic lymphadenectomy using the argon-beam coagulator: pilot data and comparison to laparotomy. *Gynecol Oncol* 2003;91(2):402–9.

[19] Magrina JF. Outcomes of laparoscopic treatment for endometrial cancer. *Curr Opin Obstet Gynecol* 2005;17(4):343–6.

[20] Ramirez PT, Slomovitz BM, Soliman PT, Coleman RL, Levenback C. Total laparoscopic radical hysterectomy and lymphadenectomy: the M.D. Anderson Cancer Center experience. *Gynecol Oncol* 2006;102:252–5.

[21] Boggess JF. Robotic surgery in gynecologic oncology: evolution of a new surgical paradigm. *J Robotic Surgery* 2007;1(1):31–7.

[22] Marchiole P, Benchaib M, Buenerd A, Lazlo E, Dargent D, Mathevet P. Oncological safety of laparoscopic assisted vaginal radical trachelectomy (LARVT or Dargent's operation): a comparative study with laparoscopic-assisted vaginal radical hysterectomy (LARVH). *Gynecol Oncol* 2007;106:132–41.

[23] Querleu D, Leblanc E, Castelain B. Pelvic lymphadenectomy under celioscopic guidance. *J Gynecol Obstet Biol Reprod (Paris)* 1990;19:576–8.

[24] Canis M, Mage G, Wattiez A, Pouly JL, Manhes H, Bruhat MA. Does endoscopic surgery have a role in radical surgery of cancer of the cervix uteri?. *J Gynecol Obstet Biol Reprod (Paris)* 1990;19:921.

[25] Querleu D. Radical hysterectomies by the Schauta-Amreich and Schauta-Stoeckel techniques assisted by celioscopy. *J Gynecol Obstet Biol Reprod (Paris)* 1991;20:747–8.

[26] Nezhat CR, Burrell MO, Nezhat FR, Benigno BB, Welander CE. Laparoscopic radical hysterectomy with paraaortic and pelvic node dissection. *Am J Obstet Gynecol* 1992;166:864–5.

[27] Nezhat CR, Nezhat FR, Burrell MO, Ramirez CE, Welander C, Carrodeguas J, Nezhat CH. Laparoscopic radical hysterectomy and laparoscopically assisted vaginal radical hysterectomy with pelvic and paraaortic node dissection. *J Gynecol Surg* 1993;9:105–20.

[28] Hatch KD, Hallum 3rd AV, Nour M. New surgical approaches to treatment of cervical cancer. *J Natl Cancer Inst Monogr* 1996:71–5.

[29] Dargent D, Martin X, Sacchetoni A, Mathevet P. Laparoscopic vaginal radical trachelectomy: a treatment to preserve the fertility of cervical carcinoma patients. *Cancer* 2000;88:1877–82.

[30] Possover M, Stober S, Plaul K, Schneider A. Identification and preservation of the motoric innervation of the bladder in radical hysterectomy type III. *Gynecol Oncol* 2000;79:154–7.

[31] Pomel C, Atallah D, Le Bouedec G, Rouzier R, Morice P, Castaigne D, Dauplat J. Laparoscopic radical hysterectomy for invasive cervical cancer: 8-year experience of a pilot study. *Gynecol Oncol* 2001; 91:534–539.

[32] Pomel C, Canis M, Mage G, Dauplat J, Le Bouëdec G, Raiga J, Pouly JL, Wattiez A, Bruhat MA. Laparoscopically extended hysterectomy for cervix cancer: technique, indications and results. Apropos of a series of 41 cases in Clermont. *Chirurgie* 1997;122:133–6.

[33] Canis M, Mage G, Pouly JL, Pomel C, Wattiez A, Glowaczover E, Bruhat MA. Laparoscopic radical hysterectomy for cervical cancer. *Baillieres Clin Obstet Gynaecol* 1995;9:675–89.

[34] Hsieh YY, Lin WC, Chang CC, Yeh LS, Hsu TY, Tsai HD. Laparoscopic radical hysterectomy with low paraaortic, subaortic and pelvic lymphadenectomy. Results of short-term follow-up. *J Reprod Med* 1998;43:528–34.

[35] Krause N, Schneider A. Laparoscopic radical hysterectomy with para-aortic and pelvic lymphadenectomy. *Zentralbl Gynakol* 1995;117:346–8.

[36] Sedlacek TV, Campion MJ, Hutchins RA, Reich H. Laparoscopic radical hysterectomy: a preliminary report. *J Am Assoc Gynecol Laparosc* 1994;1:S32.

[37] Kim DH, Moon JS. Laparoscopic radical hysterectomy with pelvic lymphadenectomy for early, invasive cervical carcinoma. *J Am Assoc Gynecol Laparosc* 1998;5:411–7.

[38] Spirtos NM, Schlaerth JB, Kimball RE, Leiphart VM, Ballon SC. Laparoscopic radical hysterectomy (type III) with aortic and pelvic lymphadenectomy. *Am J Obstet Gynecol* 1996;174:1763–7.

[39] Abu-Rustum NR, Hoskins WJ. Radical abdominal hysterectomy. *Surg Clin NorthAm* 2001;81:815–28.

[40] Kohler C, Tozzi R, Klemm P, Schneider A. "Schauta sine utero": technique and results of laparoscopic-vaginal radical parametrectomy. *Gynecol Oncol* 2003;91:359–68.

[41] Pomel C, Atallah D, Le Bouedec G, Rouzier R, Morice P, Castaigne D, Dauplat J.. Laparoscopic radical hysterectomy for invasive cervical cancer: 8-year experience of a pilot study. *Gynecol Oncol* 2003;91:534–9.

[42] Obermair A, Ginbey P, McCartney AJ. Feasibility and safety of total laparoscopic radical hysterectomy. *J Am Assoc Gynecol Laparosc* 2003;10:345–9.

[43] Liang Z, Xu H, Chen Y, Chang Q, Shi C.. Laparoscopic radical trachelectomy or parametrectomy and pelvic and para-aortic lymphadenectomy for cervical or vaginal stump carcinoma: report of six cases. *Int J Gynecol* Cancer 2006;16:1713–6.

[44] Malzoni M, Malzoni C, Perone C, Rotondi M, Reich H. Total laparoscopic radical hysterectomy (type III) and pelvic lymphadenectomy. *Eur J Gynaecol Oncol* 2004;25:525–7.

[45] Lecuru F, Taurelle R. Transperitoneal laparoscopic pelvic lymphadenectomy for gynecologic malignancies (II). *Indications. Surg Endosc* 1998;12:97–100.

[46] Chen Y, Xu H, Li Y, Wang D, Li J, Yuan J, Liang Z. The outcome of laparoscopic radical hysterectomy and lymphadenectomy for cervical cancer: a prospective analysis of 295 patients. *Ann Surg Oncol* 2008;15:2847–55.

[47] Possover M, Krause N, Kuhne-Heid R, Schneider A. Laparoscopic assistance for extended radicality of radical vaginal hysterectomy: description of a technique. *Gynecol Oncol* 1998;70:94–9.

[48] Decloedt J, Vergote I. Laparoscopy in gynaecologic oncology: a review. *Crit Rev Oncol Hematol* 1999;31:15–26.

[49] Zakashansky K, Chuang L, Gretz H, Nagarsheth NP, Rahaman J, Nezhat FR. A case-controlled study of total laparoscopic radical hysterectomy with pelvic lymphadenectomy versus radical abdominal hysterectomy in a fellowship training program. *Int J Gynecol Cancer* 2007;17:1–8.

[50] Spirtos NM, Eisenkop SM, Schlaerth JB, Ballon SC. Laparoscopic radical hysterectomy (type III) with aortic and pelvic lymphadenectomy in patients with stage I cervical cancer: surgical morbidity and intermediate follow-up. *Am J Obstet Gynecol* 2002;187:340–8.

[51] Gil-Moreno A, Diaz-Feijoo B, Roca I, Puig O, Pérez-Benavente MA, Aguilar I, Martínez-Palones JM, Xercavins J.. Total laparoscopic radical hysterectomy with intraoperative sentinel node identification in patients with early invasive cervical cancer. *Gynecol Oncol* 2005;96:187–93.

[52] Li G, Yan X, Shang H, Wang G, Chen L, Han Y. A comparison of laparoscopic radical hysterectomy and pelvic lymphadenectomy and laparotomy in the treatment of Ib-IIa cervical cancer. *Gynecol Oncol* 2007;105:176–80.

[53] Frumovitz M, dos Reis R, Sun CC, Milam MR, Bevers MW, Brown J, Slomovitz BM, Ramirez PT. Comparison of total laparoscopic and abdominal radical hysterectomy for patients with early-stage cervical cancer. *Obstet Gynecol* 2007;110:96–102.

[54] Frumovitz M, Ramirez PT, Greer M, Gregurich MA, Wolf J, Bodurka DC, Levenback C. Laparoscopic training and practice in gynecologic oncology among Society of Gynecologic Oncologists members and fellow-in-training. *Gynecol Oncol* 2004;94:746–53.

[55] Puntambekar SP, Palep RJ, Puntambekar SS, Wagh GN, Patil AM, Rayate NV. Laparoscopic total radical hysterectomy by the Pune technique: our experience of 248 cases. *J Minim Invasive Gynecol* 2007;14:682-9.

[56] Colombo PE, Bertrand MM, Gutowski M, Mourregot A, Fabbro M, Saint-Aubert B, Quenet F, Gourgou S, Kerr C, Rouanet P. Total laparoscopic radical hysterectomy for locally advanced cervical carcinoma (stages IIB, IIA and bulky stages IB) after concurrent chemoradiation therapy: Surgical morbidity and oncological results *Gynecol Oncol* 2009;114:404–409.

[57] Xu H, Chen Y, Li Y, Zhang Q, Wang D, Liang Z. Complications of laparoscopic radical hysterectomy and lymphadenectomy for invasive cervical cancer: Experience based on 317 procedures. *Surg Endosc* 2007; 21: 960–964.

[58] Malzoni M, Tinelli R, Cosentino F, Perone C, Iuzzolino D, Rasile M, Tinelli A. Laparoscopic radical hysterectomy with lymphadenectomy in patients with early cervical cancer: Our instruments and technique. *Surg Oncol* 2009;18:289-297.

[59] Magrina JF. Robotic surgery in gynecology. *Eur J Gynaecol Oncol* 2007; 28: 77–82.

[60] Magrina JF, Kho RM, Weaver AL, Montero RP, Magtibay PM. Robotic radical hysterectomy: Comparison with laparoscopy and laparotomy. *Gynecol Oncol* 2008;109:86–91.

[61] Kim YT, Kim SW, Hyung WJ, Lee SJ, Nam EJ, Lee WJ. Robotic radical hysterectomy with pelvic lymphadenectomy for cervical carcinoma: A pilot study. *Gynecol Oncol* 2008; 108:312–316.

[62] Boggess JF. Robotic-assisted radical hysterectomy for cervical cancer: National Library of Medicine Archives. 2006 [updated 2006; cited 2008 11/08/2008]; Available from: http://www.nlm.nih.gov/medlineplus/surgeryvideos.html.

[63] Nezhat F. Minimally invasive surgery in gynecologic oncology: Laparoscopy versus robotics. *Gynecol Oncol* 2008; 111(2):29-32

[64] Fanning J, Fenton B, Purohit M. Robotic radical hysterectomy. *Am J Obstet Gynecol* 2008;198:649:1-4.

[65] Ko EM, Muto MG, Berkowitz RS, Feltmate CM. Robotic versus open radical hysterectomy: A comparative study at a single institution. *Gynecol Oncol* 2008;111:425–430.

[66] Boggess JF, Gehrig PA, Cantrell L, Shafer A, Ridgway M, Skinner EN, Fowler WC. A comparative study of 3 surgical methods for hysterectomy with staging for endometrial cancer: Robotic assistance, laparoscopy, laparotomy. *Am J Obstet Gynecol* 2008;199:360:1-9.

[67] DeSimone CP, Ueland FR. Gynecologic laparoscopy. *Surg Clin N Am*. 2008;/88:/319-41.

[68] Advincula AP, Song A. The role of robotic surgery in gynecology. *Curr Opin Obstet Gynecol.* 2007;/19:/331-6.

[69] Sinha R, Hedge A, Mahajan C, Dubey N, Sundaram M. Laparoscopic myomectomy: do size, number and location of the myomas form limiting factors for laparoscopic myomectomy? *J Minim Invasive Gynecol.* 2008;15:292-300.

[70] Bandera CA, Magrina JF. Robotic surgery in gynecologic oncology. *Curr Opin Obstet Gynecol.* 2009;21:25-30.

[71] Veljovich DS, Paley PJ, Dreschner CW, Everett EN, Shah C, Peters WA. Robotic surgery in gynecologic oncology: program initiation and outcomes after the first year with comparison with laparotomy for endometrial cancer staging. *Am J Obstet Gynecol* 2008;198: 679.e1–9.

[72] Sert BM, Abeler VM. Robotic-assisted laparoscopic radical hysterectomy (Piver type III) with pelvic node dissection–case report. *Eur J Gynaecol Oncol* 2006;27:531–3.

[73] Kowalski L, Falkner CA, Wishnev SA. Robotic versus abdominal radical hysterectomy for gynaecological cancer. *Gynecol Oncol* 2008;108 (Suppl Abstract 141):S32–S155.

[74] Lowe MP, Chamberlain DH, Kamelle SA, Johnson PR, Tillmanns TD. A multi-institutional experience with robotic-assisted radical hysterectomy for early stage cervical cancer. *Gynecol Oncol.* 2009;113(2):191-4.

[75] Persson J, Reynisson P, Borgfeldt C, Kannisto P, Lindahl B, Bossmar T. Robot assisted laparoscopic radical hysterectomy and pelvic lymphadenectomy with short and long term morbidity data. *Gynecol Oncol* 2009;113:185–190.

[76] Cantrell LA, Mendevil A, Gehring PA, Boggess JF. Survival outcomes for women undergoing type III robotic radical hysterectomy for cervical cancer: A 3-year experience. *Gynecol Oncol* 2010;117:260–265.

[77] Lagasse LD, Creasman WT, Shingleton HM, Ford JH, Blessing JA. Results and complications of operative staging in cervical cancer: experience of the Gynecologic Oncology Group. *Gynecol Oncol* 1980;9:90.

[78] LaPolla JP, Schlaerth JB, Gaddis O, Morrow CP. The influence of surgical staging on the evaluation and treatment of patients with cervical carcinoma. *Gynecol Oncol* 1986;24:194.

[79] Mortier DG, Stroobants S, Amant F, Neven P, Van Lambergen E, Vergote I. Laparoscopic para-aortic lymphadenectomy and positron emission tomography scan as staging procedures in patients with cervical carcinoma stage IB2–IIIB. *Int J Gynecol Cancer* 2008;18:723–9.

[80] Possover M, Krause N, Plaul K, Kuhne-Heid R, Schneider A. Laparoscopic paraaortic and pelvic lymphadenectomy: experience with 150 patients and review of the literature. *Gynecol Oncol* 1998;71(1):19–28.

[81] Kohler C, Klemm P, Schau A, Possover M, Krause N, Tozzi R, Schneider A. Introduction of transperitoneal lymphadenectomy in a gynecologic oncology center: analysis of 650 laparoscopic pelvic and/or para-aortic transperitoneal lymphadenectomies. *Gynecol Oncol* 2004;95(1):52–61.

[82] Marnitz S, Kohler C, Roth C, Fuller J, Hinkelbein W, Schneider A. Is there a benefit of pretreatment laparoscopic transperitoneal surgical staging in patients with advanced cervical cancer? *Gynecol Oncol* 2005;99(3):536–44.

[83] Denschlag D, Gabriel B, Mueller-Lantzsch C, Tempfer C, Henne K, Gitsch G, Hasenburg A. Evaluation of patients after extraperitoneal lymph node dissection for cervical cancer. *Gynecol Oncol* 2005;96(3):658–64.

[84] Querleu D. Laparoscopic paraaortic node sampling in gynecologic oncology: a preliminary experience. *Gynecol Oncol* 1993;49(1):24–9.

[85] Spiritos NM, Schlaerth JB, Spirtos TW, Schlaerth AC, Indman PD, Kimbal RE. Laparoscopic bilateral pelvic and para-aortic lymph node sampling: an evolving technique. *Am J Obstet Gynecol* 1995;173:105–11.

[86] Dargent D, Ansquer Y, Mathevet P. Technical development and results of left extraperitoneal laparoscopic para-aortic lymphadenectomy for cervical cancer. *Gynecol Oncol* 2000;77:87–92.

[87] Querleu D, Dargent D, Ansquer Y, Leblanc E, Narducci F. Extraperitoneal endosurgical aortic and common iliac dissection in the staging of bulky or advanced cervical carcinomas. *Cancer* 2000;88:1883–91.

[88] Vergote I, Amant F, Beterloot P, Van Gramberen M. Laparoscopic lower paraaortic staging lymphadenectomy in stage IB2, II, and III cervical cancer. *Int J Gynecol Cancer* 2002;12:22–6.

[89] Vergote I, Pouseele B, Van Gorp T, Vanacker B, Leunen K, Cadron I, Neven P, Amant F. Robotic retroperitoneal lower para-aortic lymphadenectomy in cervical carcinoma: first report on the technique used in 5 patients. *Acta Obstet Gynecol Scand* 2008;87:783–7.

[90] Fastrez M, Vandromme J, George P, Rozenberg S, Degueldre M. Robot assisted laparoscopic transperitoneal para-aortic lymphadenectomy in the management of advanced cervical carcinoma. *Eur J Obstet Gynaecol Reprod Biol* 2009; 147:226–229.

[91] Ramirez PT, Schmeler KM, Wolf JK, Brown J, Soliman PT. Robotic radical parametrectomy and pelvic lymphadenectomy in patients with invasive cervical cancer. *Gynecol Oncol* 2008;111: 18–21.

[92] Dargent D, Brun JL, Roy M, Remy I. Pregnancies following radical trachelectomy for invasive cervical cancer (Abstract). *Gynecol Oncol* 1994;52:105.

[93] Plante M. Radical vaginal trachelectomy: an update. Gynecol Oncol 2008;111:S105–110.

[94] Beiner ME, Hauspy J, Rosen B, Murphy J, Laframboise S, Nofech-Mozes S, Ismiil N, Rasty G, Khalifa MA, Covens A. Radical vaginal trachelectomy vs. radical hysterectomy for small early stage cervical cancer: a matched case-control study. *Gynecol Oncol* 2008;110:168–71.

[95] Dursun P, LeBlanc E, Nogueira MC. Radical vaginal trachelectomy (Dargent's operation): a critical review of the literature. *Eur J Surg Oncol* 2007;33:933–41.

[96] Hertel H, Kohler C, Grund D, Hillemanns P, Possover M, Michels W, Schneider A; German Association of Gynecologic Oncologists (AGO). Radical vaginal trachelectomy (RVT) combined with laparoscopic pelvic lymphadenectomy: prospective multicenter study of 100 patients with early cervical cancer. Gynecol *Oncol* 2006;103:506–11.

[97] Ungar L, Palfalvi L, Hogg R, Siklós P, Boyle DC, Del Priore G, Smith JR. Abdominal radical trachelectomy: a fertilitypreserving option for women with early cervical cancer. *Br J Obstet Gynaecol* 2005;112:366–9.

[98] Pareja RF, Ramirez PT, Borrero MF, Angel CG. Abdominal radical trachelectomy: a case series and literature review. *Gynecol Oncol* 2008;111:555–60.

[99] Geisler JP, Orr CJ, Manahan KJ. Robotically assisted total laparoscopic radical trachelectomy for fertility sparing in stage IB1 adenosarcoma of the cervix. *J Laparoendosc Adv Surg Tech* 2008;18:727–9.

[100] Persson J, Kannisto P, Bossmar T. Robot-assisted abdominal laparoscopic radical trachelectomy. *Gynecol Oncol.* 2008;111:564–567.

[101] Chuang LT, Lerner DL, Liu CS, Nezhat FR. Fertility-sparing robotic assisted radical trachelectomy and bilateral pelvic lymphadenectomy in early-stage cervical cancer. *J Minim Invasive Gynecol.* 2008;15:767–770.

[102] Ramirez PT, Schmeler KM, Malpica A, Soliman PT. Safety and feasibility of robotic radical trachelectomy in patients with early-stage cervical cancer. *Gynecol Oncol* 2010; 116:512–515.

[103] Diaz JP, Sonoda Y, Leitao MM, Zivanovic O, Brown CL, Chi DS, Barakat RR, Abu-Rustum NR. Oncologic outcomes of fertility-sparing radical trachelectomy versus radical hysterectomy for stage IB1 cervical carcinoma. *Gynecol Oncol* 2008;111:255–60.

[104] Chavin G. Robotic-assisted laparoscopy for the excision of a pelvic leiomyosarcoma. *J Robotic Surg* 2008; 1: 315–317.

[105] van Dam PA, van Dam PJ, Verkinderen L, Vermeulen P, Deckers F, Dirix LY. Robotic-assisted laparoscopic cytoreductive surgery for lobular carcinoma of the breast metastatic to the ovaries. *J Minim Invasive Gynecol* 2007; 14: 746–749.

[106] Cho JE, Nezhat FR. Robotics and Gynecologic Oncology: Review of the Literature. *J Minim Invasive Gynecol* 2009; 16 (6):669-681.

[107] Lambaudie E, Narducci F, Bannier M, Jauffret C, Pouget N, Leblanc E, Houvenaeghel G. Role of robot-assisted laparoscopy in adjuvant surgery for locally advanced cervical cancer. *Eur J Surg Oncol* 2010;36: 409-413.

[108] Marchal F, Rauch P, Vandromme J, Laurent I, Lobontiu A, Ahcel B, Verhaeghe JL, Meistelman C, Degueldre M, Villemot JP, Guillemin F. Telerobotic-assisted laparoscopic hysterectomy for benign and oncologic pathologies: initial clinical experience with 30 patients. *Surg Endosc* 2005;19:826–31.

[109] Kim YT, Kim SW, Hyung WJ, Lee SJ, Nam EJ, Lee WJ. Robotic radical hysterectomy with pelvic lymphadenectomy for cervical carcinoma: a pilot study. *Gynecol Oncol* 2008;108: 312–6.

[110] Zapardiel I. Is robotic surgery suitable for all gynecologic procedures? *Acta Obst Gynecol Scandinavica* 2009;88:10,1176.

[111] Estape R, Lambrou N, Diaz R, Estape E, Dunkin N, Rivera A. A case matched analysis of robotic radical hysterectomy with lymphadenectomy compared with laparoscopy and laparotomy. *Gynecol Oncol.* 2009;113(3):357-61.

[112] DeNardis SA, Holloway RW, Bigsby GE, Pikaart DP, Ahmad S, Finkler NJ. Robotically assisted laparoscopic hysterectomy versus total abdominal hysterectomy and lymphadenectomy for endometrial cancer. *Gynecol Oncol.* 2008;111:412–417.

In: Brain Cancer, Tumor Targeting and Cervical Cancer
Editor: Elena K. Salvatti
ISBN: 978-1-61122-738-3

Chapter V

Malignant Glioma: Challenges, Advances and Perspectives for Treatment

A. Nimer Amr,[1] Sven Kantelhardt,[2] Alf Giese[2,] and Ella L. Kim[2,*]*

[1]Department of Neurosurgery, Klinikum Bremenhaven Reinkenheide, Bremenhaven, Germany

[2]The Translational Neurooncology Research Group, Department of Neurosurgery, University Medical Centre Göttingen, Göttingen, Germany

Abstract

From the earliest records of surgical resection of a cerebral lesion in 1885 [11,12] till the present time, neoplasms of the CNS have posed an interdisciplinary therapeutic challenge. While benign brain tumours have been treated with relatively satisfactory results since the early decades of the last century, [76,85,124] malignant neoplasms continue to elude cure despite advances in diagnostics and steadily evolving neurosurgical techniques. In fact, mean survival rates for the largest group of malignant primary brain tumours, glioblastoma multiforme (GBM), have only slowly improved in the last four decades. [87] The major underlying reason for the limited clinical success of therapeutic surgical options is that GBMs, while rarely capable of metastasizing outside of the central nervous system (CNS), possess the propensity for infiltrative spread within the CNS, a phenomenon known as "local glioma invasion". [40,42,43,93,108] Another

* Correspondence may be addressed to:
Dr. Ella L. Kim, The Translational Neurooncology Research Group, Department of Neurosurgery, University Medical Centre Göttingen, Robert-Koch-Strasse 40, 37075 Göttingen, Germany.; tel +49-(0)551-3922966; ella.kim@med.uni-goettingen.de
Dr. Alf Giese, The Translational Neurooncology Research Group, Department of Neurosurgery, University Medical Centre Göttingen, Robert-Koch-Strasse 40, 37075 Göttingen, Germany.; tel +49-(0)551-396036; alf.giese@med.uni-goettingen.de

prominent feature of gliomas is their heterogeneity with respect to the cell type composition, biological phenotypes and responses to cytotoxic treatments. [75,120] Over the last two decades, much progress has been made towards understanding of the GBM biology, including the realisation that malignant gliomas possess an hierarchical structure, whereby various types of cells might exist within the same tumour to exert different functions [35]. Recently, the potential research breakthrough in understanding the biological basis for the less than satisfying response of malignant gliomas to current therapies has been made through the identification of a specialized subpopulation of cells generally termed brain tumour initiating cells (BTICs) that might be largely responsible for post-treatment recurrence of malignant glioma. [51,100] In this chapter, we discuss the history and current status of glioma therapy as well as some of the newly opened research avenues with special reference to BTICs.

Keywords: Glioma; neurosurgery; neurooncology; glioma therapy Abstract

I. Classification and Grading of Brain Tumours

The classification of brain tumours plays a key role in determining the therapeutic approaches to be utilised. Broadly speaking, neoplasms of the CNS can be divided into two major groups: those arising from constituent cells of the CNS (primary brain tumours) and those originating from metastatic spread of non-CNS cancers (secondary brain tumours). The arguably first classification system proposed for brain tumours was that of Cushing and Bailey, first published in 1926, wherein glial tumours were first subcategorised into thirteen groups according to their cellular lineage. [7,34] Even in that early period, it was clear that not all CNS neoplasms grew at the same rate or had the same level of invasiveness or prognosis, thus necessitating a grading system that accommodated those facts. The most widely accepted tumour classification system today, the WHO classification, further categorizes primary brain tumours into various subcategories according to their resemblance to CNS cell type and the degree of aggressiveness. The WHO grades brain tumours on a scale of I (benign tumours) to IV (malignant tumours). [80]

GBMs classified as WHO IV tumours of astrocytic origin comprise the largest group of malignant CNS neoplasms. According to the WHO criteria GBMs are divided into two main groups: de novo (primary) GBM that display a sudden onset and which constitute most GBM cases, as well as the so-called progressive pathway, where lower-grade gliomas progress with time to higher-grade gliomas, ultimately progressing to secondary GBM. Primary and secondary GBMs apparently evolve through different molecular mechanisms as evidenced by distinct patterns of genetic alterations characteristic of each group. [37]

It is becoming increasingly clear that a simplistic classification and grading system based solely on histopathological features is insufficient in the case of GBM and that consideration of molecular profiles in relation to clinical outcome might be more accurate at stratifying gliomas into clinically distinct entities. [79,122] Recent research findings indicate that similar histopathological characteristics may not necessarily reflect similarity in clinical outcomes whereas stratification schemes based upon similarity of gene expression patterns have a higher predictive power for glioma prognosis. [131]

Table 1. Genetic aberrations associated with GBM. From Furnari et al, 2007 [37]

	Primary GBM	Secondary GBM
Genetic aberration	INK4aARF mutation EGFR amplification/mutation Cyclin D1/3 amplification/overexpression MDM2 or 4 amplification/overexpression LOH 10p/ 10q PTEN mutation Bcl2L Overexpression	10q loss PTEN mutation

The status of individual molecular markers plays an increasingly important role in predicting the therapeutic response in glioma patients. [83] For instance, promoter methylation of the MGMT (O6-methylguanine-DNA methyltransferase) DNA-repair gene has proven to be a positive prognostic factor for the response to RT and CT regardless of the therapeutic modality utilised. [50,103] Over-expression of MGMT inversely leads to increased chemoresistance. [21]

Even though the utilisation of a molecular classification of GBM is not yet well established in routine clinical practice, such a classification may serve to improve the accuracy of determining the prognosis of glioma patients and lay basis for an individualised tailor-made therapy in the future.

II. Current Status of Diagnostic and Therapeutic Modalities

II. a. Diagnostics

Prior to the era of radiological imaging, the location of cerebral lesions was solely the domain of meticulous neurological examination and history taking. [102,107] The bedside neurological assessment remains the cornerstone of and initial step in brain tumour diagnosis. It was not until Dandy introduced air ventriculography in 1918 and pneumoencephalography in 1919 that CNS lesions were (indirectly) radiologically detectable. [44,102] Although this method had several limitations and risks such as cardiovascular complications, [125] vomiting and neurological deficits [129], it remained a key diagnostic tool till the introduction of angiography [99] and computerised tomography (CT) in the practice of clinical neuro-oncology starting from the 1950s in the 1970s, respectively. [54,67] The introduction of CTs heralded a new era in the field of neurosurgery. For the first time, non-invasive imaging of brain tissue with an acceptable picture resolution was possible. No longer were locations of lesions of the CNS the subject of an educated guess – CNS neoplasms and other brain lesions could be located with precision through a simple diagnostic test. The main drawbacks in the early era of CT-based diagnostics were the relatively high radiation dose emitted by CT-scans and that the display of certain anatomical areas, particularly those with a higher bone density such as the posterior fossa, was artefact laden. The introduction of Magnetic Resonance Imaging (MRI) in the 1980s had been an essential step towards alleviating these problems.

Further developments such as coupling MR imaging with cerebral function mapping in the 1990s (functional MRI, fMRI) provided the possibility to identify and avoid eloquent areas in the brain. [62,92] Diffusion tensor imaging (DTI), a complex process whereby cortical tracts are visualised, helps sparing cortical pathways from resection. [62] Metabolic imaging with Positron Emission Tomography (PET) defines a larger area of metabolic abnormality as a target for resection compared to T1 Gadolineum MRT. Recently, Diffusion and Perfusion MRI as well as MR-Spectroscopy have seen more widespread utilisation to further differentiate between various types of CNS lesions. [132,135]

When an unidentifiable CNS lesion is detected in a CT-scan with contrast medium, and a malignant tumour is suspected, MR imaging is performed to help differentiate the nature of the lesion. In particular, dynamic contrast T2-weighted perfusion MRI may help differentiate between the different types of malignancies and other lesions such as abscesses and metastatic lesions pre-operatively. [62] As non-invasive procedures with minimal side effects and extremely high image resolution, MRI is today the golden standard in neuro-oncological diagnostics, whereby CTs often form the first line of diagnostics due to their ready availability.

The radiologic appearance of GBM in MR- and CT-scans is illustrated and described in Figure 1.

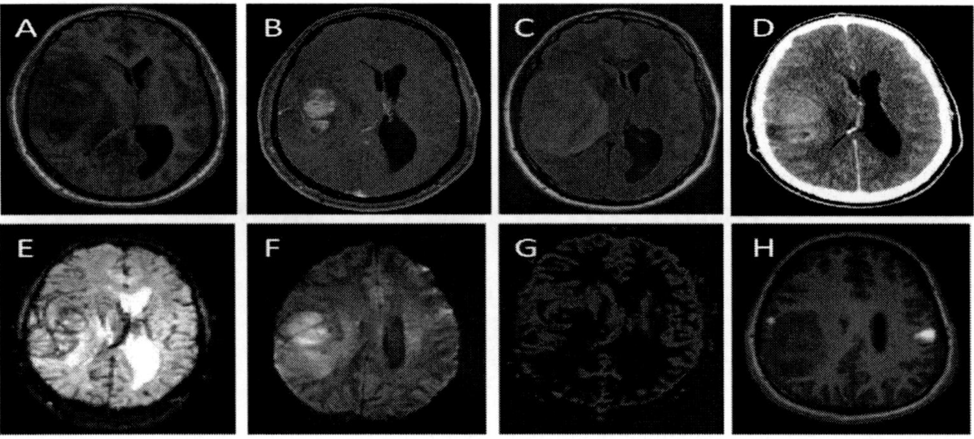

Figure 1. MKultimodal imaging of an Glioblastoma multiforme WHO oIV. A shows a T1 weighted MRI without aand B with Gadolinium-contrast enhancement. The tumor shows strong contrast enhancement, revealing a downbreak of the blood-brain barrier. C is a T2 FLAIR weighted image, showing the tumor with surrounding edema. D is a conventional CT=scan. E and F are diffusion and ADC weighted MRIs, E is a MRI-perfusion, highlighting the blood supply to different regions of the brain, while H shows a functional MRI. Motor-cortical areas activated during finger-tapping are gighlighted in bright colours.

II. b. Surgery

Historical Review and Current Status

The development of neurosurgery as a medical discipline is closely intertwined with neuro-oncology. In fact, it could be argued that the strive to find a "cure" for brain tumours spurned the birth of neurosurgery as we know it today. While accounts of ancient operations on the CNS are well known, [18,78,94,118] the first recorded post-medieval operation of a brain tumour, presumably a glioma, was performed by the London surgeon R. Godlee in 1885, [11,12] whereby the tumour was primitively removed by means of a spoon. [130] This was closely followed by the 1886 glioma operation by the surgeons Hirschfelder and Morse of San Francisco. [63] At that time, the surgical therapy of neuro-oncological disorders suffered from the same problems that plagued all surgical fields: namely infection, haemostasis and oedema. Resection of brain tumours was further complicated by the fact that the CNS and its functions were poorly understood. These were the key factors holding back the development of surgery as a therapeutic modality in the treatment of brain tumours till the early years of the 20th century when it became possible to overcome life threatening hurdles due to infection, bleeding and neurological deficits associated with brain surgery. The introduction of Listerian methods, aseptic surgical techniques, and discovery of antibiotics helped minimise infection. Invention of the so called “silver clips” to ligate small vessels by Harvey Cushing of Baltimore, the pioneering neurosurgeon, was an important first step in controlling bleeding in tumour operations. [76,124] The introduction of electric currents to heat tissues (electrocautery) helped further control bleeding. William Bovie developed Monopolar electrocautery for neurosurgical use in 1926 as requested by Cushing. [121] Bipolar instruments, which made a more limited and refined electrocoagulation possible, were designed by Leonard Malis of Yale in 1951. [31]

In its early days, glioma surgery aimed primarily at relieving the increased intracranial pressure by the means of decompression, (subtemporal in hemispheric tumours [105] or suboccipital decompression in the case of tumours of the posterior fossa [25]). Cushing is credited with developing this approach with respect to brain tumours. [76,105,124] Not surprisingly, this palliative approach did little in the way of significantly increasing survival times leading neurosurgeons realise that the extent of tumour resection is a crucial factor in achieving a meaningful prolongation of survival time in patients with brain tumours. Cushing’s former pupil Walter Dandy at the Stanford University Hospital in San Fransisco pursued the idea that radical resection might be the approach to improve the clinical outcome in brain tumour patients in terms of increased survival. Dandy performed series of hemispherectomies (an operation where one cerebral hemisphere is removed) in patients that had brain tumours on the right hemisphere and were paralized on the left side [27]. The outcome was that Dandy’s patients who had undergone hemispherectomy eventually developed and succumbed to recurrent tumours [27]. Though disappointing, the results of Dandy’s hemispherectomy approach provided first clear evidence that even the most radical surgery does not allow the complete removal of all brain tumour cells and that few tumour cells left behind are capable of repopulating the tumour, the phenomenon known as tumour recurrence. A middle ground was to be found: radical resection of the tumour while sparing healthy brain tissue.

The introduction of the operating microscope is universally acknowledged as a major breakthrough in the surgical therapy of brain tumours. [58] Although the first operation involving a tumour where a microscope was utilised took place in 1957, [119] it took many refinements, spanning decades, [119,133] till the microscope has firmly taken the central place it occupies today in the operating theatre. The benefits of microneurosurgery (the practice of using microscopes) and specialised surgical instruments came to be known, were manifold. The illumination of the operation field yielded better visual access to dim structures within the brain. This allowed finer incisions, which in turn led to shortened operation durations, a smaller risk of damage to healthy tissue, and better wound healing. The magnification, coupled with illumination, aids the differentiation between normal brain and malignant tissue in the cases where this is not possible with the naked eye under dim light conditions, such as the borders of glioma tissue.

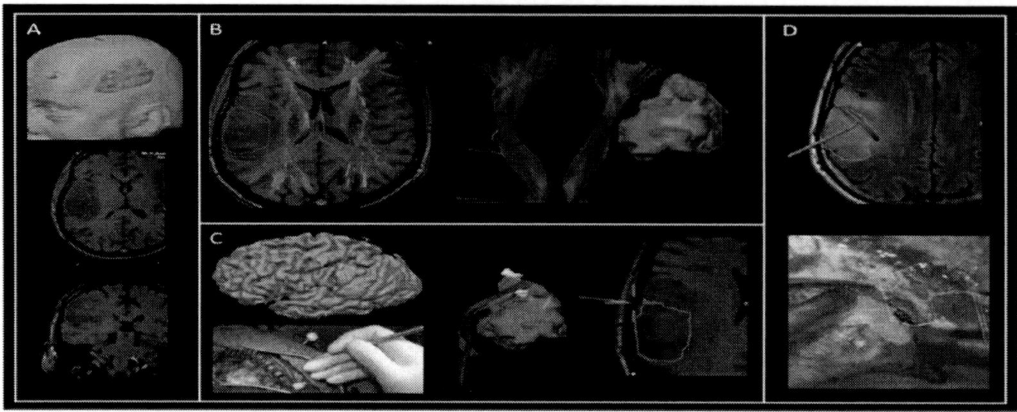

Figure 2. Image-guided resection of brain tomors. A: Three-dimensional segmentation and display of a glioma in a neuro-bavugatuib ststen, B: Diffusion tensor imaging (DTI) of the pyramidal tracks in relation to the tumor volume. C: Functional MRI demonstratinhg activation of the cortical tongue area, segmentation of the functional area in three-dimensional navigation data and navigation-guided direct cortical electrical stimulation for functional mapping.l D: Display of the trajectory and field of view of the operating microscope in navigation data and image injection of the segmented tumor volume into the light router of the operating microscope.

The amalgamation of imaging techniques and intraoperative computers gave rise to image guided neurosurgery or neuronavigation. [32,67,77] The principle behind neuronavigation is straightforward: the position of the tumour is calculated relative to fixed points on the operating theatre. By determining the position in 3D space of a mobile object (pointer or microscope) using sensors, its distance from the fixed points (and hence, from the tumour) is calculated. This data is relayed to the computer, which then displays 3 dimensional pictures of the tumour relative to the position of the mobile object. Neuronavigation allows relatively precise pre-operative localisation of the tumour in question, thereby greatly aiding the surgeon in the choice of surgical approach to be used (operative planning). [10] Great strides have been made with respect to neuronavigation – it is possible today to combine fMRI [16,48] and diffusion tensor imaging (DTI) data and metabolic imaging with neuronavigation, thus displaying this data intraoperatively (Figure 2). Integrating navigation of the operating microscope itself into navigation systems and adding image injection of spatial and functional data into the microscopes light router today provides the surgeon with the information needed to identify the tumour tissue to a greater extent while sparing healthy tissue from resection.

All the aforementioned tools help achieve the main aim of surgery, which is the maximal resection of GBM tissue while sparing cortical pathways and eloquent areas. As mentioned earlier, the migration of glioma cells and their invasion of healthy brain tissue render complete microscopic surgical resection impossible, which dictates the need for adjuvant therapeutic modalities – chemotherapy and radiotherapy.

II.c. Radiotherapy

Radiation has been used to treat malignancies of the CNS since the early decades of the 20th century. [104,127] The type of radiation itself has not changed since decades, [120] what has changed is the method of delivery, the dosage used and the extent of radiation. Cushing experimented with interstitial irradiation of gliomas by placing a 'radium bomb' in the surgical cavity – a crude cluster of "radium needles enclosed in a rubber sponge and wrapped in thin rubber tissue" [104] as he described. The discouraging results put an end to this method of radiation delivery. The use of radiotherapy in the following period, though widespread, was not without controversy. [106] It was not until Walker et al published their findings [127] about the correlation between survival time and radiation dose was a the benefit of radiation in increasing survival times in gliomas established without doubt. The initially used whole brain irradiation (WBRT) gradually gave way to the specific targeting of the tumour site (alongside a small margin around the site) due to the fact that no significant advantage of WBRT vs. specific targeting could be proven in trials. [109] Specific targeting had the further advantage of sparing radiosensitive tissues such as the eye.

The current standard dosage of radiotherapy is 60 Gy (fractioned at 2 Gy/d) delivered over a 6 week period, with the focus of the radiation on the site of the lesion itself, while a so-called 'boost volume' radiation focusing on a 2 cm margin around the tumour site.

The adjuvant treatment of GBM with radiotherapy alone, even after aggressive and radical resection, does not prevent recurrence. This is thought to be due to the proven radioresistance of GBM cells. To increase the survival rate among GBM patients, a third modality is utilised: concomitant chemotherapy.

II.d. Chemotherapy

While surgery and later radiotherapy where developing as feasible, if insufficient, therapy modalities for gliomas since the early years of the last century, chemotherapy (CTx) came to be discovered much later. Like many other great discoveries in medicine, the novel idea of using what is essentially poison to treat cancerous lesions was the product of a chance observation. Physicians associated with the US Department of Defence noticed the depletion of the lymphatic organs in soldiers returning from the battlefields of WWII when mustard gas was deployed. [26] This has led to the study of the effect of mustard gas on lymphoma. The positive results spurned numerous studies, developments and trials of scores of chemotherapeutic agents and their effect on virtually every existing neoplasm type. A new era in cancer therapy was heralded.

In the specific case of malignant gliomas, systematic administration of the nitrosoureas CCNU (lomustine) and BCNU (carmustine) has long been the mainstay of brain tumour CTx despite their side effects. Although these agents are able to pass the blood-brain barrier (BBB), relatively high systematic doses are needed to achieve an effective dose in the CNS. The on-going debate as to the efficacy of CCNU, which was usually administered with procarbazine and vincristine (the PCV regimen), [49] vs. BCNU now plays a secondary role after the discovery of the alkalyting agent temozolomide (TMZ). TMZ, which is administered orally, has proven to increase survival time in various studies when compared to surgery and radiotherapy alone, [6,114] and is therefore today the standard chemotherapeutic agent in glioma therapy.

BCNU has made a comeback as a therapeutic agent in the form of biodegradable wafers (Gliadel), which are to be implanted in the surgical cavity after glioma resection. The rationale for local CTx delivery is the fact that most GBM recur in the immediate vicinity of the original tumour, thus necessitating a concentrated delivery of the drugs to the area around the surgical cavity. This novel form of chemotherapy delivery addresses the key issues of systemic administration of BCNU, namely BCNU's toxicity in cells outside the CNS in high doses and its short half-life. [128] The local administration of the drug effectively lowers its peripheral dose levels and hence its systemic toxic potential while simultaneously increasing the dose in the tumour site. [17,91] Notwithstanding its limitations, specifically the contra-indication of implanting Gliadel in surgical cavities communicating with the cerebrospinal fluid, and its side effects, namely the increased risk of infection and cerebral oedema in the implantation site, [41,98] Gliadel, and local delivery of CTx agents in general, may play a pivotal role in the therapy of gliomas in the future. Currently, the focus is on integration of local and systemic chemotherapy to yield longer survival times. [1,81,89]

Despite the aforementioned advances in surgery, radiotherapy and concomitant chemotherapy, the prognosis of GBM remains dismal (See Figure 3). The need for a new approach to treat these neoplasms is therefore clear. Promising novel therapeutic modalities are discussed in the following section.

III. Experimental Therapies

The inevitable escape of a portion of tumour cells from surgical dissection due to anatomical inaccessibility and cytotoxic resistance are the major challenges facing the current treatment of malignant brain tumours. Different strategies have been proposed to overcome these limitations and complement the established therapeutic modalities. We discuss some directions in experimental therapy of malignant brain tumours.

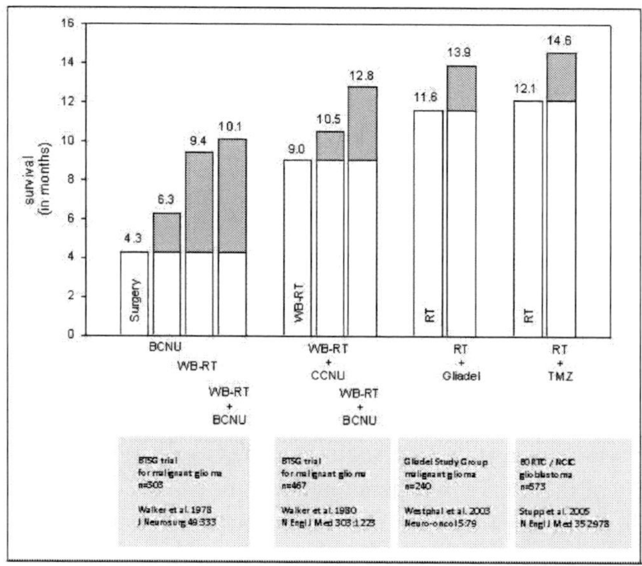

Figure 3. Clinical Phase III trials: Evolution of survival times in patients with malignant gliomas.

III. a. Virus-based Therapies

Selective killing of tumour cells by oncolytic virus infection is an actively developing area. Although the oncolytic properties of certain viruses were observed and experimented upon since the first decades of the last century, [71] the application of viruses in the case of gliomas is comparatively recent. [2] Therapeutic approaches based on using oncolytic viruses including herpex simplex virus, influenza virus, poliovirus and Newcastle disease virus are being developed.

The purpose of utilising viruses as vehicles is to induce cytotoxicity in tumour cells while sparing healthy tissue. Cytotoxic response may be triggered by means of delivery of suicidal genes or genes that encode enzymes activating pro-drugs into cytotoxic agents or inducing an immune response to viral infection in the affected cells. The fundamental concern in the application of viruses in glioma therapy is the safety factor. The virus should ideally target and kill only the neoplastic cells. A further concern is the effectiveness of the used viral dosage, which highlights the dilemma of safety vs. effective dosis. In recent studies, no maximal tolerated dose (MTD) was reached even with high virus titers, thus demonstrating the safety of viral administration. [65,134]

The pro-drug activation strategy relies on the principle of converting a harmless pro-drug into a potent cytotoxic agent. HSV-1 thymidine kinase (HSVtk) is an enzyme, which converts the 2'-deoxy-guanosine analogue Ganciclovir into its extremely toxic derivate, deoxyguanosine triphosphate. [5,90] Expression of HSVtk in glioma cells can be achieved by means of a non-replicating Adenovirus serotype 5 (Ad5) (Cerepro®). Phase III trials of Cerepro® delivered results that were deemed insufficient and ineffective by the pharmaceutical regulatory body of the European Union, the European Medicines Agency (EMA), [84] and subsequent efforts to market the drug have been put on hold as of March 2010. [30] Another promising pro-drug strategy is the conversion of the non-toxic (for humans) 5-fluorocytosine (5-FC) to the highly toxic 5-fluorouracil (5-FU) by the enzyme cytosine deaminase (CD). [111] CD does not occur naturally in humans; hence the only cells affected by 5-FC adminstration are those targeted by the carrier adenoviruses, oncolytic viruses or retroviruses. There have been 5 pre-clinical trials so far concerning 5-FC/CD, whereby a reduction of tumour mass in animal models was achieved. [5,36,39,56,111] Whether or not 5-FC/CD will live up to its promise in clinical trials remains to be seen.

A different gene therapy approach aiming at functional reconstitution of the p53 tumour suppressor pathway. The tumour suppressor protein p53 is an exceptionally complex protein involved in the regulation of major cellular processes including DNA repair, cell cycle control and programmed cell death (apoptosis). TP53 is one of the most frequently mutated genes in different types of human cancers including gliomas. The rationale behind restoring p53 function was that exogenous expression of wild type p53 protein in glioma cells should lead to the activation of the apoptotic response. A phase I/II clinical study evaluated the therapeutic efficacy of p53 restoration by means of intratumoral delivery of replication-deficient adenovirus encoding wild type p53 (Adp53). [73] Although p53 transduction was evident, apoptosis induction was low and restricted to a small number of p53 expressing cells. Furthermore, p53 expressing cells were found only 5-8 mm around the injection site. Selective targeting of tumour cells lacking functional p53 is another approach evaluated in clinical studies using an oncolytic virus, ONYX-15, which is thought to specifically target cells with a defective p53 pathway while sparing 'normal' cells [24]. While being well tolerated, none of the 24 patients involved in the study showed significant tumour regression, [24] although earlier studies have demonstrated oncolytic properties of ONYX-15 in rat models.

III. b. Stem-cell Based Therapies

Pluripotent neural stem cells (NSC) are capable of self-regeneration and differentiation into any of the basic CNS cell types. [115] An interesting phenomenon was observed in animal models: NSC, once injected to a tumour-infested CNS, flocked to the tumour site, even if injected to the contralateral hemisphere. [64] This led to the novel idea of utilising genetically modified NSC to directly deliver therapeutic agents to tumour cells. Several agents were tested in pre-clinical and phase I models, including interleukins (most notably IL-12) and Tumour necrosis factor-alpha (TNF-α). [28,29,64,88]

While being promising in pre-clinical experiments, the use of stem cells in glioma therapy failed to live up to their expectations. Among the problems facing stem-cell based therapies are a) the legal and ethical issues with regards to the harvest of such cells from embryonic and fetal tissue, and the technical issues in harvesting NSC from adult bone marrow; b) the disappointing results in clinical trials, and c) the concern that NSC possess a prototumourigenic potential. It has now been suggested that NSCs may undergo transformation, [123] and induce tumours under certain experimental conditions. [19] Hence, the potential risks and benefits of using NSC for targeting tumour cells need careful evaluation and require further investigation.

III. c. Antiangiogenesis Therapy

Malignant gliomas are, in general, highly vascularised lesions. The formation of new blood vessels is essential for tumour growth. Inhibiting angiogenesis-promoting factors as a therapeutic strategy is a subject of intense investigations.

The main angiogenetic factors that have been identified as playing a pivotal role in blood vessel formation in GBM are VEGF, FGF, PDGF, and HGF. [23,120] Agents targeting the VEGF pathway do so through different mechanisms; namely blocking the VEGF receptor or directly inhibiting VEGF. An example of the first group is Cediranib, where phase II trials showed a 56% response rate, leading to approval for phase III trials and phase II trials in combination with radiotherapy and chemtherapy. [23] The prime example of inhibition of the VEGF pathway is however the VEGF inhibiting monoclonal antibody Bevacizumab (Avastin). Several phase II trial showed that, in combination with the topoisomerase I inhibitor irinotecan, Bevacizumab prolonged the survival time in patients with recurrent GBM, [69,126] although studies with Bevacizumab as a monotherapy agent showed similar results, [95] thus shedding doubt on the role of irinotecan in the aforementioned trials. A further concern in the administration of Bevacizumab is toxicity witnessed in a small portion of patients partaking in the trials. Side effects included pulmonary embolism, myocardial infarction and stroke. [23] A cluster of drugs that target the non-VEGF angiogenesis pathway, such as the tyrosine kinase inhibitor Imatinib (Gleevec, targeting PDGFR) and Celecoxib (targeting COX-2 to suppress FGF and VEGF) are currently undergoing clinical trials. Preliminary results are disappointing. [23]

In practically all tested agents, tumour resistance was noticed in the subsequent cycles, underscoring the adaptability of GBM to their surroundings. New agents being tested, as well as current agents tested in combination with chemoradiotherpy, need to address this shortcoming. In short, angiogenesis inhibitors, with the exception of Bevacizumab, have yet to demonstrate a substantial prolongation of survival times and tumour regression in the therapy of GBM.

IV. New Avenues to Glioma Cell Therapy

The necessity of novel approaches in glioma therapy, in light of the insufficiency of current therapeutic regimen, has been highlighted several times in this chapter. The numerous challenges facing neurosurgeons include diffuse tumour cell infiltration, the intra-operative differentiation between tumour and normal cells, and the risk of operative damage of eloquent areas. On the adjuvant therapeutic front identifying and targeting tumour cells while sparing healthy brain tissue was and remains a major obstacle. New discoveries in the underlying tumour biology may build the cornerstone of novel therapeutic approaches. In this section, future avenues in glioma therapy that aim to address the aforementioned problems are discussed.

IV. a. Understanding Glioma Biology

The glioma signalling network, hierarchical organization of brain tumours and implications for therapy.

The underlying principle of the so-called "molecularly targeted therapies" is to restore or inhibit cellular functions that have been affected by genetic changes in tumour cells. The acquisition and accumulation of genetic changes is the driving force for malignant transformation [55,68]. Accordingly, the extent and type of genetic alterations are the prime molecular measures of tumour growth potential and of the ability to resist internal and external counteracting forces such as antitumor immune defence or cytotoxic treatments, respectively. Global analyses of patterns of genomic alterations in human GBMs revealed the existence of the glioma signalling network, which is characterised by a combination of genetic changes affecting functions of certain cellular pathways. [116] It turns out that genetic alterations that disable the proper functioning of RB, p53 and RTK pathways, which are all involved in the regulation of cell proliferation, growth factor signaling, DNA repair and apoptosis is an obligatory event in GBMs. [116] Within signaling cascades, genes are organized in a functional hierarchy. An important finding from genetic analyses in GBMs is that disregulation of the same pathway may involve distinct genetic events, through affecting different genes or groups of genes within the cascade. Hence, the hierarchical position of a particular gene in the signaling cascade is pivotal in predicting the efficacy of therapies that target a specific step in the pathway. One of the challenges to achieving satisfactory therapeutic efficacy of molecularly targeted therapies is the functional redundancy of glioma core pathways, which all are involved in the regulation of cellular processess essential for glioma cell proliferation and survival. Furthermore, it turns out that the same genetic alterations may have different impact on the tumorigenic potential in different types of brain cells. A compelling evidence that the cancer-propelling effects of oncogenic mutations may vary depending on the cell context was provided by the pivotal study by Holland et al reporting that the combination of activated Ras and Akt, which comprise two of the core glioma pathways, renders neural progenitors but not differentiated astrocytes tumorigenic [53]. These findings provided a strong impetus for further investigations of the possibility that gliomagenesis and adult neurogenesis may share some traits or even driven by similar types of cells, which have common origin. In this regard, adult neural stem cells (NSCs) or neural progenitor cells (NPCs) would suit well the requirements to be a cell of origin for the so-called cancer stem cells. However, testing the possibility that brain cancers may derive from cancer stem cells was hampered by the fact that the very existence of NSCs was questioned until the late 90's. For decades, the no-new-neuron theory dominated in the field of neuroscience. [9] Its basic premise was that no new cells could be formed in the mature central nervous system (CNS). Once damaged, the cells of the nervous system were thought to be incapable of regeneration. Starting in the late 60's, the experimental data contradicting the no-new-neuron dogma began to accumulate [3,45-47,86] with the unequivoval evidence for cell genesis taking place in the adult human brain being provided at the end of 90's. [33,60,66,70,97]

A paradigm shift in the understanding of neurogenesis had important implications for further development of the concept of genesis and malignant progression of adult brain neoplasia. Recent years have been marked by dramatic advances in the field of neurooncology, with one of the most remarkable discoveries being that malignant brain tumours contain a population of cells possessing the cardinal properties of neural stem cells (NSCs). It appears that malignant brain tumours possess a hierarchical organization whereby distinct cell populations from the same tumour differ with respect to the key properties such as the tumorigenic potential and ability to reconstitute the tumour. [38,51,110] Among a diverse cell population comprising the tumour, the so-called "brain tumour initiating cells" (BTICs) or "brain tumour stem cells" (BTSCs) share several characteristics with NSCs such as the self-renewing capability and multipotency, allowing for multilineage differentiation. Intriguingly, experimental evidence suggests that BTICs represent the ultimate tumorigenic population and exhibit the enhanced resistance to radio- and chemotherapy. [8,13,20] The emerging view is that glioma recurrence is primarily driven by a functionally distinct population of cells, which possess stem-like properties and differ from the majority of cells comprising the tumour bulk in terms of their responses to cytotoxic treatments.

Although considerable controversy still exists regarding the identity of BTICs, [52] the realization that malignant brain tumours comprise a functionally diverse population has enormous implications for the future development of anti-glioma therapeutic strategies. There have been some encouraging results from pre-clinical experimental studies on the use of the oncolytic adenovirus Delta-24-RGD to target BTICs. [59] There is a tremendous effort currently to elucidate molecular pathways operating in BTICs. [22,74] Elucidating the molecular and cellular characteristics of BTICs will certainly provide important novel insights into the glioma biology and, hopefully, open new avenues in therapy of malignant brain tumours.

IV. b. Future Perspectives for Intraoperative Imaging in Brain Tumour Surgery

Although surgical therapy has progressed immensely since the time of Godlee, Cushing and other pioneers, vast avenues for improvement remain. One is the aforementioned intra-operative differentiation between tumour and normal cells. Although an intra-operative visualisation of individual tumour cells still remains a challenge to overcome, several promising strategies aiming to allow distinguishing between healthy tissue and individual tumour cells are currently explored: namely 5-ALA, OCT, and Q-Dot.

5-aminolevulinic acid (5-ALA) is a naturally occurring amino-acid that metabolises a fluorescent porphyrin, protoporphyrin IX (PpIX) in glioma cells. [61,112] PpIX can be excited by UV-light illumination and is utilised to identify residual tumour tissue unidentifiable as such under normal light conditions by means of specialised operative microscope fitted with an UV-light excitation, thus enabling guided resection of residual tissue. It has been proven that maximal reduction of the tumour mass correlates positively with the outcome. [72,101] Fluorescent guided resection helps achieve this goal, thereby increasing short-term survival rates. [113,117]

Optical coherence tomography (OCT) is a complex imaging modality that utilises light waves possessing a long wavelength, that are able to infiltrate several millimetres into biological tissue. [14] The reflected light is registered and displayed optically. Tumour tissue in the case of GBM has a different light-reflecting and scattering properties than normal brain tissue. Using OCT, the surgeon can differentiate between various types of tissue. [15] The main advantages of OCT are that it produces intra-operative images in real-time, in a non-invasive manner with a high resolution and it's comparatively deep tissue penetration.

Quantum dots (Q-Dots) are semiconductor nanoparticles capable of fluorescent emission. [82,96] When coupled to biomolecules that specificly bind to molecular targets on glioma cells, such as the epidermal growth factor receptor (EGFR), they are able to point to these cells specifically, thus enabling the surgeon to identify and specifically target fluorescent cells. [4] This imaging modality has shown promising results in cell cultures, animal models [57] and human glioma biopsies. [4] Since Q-Dots can be coupled with a variety of targeting molecules, the possibilities of Q-Dot identification of glioma and other CNS tumour cells using the aforementioned structures are boundless and warrant further investigation.

Conclusion

The evolution of glioma therapy from the early days of neuro-oncology till the present has seen astounding advances in every therapeutic modality. On the surgical front, we have progressed from Godlee's indiscriminate resection utilising crude instruments to precise image-guided resection with the aid of state-of-the-art intra-operative imaging modalities and microinstruments. Radiation Therapy progressed from the use of arbitrary doses of whole brain radiation and Cushing's "radium bomb" to the use of anatomically precise, well-defined protocols, while chemotherapy is now much better tolerated and offers a limited but significant survival advantage, and the tailor-made, individualised chemotherapies of the future may prove advantageous over the current one-size-fits-all regimen. The better understanding of tumour biology in GBM may open yet unrealised therapeutic avenues. Despite the dismal prognosis associated with GBM, advances on the surgical, perioperative and biological fronts give hope that a significant prolongation of survival times in GBM patients may be achieved.

References

[1] Affronti, M.L., Heery, C.R., Herndon, J.E., 2nd, Rich, J.N., Reardon, D.A., Desjardins, A., Vredenburgh, J.J., Friedman, A.H., Bigner, D.D. & Friedman, H.S. Overall survival of newly diagnosed glioblastoma patients receiving carmustine wafers followed by radiation and concurrent temozolomide plus rotational multiagent chemotherapy. *Cancer* 115, 3501-3511 (2009).

[2] Aksel, I.S. & Aykan, T.B. Growth behavior of the rabies virus in a glioblastomatous tumor induced with methylcholanthrene in mice. *World Neurol* 2, 398-405 (1961).

[3] Altman, J. & Das, G.D. Postnatal neurogenesis in the guinea-pig. *Nature* 214, 1098-1101 (1967).

[4] Arndt-Jovin, D.J., Kantelhardt, S.R., Caarls, W., de Vries, A.H., Giese, A. & Jovin Ast, T.M. Tumor-targeted quantum dots can help surgeons find tumor boundaries. *IEEE Trans Nanobioscience* 8, 65-71 (2009).

[5] Asadi-Moghaddam, K. & Chiocca, E.A. Gene- and viral-based therapies for brain tumors. *Neurotherapeutics* 6, 547-557 (2009).

[6] Athanassiou, H., Synodinou, M., Maragoudakis, E., Paraskevaidis, M., Verigos, C., Misailidou, D., Antonadou, D., Saris, G., Beroukas, K. & Karageorgis, P. Randomized phase II study of temozolomide and radiotherapy compared with radiotherapy alone in newly diagnosed glioblastoma multiforme. *J Clin Oncol* 23, 2372-2377 (2005).

[7] Bailey, P. & Cushing, H. A classification of the tumors of the glioma group on a histogenetic basis with a correlated study of prognosis, (J.B. Lippincott Company, Philadelphia, London etc., 1926).

[8] Bao, S., Wu, Q., McLendon, R.E., Hao, Y., Shi, Q., Hjelmeland, A.B., Dewhirst, M.W., Bigner, D.D. & Rich, J.N. Glioma stem cells promote radioresistance by preferential activation of the DNA damage response. *Nature* 444, 756-760 (2006).

[9] Barinaga, M. No-new-neurons dogma loses ground. *Science* 279, 2041-2042 (1998).

[10] Barnett, G.H. The role of image-guided technology in the surgical planning and resection of gliomas. *J Neurooncol* 42, 247-258 (1999).

[11] Bennett, A.H. Case of Cerebral Tumor. *J Nerv Ment Dis* 13, 255 (1886).

[12] Bennett, A.H. & Godlee, R.J. A case of cerebral tumor-the surgical treatment. (Reprint of the 1885 article in the British Medical Journal). *CA Cancer J Clin* 24, 171-181 (1974).

[13] Blough, M.D., Westgate, M.R., Beauchamp, D., Kelly, J.J., Stechishin, O., Ramirez, A.L., Weiss, S. & Cairncross, J.G. Sensitivity to temozolomide in brain tumor initiating cells. *Neuro Oncol* (2010).

[14] Boehringer, H.J., Boller, D., Leppert, J., Knopp, U., Lankenau, E., Reusche, E., Huettmann, G. & Giese, A. Time-domain and spectral-domain optical coherence

tomography in the analysis of brain tumor tissue. *Lasers in Surgery and Medicine* 38, 588-597 (2006).

[15] Bohringer, H.J., Lankenau, E., Stellmacher, F., Reusche, E., Huttmann, G. & Giese, A. Imaging of human brain tumor tissue by near-infrared laser coherence tomography. *Acta Neurochir (Wien)* 151, 507-517; discussion 517 (2009).

[16] Braun, V., Dempf, S., Tomczak, R., Wunderlich, A., Weller, R. & Richter, H.P. Functional cranial neuronavigation. Direct integration of fMRI and PET data. *J Neuroradiol* 27, 157-163 (2000).

[17] Brem, H., Piantadosi, S., Burger, P.C., Walker, M., Selker, R., Vick, N.A., Black, K., Sisti, M., Brem, S., Mohr, G. & et al. Placebo-controlled trial of safety and efficacy of intraoperative controlled delivery by biodegradable polymers of chemotherapy for recurrent gliomas. The Polymer-brain Tumor Treatment Group. *Lancet* 345, 1008-1012 (1995).

[18] Cappabianca, P. & de Divitiis, E. Back to the Egyptians: neurosurgery via the nose. A five-thousand year history and the recent contribution of the endoscope. *Neurosurg Rev* 30, 1-7; discussion 7 (2007).

[19] Casalbore, P., Budoni, M., Ricci-Vitiani, L., Cenciarelli, C., Petrucci, G., Milazzo, L., Montano, N., Tabolacci, E., Maira, G., Larocca, L.M. & Pallini, R. Tumorigenic potential of olfactory bulb-derived human adult neural stem cells associates with activation of TERT and NOTCH1. *PLoS One* 4, e4434 (2009).

[20] Chalmers, A.J. Radioresistant glioma stem cells--therapeutic obstacle or promising target? *DNA Repair (Amst)* 6, 1391-1394 (2007).

[21] Charles, N. & Holland, E.C. Brain tumor treatment increases the number of cancer stem-like cells. *Expert Rev Neurother* 9, 1447-1449 (2009).

[22] Chen, R., Nishimura, M.C., Bumbaca, S.M., Kharbanda, S., Forrest, W.F., Kasman, I.M., Greve, J.M., Soriano, R.H., Gilmour, L.L., Rivers, C.S., Modrusan, Z., Nacu, S., Guerrero, S., Edgar, K.A., Wallin, J.J., Lamszus, K., Westphal, M., Heim, S., James, C.D., VandenBerg, S.R., Costello, J.F., Moorefield, S., Cowdrey, C.J., Prados, M. & Phillips, H.S. A hierarchy of self-renewing tumor-initiating cell types in glioblastoma. *Cancer Cell* 17, 362-375 (2010).

[23] Chi, A.S., Norden, A.D. & Wen, P.Y. Antiangiogenic strategies for treatment of malignant gliomas. *Neurotherapeutics* 6, 513-526 (2009).

[24] Chiocca, E.A., Abbed, K.M., Tatter, S., Louis, D.N., Hochberg, F.H., Barker, F., Kracher, J., Grossman, S.A., Fisher, J.D., Carson, K., Rosenblum, M., Mikkelsen, T., Olson, J., Markert, J., Rosenfeld, S., Nabors, L.B., Brem, S., Phuphanich, S., Freeman, S., Kaplan, R. & Zwiebel, J. A phase I open-label, dose-escalation, multi-institutional trial of injection with an E1B-Attenuated adenovirus, ONYX-015, into the peritumoral region of recurrent malignant gliomas, in the adjuvant setting. *Mol Ther* 10, 958-966 (2004).

[25] Cohen-Gadol, A.A. & Spencer, D.D. Inauguration of pediatric neurosurgery by Harvey W. Cushing: his contributions to the surgery of posterior fossa tumors in children. Historical vignette. *J Neurosurg* 100, 225-231 (2004).

[26] DeVita, V.T., Jr. & Chu, E. A history of cancer chemotherapy. *Cancer Res* 68, 8643-8653 (2008).

[27] Diven, K. Dandy's Radical Extirpations of Brain Tissue in Man. *Am J Psych* 46, 500-503 (1934).

[28] Ehtesham, M., Kabos, P., Gutierrez, M.A., Chung, N.H., Griffith, T.S., Black, K.L. & Yu, J.S. Induction of glioblastoma apoptosis using neural stem cell-mediated delivery of tumor necrosis factor-related apoptosis-inducing ligand. *Cancer Res* 62, 7170-7174 (2002).

[29] Ehtesham, M., Kabos, P., Kabosova, A., Neuman, T., Black, K.L. & Yu, J.S. The use of interleukin 12-secreting neural stem cells for the treatment of intracranial glioma. *Cancer Res* 62, 5657-5663 (2002).

[30] EMA. Press release: Ark Therapeutics Ltd withdraws its marketing authorisation application for Cerepro (sitimagene ceradenovec). in European Union Medicines Agency (http://www.ema.europa.eu/docs/en_GB/document_library/Press_release/2010/03/WC500075457.pdf, 11.03.2010).

[31] Enam, S.A. Neurosurgeons and Electrocautery. *Pak J Neurol Sci* 2, 59 (2007).

[32] Enchev, Y. Neuronavigation: geneology, reality, and prospects. *Neurosurg Focus* 27, E11 (2009).

[33] Eriksson, P.S., Perfilieva, E., Bjork-Eriksson, T., Alborn, A.M., Nordborg, C., Peterson, D.A. & Gage, F.H. Neurogenesis in the adult human hippocampus. *Nat Med* 4, 1313-1317 (1998).

[34] Ferguson, S. & Lesniak, M.S. Percival Bailey and the classification of brain tumors. *Neurosurg Focus* 18, e7 (2005).

[35] Fomchenko, E.I. & Holland, E.C. Stem cells and brain cancer. *Exp Cell Res* 306, 323-329 (2005).

[36] Freeman, S.M., Abboud, C.N., Whartenby, K.A., Packman, C.H., Koeplin, D.S., Moolten, F.L. & Abraham, G.N. The "bystander effect": tumor regression when a fraction of the tumor mass is genetically modified. *Cancer Res* 53, 5274-5283 (1993).

[37] Furnari, F.B., Fenton, T., Bachoo, R.M., Mukasa, A., Stommel, J.M., Stegh, A., Hahn, W.C., Ligon, K.L., Louis, D.N., Brennan, C., Chin, L., DePinho, R.A. & Cavenee, W.K. Malignant astrocytic glioma: genetics, biology, and paths to treatment. *Genes Dev* 21, 2683-2710 (2007).

[38] Galli, R., Binda, E., Orfanelli, U., Cipelletti, B., Gritti, A., De Vitis, S., Fiocco, R., Foroni, C., Dimeco, F. & Vescovi, A. Isolation and characterization of tumorigenic, stem-like neural precursors from human glioblastoma. *Cancer Res* 64, 7011-7021 (2004).

[39] Ge, K., Xu, L., Zheng, Z., Xu, D., Sun, L. & Liu, X. Transduction of cytosine deaminase gene makes rat glioma cells highly sensitive to 5-fluorocytosine. *Int J Cancer* 71, 675-679 (1997).

[40] Giese, A., Bjerkvig, R., Berens, M.E. & Westphal, M. Cost of migration: invasion of malignant gliomas and implications for treatment. *J Clin Oncol* 21, 1624-1636 (2003).

[41] Giese, A., Bock, H.C., Kantelhardt, S.R. & Rohde, V. Risk Management in the Treatment of Malignant Gliomas with BCNU Wafer Implants. *Cen Eur Neurosurg* (2010).

[42] Giese, A., Rief, M.D., Loo, M.A. & Berens, M.E. Determinants of human astrocytoma migration. *Cancer Res* 54, 3897-3904 (1994).

[43] Giese, A. & Westphal, M. Glioma invasion in the central nervous system. *Neurosurgery* 39, 235-250; discussion 250-232 (1996).

[44] Gobo, D.J. Localization Techniques: Neuroimaging and Electroencephalography. in A history of neurosurgery: in its scientific and professional contexts (eds. Greenblatt, S.H., Dagi, T.F. & Epstein, M.) 226-228 (American Association of Neurological Surgeons, Park Ridge, IL, 1997).

[45] Gould, E., Beylin, A., Tanapat, P., Reeves, A. & Shors, T.J. Learning enhances adult neurogenesis in the hippocampal formation. *Nat Neurosci* 2, 260-265 (1999).

[46] Gould, E., McEwen, B.S., Tanapat, P., Galea, L.A. & Fuchs, E. Neurogenesis in the dentate gyrus of the adult tree shrew is regulated by psychosocial stress and NMDA receptor activation. *J Neurosci* 17, 2492-2498 (1997).

[47] Gould, E., Tanapat, P., McEwen, B.S., Flugge, G. & Fuchs, E. Proliferation of granule cell precursors in the dentate gyrus of adult monkeys is diminished by stress. *Proc Natl Acad Sci U S A* 95, 3168-3171 (1998).

[48] Gumprecht, H., Ebel, G.K., Auer, D.P. & Lumenta, C.B. Neuronavigation and functional MRI for surgery in patients with lesion in eloquent brain areas. *Minim Invasive Neurosurg* 45, 151-153 (2002).

[49] Gutin, P.H., Wilson, C.B., Kumar, A.R., Boldrey, E.B., Levin, V., Powell, M. & Enot, K.J. Phase II study of procarbazine, CCNU, and vincristine combination chemotherapy in the treatment of malignant brain tumors. *Cancer* 35, 1398-1404 (1975).

[50] Hegi, M.E., Diserens, A.C., Gorlia, T., Hamou, M.F., de Tribolet, N., Weller, M., Kros, J.M., Hainfellner, J.A., Mason, W., Mariani, L., Bromberg, J.E., Hau, P., Mirimanoff, R.O., Cairncross, J.G., Janzer, R.C. & Stupp, R. MGMT gene silencing and benefit from temozolomide in glioblastoma. *N Engl J Med* 352, 997-1003 (2005).

[51] Hemmati, H.D., Nakano, I., Lazareff, J.A., Masterman-Smith, M., Geschwind, D.H., Bronner-Fraser, M. & Kornblum, H.I. Cancerous stem cells can arise from pediatric brain tumors. *Proc Natl Acad Sci USA* 100, 15178-15183 (2003).

[52] Hill, R.P. Identifying cancer stem cells in solid tumors: case not proven. *Cancer Res* 66, 1891-1895; discussion 1890 (2006).

[53] Holland, E.C., Celestino, J., Dai, C., Schaefer, L., Sawaya, R.E. & Fuller, G.N. Combined activation of Ras and Akt in neural progenitors induces glioblastoma formation in mice. *Nat Genet* 25, 55-57 (2000).

[54] Hounsfield, G.N. Computerized transverse axial scanning (tomography): Part I. Description of system. *Br J Radiol* 46, 1016-1022 (1973).

[55] Hutchinson, E. Alfred Knudson and his two-hit hypothesis. *Lancet Oncol* 2, 642-645 (2001).

[56] Ichikawa, T., Tamiya, T., Adachi, Y., Ono, Y., Matsumoto, K., Furuta, T., Yoshida, Y., Hamada, H. & Ohmoto, T. In vivo efficacy and toxicity of 5-fluorocytosine/cytosine deaminase gene therapy for malignant gliomas mediated by adenovirus. *Cancer Gene Ther* 7, 74-82 (2000).

[57] Jackson, H., Muhammad, O., Daneshvar, H., Nelms, J., Popescu, A., Vogelbaum, M.A., Bruchez, M. & Toms, S.A. Quantum dots are phagocytized by macrophages and colocalize with experimental gliomas. *Neurosurgery* 60, 524-529; discussion 529-530 (2007).

[58] Jannetta, P.J. Developments in neurosurgery: "the 4 factors". *Neurosurgery* 65, A9-10 (2009).

[59] Jiang, H., Gomez-Manzano, C., Aoki, H., Alonso, M.M., Kondo, S., McCormick, F., Xu, J., Kondo, Y., Bekele, B.N., Colman, H., Lang, F.F. & Fueyo, J. Examination of

the therapeutic potential of Delta-24-RGD in brain tumor stem cells: role of autophagic cell death. *J Natl Cancer Inst* 99, 1410-1414 (2007).

[60] Johansson, C.B., Svensson, M., Wallstedt, L., Janson, A.M. & Frisen, J. Neural stem cells in the adult human brain. *Exp Cell Res* 253, 733-736 (1999).

[61] Kantelhardt, S.R., Diddens, H., Leppert, J., Rohde, V., Huttmann, G. & Giese, A. Multiphoton excitation fluorescence microscopy of 5-aminolevulinic acid induced fluorescence in experimental gliomas. *Lasers Surg Med* 40, 273-281 (2008).

[62] Keles, G.E. & Berger, M.S. Advances in neurosurgical technique in the current management of brain tumors. *Semin Oncol* 31, 659-665 (2004).

[63] Keller, T. The first primary brain tumor operation in America. *Surg Neurol* 45, 463-466 (1996).

[64] Kim, S.U. Neural Stem Cell-based Gene Therapy for Brain Tumors. *Stem Cell Rev* in print, DOI 10.1007/s12015-12010-19154-12011 (2010).

[65] King, G.D., Curtin, J.F., Candolfi, M., Kroeger, K., Lowenstein, P.R. & Castro, M.G. Gene therapy and targeted toxins for glioma. *Curr Gene Ther* 5, 535-557 (2005).

[66] Kirschenbaum, B., Nedergaard, M., Preuss, A., Barami, K., Fraser, R.A. & Goldman, S.A. In vitro neuronal production and differentiation by precursor cells derived from the adult human forebrain. *Cereb Cortex* 4, 576-589 (1994).

[67] Kitchen, N.D. Image-directed neurosurgery. *Ann R Coll Surg Engl* 79, 148-153 (1997).

[68] Knudson, A.G., Jr. Mutation and cancer: statistical study of retinoblastoma. *Proc Natl Acad Sci U S A* 68, 820-823 (1971).

[69] Kreisl, T.N., Kim, L., Moore, K., Duic, P., Royce, C., Stroud, I., Garren, N., Mackey, M., Butman, J.A., Camphausen, K., Park, J., Albert, P.S. & Fine, H.A. Phase II trial of single-agent bevacizumab followed by bevacizumab plus irinotecan at tumor progression in recurrent glioblastoma. *J Clin Oncol* 27, 740-745 (2009).

[70] Kukekov, V.G., Laywell, E.D., Suslov, O., Davies, K., Scheffler, B., Thomas, L.B., O'Brien, T.F., Kusakabe, M. & Steindler, D.A. Multipotent stem/progenitor cells with similar properties arise from two neurogenic regions of adult human brain. *Exp Neurol* 156, 333-344 (1999).

[71] Kuruppu, D. & Tanabe, K.K. Viral oncolysis by herpes simplex virus and other viruses. *Cancer Biol Ther* 4, 524-531 (2005).

[72] Lacroix, M., Abi-Said, D., Fourney, D.R., Gokaslan, Z.L., Shi, W., DeMonte, F., Lang, F.F., McCutcheon, I.E., Hassenbusch, S.J., Holland, E., Hess, K., Michael, C., Miller, D. & Sawaya, R. A multivariate analysis of 416 patients with glioblastoma multiforme: prognosis, extent of resection, and survival. *J Neurosurg* 95, 190-198 (2001).

[73] Lang, F.F., Bruner, J.M., Fuller, G.N., Aldape, K., Prados, M.D., Chang, S., Berger, M.S., McDermott, M.W., Kunwar, S.M., Junck, L.R., Chandler, W., Zwiebel, J.A., Kaplan, R.S. & Yung, W.K. Phase I trial of adenovirus-mediated p53 gene therapy for recurrent glioma: biological and clinical results. *J Clin Oncol* 21, 2508-2518 (2003).

[74] Lee, J., Kotliarova, S., Kotliarov, Y., Li, A., Su, Q., Donin, N.M., Pastorino, S., Purow, B.W., Christopher, N., Zhang, W., Park, J.K. & Fine, H.A. Tumor stem cells derived from glioblastomas cultured in bFGF and EGF more closely mirror the phenotype and genotype of primary tumors than do serum-cultured cell lines. *Cancer Cell* 9, 391-403 (2006).

[75] Lefranc, F., Brotchi, J. & Kiss, R. Possible future issues in the treatment of glioblastomas: special emphasis on cell migration and the resistance of migrating glioblastoma cells to apoptosis. *J Clin Oncol* 23, 2411-2422 (2005).

[76] Light, R.U. The contributions of Harvey Cushing to the techniques of neurosurgery. *Surg Neurol* 35, 69-73 (1991).

[77] Litofsky, N.S., Bauer, A.M., Kasper, R.S., Sullivan, C.M. & Dabbous, O.H. Image-guided resection of high-grade glioma: patient selection factors and outcome. *Neurosurg Focus* 20, E16 (2006).

[78] Liu, C.Y. & Apuzzo, M.L. The genesis of neurosurgery and the evolution of the neurosurgical operative environment: part I-prehistory to 2003. *Neurosurgery* 52, 3-19; discussion 19 (2003).

[79] Louis, D.N., Holland, E.C. & Cairncross, J.G. Glioma classification: a molecular reappraisal. *Am J Pathol* 159, 779-786 (2001).

[80] Louis, D.N., Ohgaki, H., Wiestler, O.D., Cavenee, W.K., Burger, P.C., Jouvet, A., Scheithauer, B.W. & Kleihues, P. The 2007 WHO classification of tumours of the central nervous system. *Acta Neuropathol* 114, 97-109 (2007).

[81] McGirt, M.J., Than, K.D., Weingart, J.D., Chaichana, K.L., Attenello, F.J., Olivi, A., Laterra, J., Kleinberg, L.R., Grossman, S.A., Brem, H. & Quinones-Hinojosa, A. Gliadel (BCNU) wafer plus concomitant temozolomide therapy after primary resection of glioblastoma multiforme. *J Neurosurg* 110, 583-588 (2009).

[82] Michalet, X., Pinaud, F.F., Bentolila, L.A., Tsay, J.M., Doose, S., Li, J.J., Sundaresan, G., Wu, A.M., Gambhir, S.S. & Weiss, S. Quantum dots for live cells, in vivo imaging, and diagnostics. *Science* 307, 538-544 (2005).

[83] Mischel, P.S., Cloughesy, T.F. & Nelson, S.F. DNA-microarray analysis of brain cancer: molecular classification for therapy. *Nat Rev Neurosci* 5, 782-792 (2004).

[84] Mitchell, P. Ark's gene therapy stumbles at the finish line. *Nat Biotechnol* 28, 183-184 (2010).

[85] Moore, M.R., Rossitch, E., Jr. & Black, P.M. The development of neurosurgical techniques: the postoperative notes and sketches of Dr. Harvey Cushing. *Acta Neurochir (Wien)* 101, 93-99 (1989).

[86] Nottebohm, F. The road we travelled: discovery, choreography, and significance of brain replaceable neurons. *Ann N Y Acad Sci* 1016, 628-658 (2004).

[87] Oertel, J., von Buttlar, E., Schroeder, H.W. & Gaab, M.R. Prognosis of gliomas in the 1970s and today. *Neurosurg Focus* 18, e12 (2005).

[88] Oh, M.C. & Lim, D.A. Novel treatment strategies for malignant gliomas using neural stem cells. *Neurotherapeutics* 6, 458-464 (2009).

[89] Pan, E., Mitchell, S.B. & Tsai, J.S. A retrospective study of the safety of BCNU wafers with concurrent temozolomide and radiotherapy and adjuvant temozolomide for newly diagnosed glioblastoma patients. *J Neurooncol* 88, 353-357 (2008).

[90] Parker, J.N., Bauer, D.F., Cody, J.J. & Markert, J.M. Oncolytic viral therapy of malignant glioma. *Neurotherapeutics* 6, 558-569 (2009).

[91] Perry, J., Chambers, A., Spithoff, K. & Laperriere, N. Gliadel wafers in the treatment of malignant glioma: a systematic review. *Curr Oncol* 14, 189-194 (2007).

[92] Pillai, J.J. The evolution of clinical functional imaging during the past 2 decades and its current impact on neurosurgical planning. *AJNR Am J Neuroradiol* 31, 219-225 (2010).

[93] Poonnoose, S.I. & Daniel, R.T. Radiological evidence of glioma invasion of the central nervous system along tracts. *Surg Neurol* 54, 194-196 (2000).

[94] Rahimi, S.Y., McDonnell, D.E., Ahmadian, A. & Vender, J.R. Medieval neurosurgery: contributions from the Middle East, Spain, and Persia. *Neurosurg Focus* 23, E14 (2007).

[95] Raizer, J.J., Grimm, S., Chamberlain, M.C., Nicholas, M.K., Chandler, J.P., Muro, K., Dubner, S., Rademaker, A.W., Renfrow, J. & Bredel, M. A phase 2 trial of single-agent bevacizumab given in an every-3-week schedule for patients with recurrent high-grade gliomas. *Cancer* (2010).

[96] Resch-Genger, U., Grabolle, M., Cavaliere-Jaricot, S., Nitschke, R. & Nann, T. Quantum dots versus organic dyes as fluorescent labels. *Nat Methods* 5, 763-775 (2008).

[97] Roy, N.S., Wang, S., Jiang, L., Kang, J., Benraiss, A., Harrison-Restelli, C., Fraser, R.A., Couldwell, W.T., Kawaguchi, A., Okano, H., Nedergaard, M. & Goldman, S.A. In vitro neurogenesis by progenitor cells isolated from the adult human hippocampus. *Nat Med* 6, 271-277 (2000).

[98] Sabel, M. & Giese, A. Safety profile of carmustine wafers in malignant glioma: a review of controlled trials and a decade of clinical experience. *Curr Med Res Opin* (2008).

[99] Salcman, M. Historical development of surgery for glial tumors. *J Neurooncol* 42, 195-204 (1999).

[100] Sanai, N., Alvarez-Buylla, A. & Berger, M.S. Neural stem cells and the origin of gliomas. *N Engl J Med* 353, 811-822 (2005).

[101] Sanai, N. & Berger, M.S. Operative techniques for gliomas and the value of extent of resection. *Neurotherapeutics* 6, 478-486 (2009).

[102] Saris, S. Intracranial tumors: the evolution of treatment. in A history of neurosurgery: in its scientific and professional contexts (eds. Greenblatt, S.H., Dagi, T.F. & Epstein, M.) 247-258 (American Association of Neurological Surgeons, Park Ridge, IL, 1997).

[103] Sarkaria, J.N., Kitange, G.J., James, C.D., Plummer, R., Calvert, H., Weller, M. & Wick, W. Mechanisms of chemoresistance to alkylating agents in malignant glioma. *Clin Cancer Res* 14, 2900-2908 (2008).

[104] Schulder, M., Loeffler, J.S., Howes, A.E., Alexander, E., 3rd & Black, P.M. The radium bomb: Harvey Cushing and the interstitial irradiation of gliomas. *J Neurosurg* 84, 530-532 (1996).

[105] Selby, R. The surgical treatment of cerebral glioblastoma multiforme: an historical review. *J Neurooncol* 18, 175-182 (1994).

[106] Seymour, Z.A. & Cohen-Gadol, A.A. Cushing's lost cases of "radium bomb" brachytherapy for gliomas. *J Neurosurg*, published online December 18, 2009; DOI: 2010.3171/2009.2011.JNS091393 (2010).

[107] Shafqat, S. The long shadow of cerebral localization. *J R Soc Med* 98, 549 (2005).

[108] Shah, K., Bureau, E., Kim, D.E., Yang, K., Tang, Y., Weissleder, R. & Breakefield, X.O. Glioma therapy and real-time imaging of neural precursor cell migration and tumor regression. *Ann Neurol* 57, 34-41 (2005).

[109] Shapiro, W.R., Green, S.B., Burger, P.C., Mahaley, M.S., Jr., Selker, R.G., VanGilder, J.C., Robertson, J.T., Ransohoff, J., Mealey, J., Jr., Strike, T.A. & et al. Randomized trial of three chemotherapy regimens and two radiotherapy regimens and two

radiotherapy regimens in postoperative treatment of malignant glioma. Brain Tumor Cooperative Group Trial 8001. *J Neurosurg* 71, 1-9 (1989).

[110] Singh, S.K., Clarke, I.D., Terasaki, M., Bonn, V.E., Hawkins, C., Squire, J. & Dirks, P.B. Identification of a cancer stem cell in human brain tumors. *Cancer Res* 63, 5821-5828 (2003).

[111] Sonabend, A.M., Ulasov, I.V., Han, Y. & Lesniak, M.S. Oncolytic adenoviral therapy for glioblastoma multiforme. *Neurosurg Focus* 20, E19 (2006).

[112] Stummer, W., Pichlmeier, U., Meinel, T., Wiestler, O.D., Zanella, F. & Reulen, H.J. Fluorescence-guided surgery with 5-aminolevulinic acid for resection of malignant glioma: a randomised controlled multicentre phase III trial. *Lancet Oncol* 7, 392-401 (2006).

[113] Stummer, W., Tonn, J.C., Mehdorn, H.M., Nestler, U., Franz, K., Goetz, C., Bink, A. & Pichlmeier, U. Counterbalancing risks and gains from extended resections in malignant glioma surgery: a supplemental analysis from the randomized 5-aminolevulinic acid glioma resection study. *J Neurosurg* (2010).

[114] Stupp, R., Mason, W.P., van den Bent, M.J., Weller, M., Fisher, B., Taphoorn, M.J., Belanger, K., Brandes, A.A., Marosi, C., Bogdahn, U., Curschmann, J., Janzer, R.C., Ludwin, S.K., Gorlia, T., Allgeier, A., Lacombe, D., Cairncross, J.G., Eisenhauer, E. & Mirimanoff, R.O. Radiotherapy plus concomitant and adjuvant temozolomide for glioblastoma. *N Engl J Med* 352, 987-996 (2005).

[115] Svendsen, C.N., Caldwell, M.A. & Ostenfeld, T. Human neural stem cells: isolation, expansion and transplantation. *Brain Pathol* 9, 499-513 (1999).

[116] TCGA. Comprehensive genomic characterization defines human glioblastoma genes and core pathways - The Cancer Genome Atlas Research Network. *Nature* 455, 1061-1068 (2008).

[117] Tonn, J.C. & Stummer, W. Fluorescence-guided resection of malignant gliomas using 5-aminolevulinic acid: practical use, risks, and pitfalls. *Clin Neurosurg* 55, 20-26 (2008).

[118] Tubbs, R.S., Loukas, M., Shoja, M.M., Ardalan, M. & Oakes, W.J. Ibn Jazlah and his 11th century accounts (Taqwim al-abdan fi tadbir al-insan) of disease of the brain and spinal cord. Historical vignette. *J Neurosurg Spine* 9, 314-317 (2008).

[119] Uluc, K., Kujoth, G.C. & Baskaya, M.K. Operating microscopes: past, present, and future. *Neurosurg Focus* 27, E4 (2009).

[120] Van Meir, E.G., Hadjipanayis, C.G., Norden, A.D., Shu, H.K., Wen, P.Y. & Olson, J.J. Exciting new advances in neuro-oncology: the avenue to a cure for malignant glioma. *CA Cancer J Clin* 60, 166-193 (2010).

[121] Vender, J.R., Miller, J., Rekito, A. & McDonnell, D.E. Effect of hemostasis and electrosurgery on the development and evolution of brain tumor surgery in the late 19th and early 20th centuries. *Neurosurg Focus* 18, e3 (2005).

[122] Verhaak, R.G., Hoadley, K.A., Purdom, E., Wang, V., Qi, Y., Wilkerson, M.D., Miller, C.R., Ding, L., Golub, T., Mesirov, J.P., Alexe, G., Lawrence, M., O'Kelly, M., Tamayo, P., Weir, B.A., Gabriel, S., Winckler, W., Gupta, S., Jakkula, L., Feiler, H.S., Hodgson, J.G., James, C.D., Sarkaria, J.N., Brennan, C., Kahn, A., Spellman, P.T., Wilson, R.K., Speed, T.P., Gray, J.W., Meyerson, M., Getz, G., Perou, C.M. & Hayes, D.N. Integrated genomic analysis identifies clinically relevant subtypes of glioblastoma

characterized by abnormalities in PDGFRA, IDH1, EGFR, and NF1. *Cancer Cell* 17, 98-110 (2010).

[123] Vescovi, A.L., Galli, R. & Reynolds, B.A. Brain tumour stem cells. *Nat Rev Cancer* 6, 425-436 (2006).

[124] Voorhees, J.R., Cohen-Gadol, A.A. & Spencer, D.D. Early evolution of neurological surgery: conquering increased intracranial pressure, infection, and blood loss. *Neurosurg Focus* 18, e2 (2005).

[125] Vourc'h, G. & Tannieres, M.L. Cardiac arrhythmia induced by pneumo encephalography. *Br J Anaesth* 50, 833-839 (1978).

[126] Vredenburgh, J.J., Desjardins, A., Herndon, J.E., 2nd, Marcello, J., Reardon, D.A., Quinn, J.A., Rich, J.N., Sathornsumetee, S., Gururangan, S., Sampson, J., Wagner, M., Bailey, L., Bigner, D.D., Friedman, A.H. & Friedman, H.S. Bevacizumab plus irinotecan in recurrent glioblastoma multiforme. *J Clin Oncol* 25, 4722-4729 (2007).

[127] Walker, M.D., Strike, T.A. & Sheline, G.E. An analysis of dose-effect relationship in the radiotherapy of malignant gliomas. *Int J Radiat Oncol Biol Phys* 5, 1725-1731 (1979).

[128] Westphal, M., Hilt, D.C., Bortey, E., Delavault, P., Olivares, R., Warnke, P.C., Whittle, I.R., Jaaskelainen, J. & Ram, Z. A phase 3 trial of local chemotherapy with biodegradable carmustine (BCNU) wafers (Gliadel wafers) in patients with primary malignant glioma. *Neuro Oncol* 5, 79-88 (2003).

[129] White, Y.S., Bell, D.S. & Mellick, R. Sequelae to pneumoencephalography. *J Neurol Neurosurg Psychiatry* 36, 146-151 (1973).

[130] Whittle, I.R. Brain tumour surgery: triumphs and tragedies. *J R Coll Surg Edinb* 44, 72-77 (1999).

[131] Yan, H., Parsons, D.W., Jin, G., McLendon, R., Rasheed, B.A., Yuan, W., Kos, I., Batinic-Haberle, I., Jones, S., Riggins, G.J., Friedman, H., Friedman, A., Reardon, D., Herndon, J., Kinzler, K.W., Velculescu, V.E., Vogelstein, B. & Bigner, D.D. IDH1 and IDH2 mutations in gliomas. *N Engl J Med* 360, 765-773 (2009).

[132] Yang, D., Korogi, Y., Sugahara, T., Kitajima, M., Shigematsu, Y., Liang, L., Ushio, Y. & Takahashi, M. Cerebral gliomas: prospective comparison of multivoxel 2D chemical-shift imaging proton MR spectroscopy, echoplanar perfusion and diffusion-weighted MRI. *Neuroradiology* 44, 656-666 (2002).

[133] Yasargil, M.G. A legacy of microneurosurgery: memoirs, lessons, and axioms. *Neurosurgery* 45, 1025-1092 (1999).

[134] Zemp, F.J., Corredor, J.C., Lun, X., Muruve, D.A. & Forsyth, P.A. Oncolytic viruses as experimental treatments for malignant gliomas: using a scourge to treat a devil. *Cytokine Growth Factor Rev* 21, 103-117 (2010).

[135] Zonari, P., Baraldi, P. & Crisi, G. Multimodal MRI in the characterization of glial neoplasms: the combined role of single-voxel MR spectroscopy, diffusion imaging and echo-planar perfusion imaging. *Neuroradiology* 49, 795-803 (2007).

In: Brain Cancer, Tumor Targeting and Cervical Cancer
Editor: Elena K. Salvatti
ISBN: 978-1-61122-738-3

Chapter VI

Surgical Treatment of Invasive Cervical Cancer

Sabas Carlos Vieira[1], Rafael Bandeira Lages[1], Benedito de Sousa Almeida Filho[1] and Ilanna Naianny Leal Rodrigues[2]

[1]Universidade Federal do Piauí, Teresina, Piauí, Brazil
[2]Universidade Estadual do Piauí,Teresina, Piauí, Brazil

Abstract

The standard treatment of stages 0-Ia1-Ia2-Ib1 cervical carcinoma is surgery, unless the patient has no clinical conditions for this procedure. There is no standard management of stages Ib-IIa. Both radical surgery and radical radiotherapy have proven to be equally effective, but differ in associated morbidity and complications. Concurrent chemo-radiation with cisplatin-based chemotherapy is important for patients with advanced disease, lymph node invasion, as well as parametria or surgical margins involved by the tumor, because this method improve disease-free and overall survival rates. The aim of this chapter was review the current approach to surgical management of cervical cancer. The standard surgical treatment for invasive cervical carcinoma stage Ib-IIa is radical hysterectomy class III with pelvic lymphadenectomy. Radical hysterectomy was firstly described by Wetheim more than one hundred years ago and was then modified and re-popularized by Meigs in 1950s. This operation yields 5-year survival rates of 75%-90% in most cases. We also reviewed studies that discuss the use of radical hysterectomy class II as a technique with same class III survival rates. The possibility and the advantages of ovarian preservation, laparoscopic radical hysterectomy, trachelectomy, and robotic approach were also reviewed. Finally, we discussed the use of sentinel lymph node in cervical carcinoma, a promising technique for management of this cancer and an area in which we have some important researches.

Introduction

Cervical carcinoma is the second most common cancer worldwide, resulting in approximately 275,000 deaths yearly and it remains an important health problem in both developed and developing countries even though population-based screening are available. There is no standard management for patients with stage Ib to IIa carcinoma of the cervix. Most patients can be treated with surgery or radiation therapy alone. Both treatments have the same survival; however, their morbidities are distinct [1]. There is only one randomized study comparing these two treatment modalities [1]. In this study, 343 women undergoing surgery (172) or radiation therapy (171) were enrolled. With an average follow-up period of 87 months, a five-year overall and disease-free survival was observed in 83% and 75%, respectively, for both groups.

The choice between these treatments depends on the patient's general health status, biological and non-chronological age and histopathological features of the tumor. The surgical treatment is eligible for young patients owing to the possibility of ovarian preservation and because it also determines a promising better sexual performance, considering that vaginal fibrosis and stenosis are not present as they are in irradiated patients.

Class III (Wertheim-Meigs) radical hysterectomy was initially performed in 1898 by Wertheim, who published his case series in 1890 and 1910 [2]. Radical surgery may result in a short vagina; however, lubrication is usually not compromised. It is imperative that the surgeon be aware not to perform a resection beyond the length required during the colpectomy, particularly in lean patients in whom surgery is generally technically easier. A 2-cm vaginal margin is sufficient. When the vaginal margin is less than 0.5 cm, the vaginal vault brachytherapy should be performed.

Radiation therapy may reduce the caliber and length of the vaginal cavity as well as its lubrication. These unfortunate events may be improved by using local and systemic hormone replacement therapy or by carrying out a vaginal dilatation. In patients without sexual activity, a complete obliteration of the vaginal cavity may occur, which makes the follow-up of these patients difficult. In addition to cystitis, enteritis and actinic rectitis may compromise significantly the patients' quality of life. However, due the recent improvement of equipment for radiotherapy, the incidence of these complications is decreasing.

Patients who have undergone surgical treatment and present with lymph nodes, margins and parametrium compromised by the neoplasia must receive postoperative radiation therapy. The addition of platin-based chemotherapy in these patients improves survival, as it has been shown in five major randomized studies [3, 4, 5, 6, 7].

Stages: Ib and IIa

The standard surgical treatment for invasive stage Ib and IIa carcinoma of the cervix is class III radical hysterectomy with open bilateral pelvic lymph node dissection (Figures 1, 2). The surgery includes complete removal of the uterus, parametria and upper third of the vagina. The ovaries can be preserved or not depending on the tumor size, patient's age and desire.

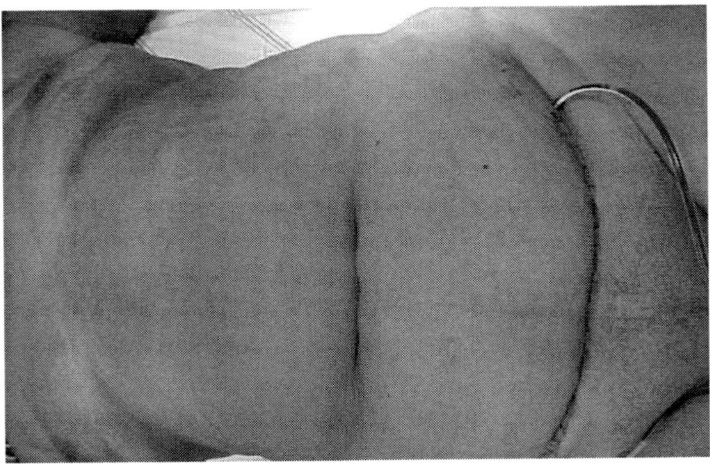

Figure 1. Transverse incision for radical hysterectomy.

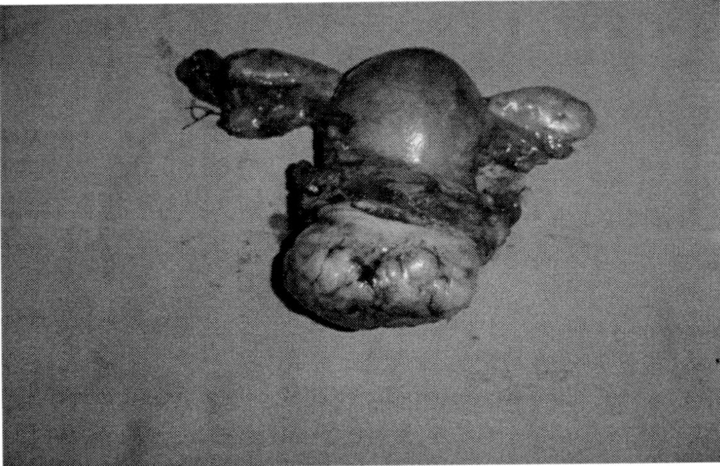

Figure 2. Stage IIa cervical carcinoma.

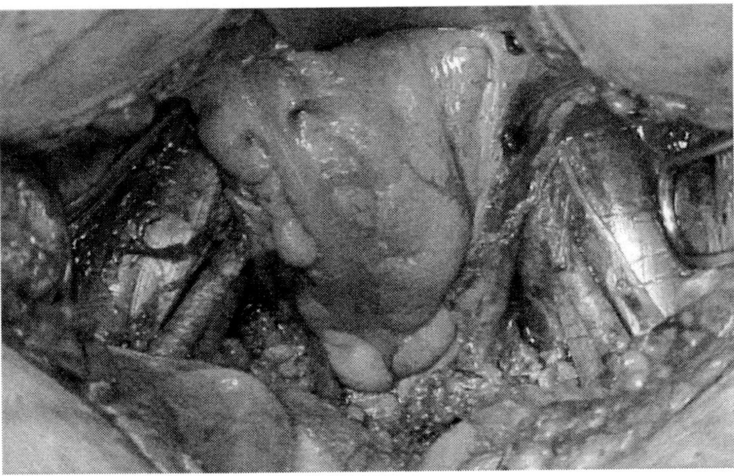

Figure 3. Pelvic lymph node dissection.

Pelvic lymph node dissection is usually performed by conforming to these limits: superiorly at the point where the ureter crosses anterior to the iliac vessels; laterally with the ileoinguinal nerve; inferiorly with the circumflex artery emergence and posteriorly with the obturator nerve (Figure 3). The amount of resected pelvic lymph nodes varies widely in the literature. However, the larger the number of resected lymph nodes, the higher is the rate of lymph node metastasis. When less of 20 pelvic lymph nodes are removed, the rate of positive lymph nodes is 10.5% and it reaches 26.5% when more than 50 lymph nodes are removed [8].

When patients present with clinically suspected lymph nodes during laparotomy or laparoscopy, frozen section examination must be performed. In presence of lymph node metastasis, the surgery must be interrupted and the patient treated with concomitant radiation and chemotherapy.

In the past, obesity and age older 65 years were contraindications to radical hysterectomy. Currently these patients can be operated on safely and with acceptable morbi-mortality.

Class II versus Class III Radical Hysterectomy

The radicality in oncology has undergone changes. The gross margins were replaced by histological margins; consequently, in many tumors the margin size has reduced without a great impact in survival. The most remarkable example has occurred in breast cancer, in which radical mastectomy was replaced by conservative treatment, without changing survival rates and improving patient's quality of life.

As for the carcinoma of the cervix, some studies have shown that patients undergoing class II radical hysterectomy have a similar survival rate to those patients undergoing class III radical hysterectomy.

In the first study [9], 243 patients with invasive stage Ib and IIa carcinoma of the cervix randomly assigned to undergo class II versus class III radical hysterectomy. The duration of surgery was significantly shorter in the group that underwent class II radical hysterectomy; however, blood transfusion and loss were similar in both groups. Mortality rate, mainly of urological origin, was significantly lower in class II (13% versus 28%). Recurrence rate was similar in both groups. Overall 5-year survival rate was 81% and 77% and 5-year disease-free survival rate was 75% and 73% for class II and class III hysterectomy, respectively. A multivariable analysis has not shown that class II versus class II radicality was significant to survival.

In an analysis of 131 patients who underwent radical hysterectomy for carcinoma of the cervix [10], only 8% had parametrial involvement. However, all these patients with parametrial involvement had other variables that advocated the need for adjuvant treatment with radiation therapy or chemotherapy, which confirms that a less radical surgery is appropriate for the treatment of early-stage cervical cancer.

Another study [11] in which 120 patients with invasive stage Ia and B1 carcinoma of the cervix, that is, with tumors smaller than 4 cm, were assessed, parametrial involvement occurred in only 5% of patients. All these patients underwent class II radical hysterectomy. In 13% of the patients, pelvic lymph node involvement was present. Of the patients with parametrial involvement, 80% had positive lymph nodes. In this study of cases only the presence of involved lymph nodes was associated with poor prognosis with a median follow-up of 48 months.

In order to decrease urinary tract dysfunctions, a new Type III radical hysterectomy with nerve-sparing technique for cervical cancer has been performed. Raspagliesi *et al.* [12] published a comparative series of 110 patients submitted to Type II Radical Hysterectomy (RH), Type III RH with nerve-sparing and Type III RH. The groups did not differ significantly in terms of urologic GI-IV morbidity, but Type II RH and Type III with nerve-sparing presented a prompt recover of bladder function, significantly different from that of Type III RH. Type III RH with nerve-sparing seemed to be comparable to Type II RH and superior to Type III RH in terms of ealry bladder dysfunctions.

Nevertheless, randomized trials with larger number of cases are needed in order to achieve definitive conclusions. The standard treatment still remains class III radical hysterectomy.

Ovarian Preservation

More recently, Bloemers et al. reported a case of preservation of ovarian function and regular normal menstrual cycles at regular months intervals by ovarian transposition prior to concurrent pelvic cisplatin-based chemoradiation for cervical cancer. This can provide an option for cervical cancer patients who would like preservation of ovarian function [13].

Ovarian preservation in patients with carcinoma of the cervix that are scheduled to undergo radical hysterectomy must be considered in association with the risk of metastasis from the carcinoma of the cervix, loss of quality of life related to surgical menopause and the risk of future need for surgery for benign or neoplasic disease in the remaining ovary.

The dread of ovarian metastasis in squamous lesions (<2%) does not advocate resection of grossly normal ovaries during the surgery of young patients undergoing radical hysterectomy, provided that these women have regular menstrual cycles and absence of signs and/or symptoms related to menopause [14, 15]. In the reports on ovarian involvement, extension to the endometrium or to the uterine corpus occurred, which suggests a possible direct extension of the cancer to the ovary [16].

The consequences of estrogen deprivation in quality of life (vasomotor and urogenital symptoms) and in the development of morbidity consequences (osteoporosis, cardiovascular disease), and considering that a significant number of patients that have this condition are members of the lower social strata (thus without resources to sustain a long-term estrogen replacement therapy), this option must be considered by the physician, as it provides a better quality of life with low morbidity rates for the patients.

The preservation of ovarian function is variable, depending mainly on the age of the patient at the moment of the surgery and the need for adjuvant pelvic radiation therapy.

Pete and Bosze [17] evaluated 65 patients with early-stage carcinoma of the cervix treated with radical hysterectomy in association to ovarian preservation and considering as diagnostic criterion of ovarian failure three measurements of FSH level higher than 30 mU/mL, they observed ovarian failure in two patients. No patient had recurrence of disease. Three patients underwent salpingo-ophorectomy 6, 11 and 24 months after radical hysterectomy because they presented with adnexal lesions, all of which were histopathologically benign.

Morice et al. [18] in a study of 107 patients with mean age of 33 years who underwent radical hysterectomy with ovarian preservation (fixation to the paracolic gutters), they achieved maintenance of ovarian function in all patients who underwent only the surgical procedure, whereas in 90% of patients treated with postoperative vaginal brachytherapy and in 60% of those patients treated with postoperative external-beam radiation therapy and brachytherapy. These data imply higher risk of ovarian failure associated with external beam radiation therapy. As for Ishii et al. [19], who evaluated 33 patients that underwent radical hysterectomy with ovarian preservation, they observed the occurrence of ovarian failure in 5 of the 21 patients younger than 39 years of age, in comparison to 10 of the 12 patients older than 40 years; and by studying these data using multiple regression analysis, they observed a significant correlation between ovarian dysfunction and patients older than 40 years.

In a study of 42 patients who underwent radical hysterectomy and ovarian sparing [20], the rate of ovarian function preservation was 78.6% with a median follow-up period of 16 months. The factor related to loss of ovarian function was postoperative pelvic radiation therapy. Four patients presented with ovarian cysts at follow-up, but none of them required surgical treatment.

A large series [21] of 3471 patients with stage I and IIb carcinoma of the cervix was analyzed in relation to presence of ovarian metastasis. The incidence of ovarian metastasis was 0.22% for stage Ib, 0.75% for stage IIa and 2.17% for stage IIb, all of which were epidermoid carcinomas. As for the cases of adenocarcinoma, the incidence of ovarian metastasis was 3.72% for stage I, 5.26% for stage IIa and 9.85% for stage IIb.

Therefore, patients that are younger than 45 years, with stage Ia and Ib1 epidermoid carcinomas and have a normal ovarian functions should have their ovaries preserved and fixated outside the pelvic cavity. The fixation must be in the parietocolic gutter above the aortic bifurcation and metallic clips should be placed in order to guide radiation therapists when adjuvant radiation therapy is required. For those women older than 45 years, all potential risks and benefits of ovarian preservation must be discussed with the patient.

Prognosis of early-stage cervical cancer patients is good. Oophorectomy at the time of hysterectomy for benign disease is associated with decreased risk of breast and ovarian cancer but an increased risk of all-cause mortality, fatal and nonfatal coronary heart disease, and lung cancer [22].

Radical Trachelectomy

In younger women diagnosed with cervical carcinoma, the desire to maintain fertility without compromising long-term survival is an important issue. As more knowledge is gained regarding the pathogenesis of cervical cancer, conservative treatment of early-stage cervical carcinoma has been proposed as a reasonable option. Nulliparous patients of childbearing age with early cervical cancer may be treated with uterus-sparing surgery. It is necessary that the tumor be small, usually smaller than 2 cm, in order to attain a satisfactory surgical margin. Options include radical trachelectomy via vaginal or abdominal approach associated with open or laparoscopic pelvic lymph node dissection. A thorough evaluation of the tumor size by means of magnetic resonance imaging is required for these cases. This procedure must be performed by an experienced and qualified staff. In a review published in 2005 [23], there were approximately 500 cases published on this subject with a number of 100 live births. Premature rupture of amniotic membranes may occur. The local recurrence rate was 4%. According to another review published in 2010 [24], more than 250 pregnancies have been reported following radical vaginal trachelectomy, resulting in 65% third trimester live births. Infertility has been described in 25-30% of patients attempting to conceive.

In another study, 100 cases of radical vaginal thrachelectomy with laparoscopic pelvic lymph node dissection were assessed. Five-year disease-free survival rate was 90.8% [25].

For those patients in whom a vaginal approach is not feasible, radical abdominal trachelectomy with uterine artery ligation is suitable to a uterus-sparing procedure. The abdominal approach is very similar to the radical hysterectomy, making it a more accessible procedure for surgeons trained in radical pelvic surgery. Recently, Abu-Rustum and al. [26] reported five cases of cervical cancer treated with this technique. Three patients presented postoperatively with normal menstrual cycles and no case of recurrence was observed. In the largest series, normal menstrual pattern was resumed within 8 weeks of surgery in all but two patients [27].

Total Laparoscopic Radical Hysterectomy

Total laparoscopic radical hysterectomy was described in 1990 by Canis *et al.* [28]. Since those initial reports, a number of other groups have published their experiences showing the feasibility and safety of this procedure for cervical cancer. However, the data available up till now suggest that laparoscopic surgery does not alter the outcome and life expectancy of the patients with carcinoma of the cervix [29].

The main advantages of the laparoscopic surgery are reduced trauma to the abdominal wall, lower infection rate in the surgical site, faster return to work, shorter hospitalization and better aesthetic result. Recently, a major randomized study involving 872 patients with colorectal carcinoma confronting laparoscopic colectomy versus open colectomy has demonstrated overall and disease-free survival curves. Recovery for the group that underwent laparoscopic surgery was faster, hospitalization was shorter and they also required less narcotic analgesic in the postoperative period [30]. A metaanalysis of prospective randomized trials has proven the benefits of advanced laparoscopic procedures compared to laparotomy which include: decreased pain, decreased surgical site infection rate, decreased length of stay, quicker return to activity and less postoperative adhesions [31].

In 2010, Naik et al. [32] performed the first attempt at a randomized controlled trial to evaluate short-term surgical outcomes after laparoscopic assisted radical vaginal hysterectomy. Fifteen women were recruited to the study over a 20-month period – 7 patients randomized to radical abdominal hysterectomy and 8 to laparoscopic assisted radical vaginal hysterectomy. It confirmed the previously suggested benefits of laparoscopic surgery in terms of a reduced duration of bladder catheterization, reduced blood loss, reduced analgesic requirement and shorter hospital stay.

Although laparoscopic surgery has many advantages, it is also associated with a number of potential drawbacks, including limited freedom of manipulation, lack of tactile sensation, amplification of tremors in prolonged cases because of the length and rigidity of the instrumentation, as well as reduced depth perception secondary to a two-dimensional vision [33].

In 2002, a study [34] evaluated 78 consecutive patients who underwent total laparoscopic radical hysterectomy. The percentage of those converted to open surgery was 6%. The average duration of the surgery was 205 minutes. The recurrence rate was 5%, with a follow up of at least three years after the surgery. Only one patient required blood transfusion. There was a large number of dissected lymph nodes, with an average of 34 lymph nodes. The rate of positive lymph nodes was 11.5%. Three patients had their margins compromised.

In another series with 50 patients [35], three of them relapsed, having an average follow up of 44 months. The patients with body mass index above 29 and/or tumors with more than 4cm of diameter were not eligible of having a laparoscopic surgery. The average duration of the surgery was 258 minutes. The average of resected lymph nodes was 13.2. Two major complications occurred: one case of urinary fistula and one of urethral stenosis. The mean follow up was 44 months. The overall survival rate in five years was 96%.

A recent study at MD Anderson [36], in which 20 people who underwent total laparoscopic radical hysterectomy were assessed, it was observed that all patients were alive and disease-free with a mean follow up of 8 months. Three patients had early complications: bladder injury, pulmonary embolism and pneumomediastinum with subcutaneous emphysema. The average number of resected lymph nodes ranged from 9 to 26, with an average of 13 lymph nodes.

Relapses in the operated area in open radical hysterectomy are rare. In laparoscopic surgery, studies have reported implants in puncture openings of the abdominal walls [37, 38]. It is believed that the implants occur by means of the contact of neoplastic cells at the moment of removal of surgical instruments with the puncture openings or by an increased intra-abdominal pressure secondary to the pneumoperitoneum, which would determine larger spacing among mesothelial cells, facilitating the incidence of implants and the development of neoplastic cells. A meticulous surgical technique which involves cautious manipulation of structures, affixation of a trocar to the operatory wound, avoiding "touching" the tumor and putting the surgical specimens in bags so they could be removed from the abdominal cavity, thus avoiding contact with patients' tissues in the puncture opening, is a measure that can decrease the incidence of implants [39].

In order to determine the long-term disease-free and overall survival outcomes of laparoscopic treatment of early-stage cervical cancer, Lee et al. [40] analyzed 139 patients undergoing laparoscopic radical hysterectomy who are followed up for 3 years or longer were enrolled. The median of follow-up duration was 92.1 (interquartile range, 54.4–127.4) months. Cumulative disease-free survival rate was 91.01% and cumulative overall survival rate was 92.78%. Eleven patients had recurrence and the recurrence rate was 7.91%, 3 of whom were in FIGO stage IA2, 7 in IB1, and 1 in IIA. The interval between surgery and recurrence ranged from 6.2 to 66.8 months. The sites of recurrence included pelvic lymph nodes, paraaortic lymph nodes, adnexa, vaginal cuff, parametria, vulva, and lungs. According to them, the laparoscopic approach has favorable long-term survival outcomes and perioperative morbidity. By the other hand, Li et al. [41] found that the recurrence rates (14% and 12%, respectively) and mortality rates (10% and 8%, respectively) were similar for patients undergoing laparoscopic radical hysterectomy and abdominal radical hysterectomy.

Robotic Radical Hysterectomy

More recently, robotic surgery is being used with increasing frequency in gynecologic oncology. Robotic technology offers several advantages over conventional laparoscopy as demonstrated in laboratory drills, including three-dimensional vision, tremor reduction, 7° of intra-abdominal articulation, more ergonomic surgeon position, the potential for a shorter learning curve, and motion scaling. Among the disadvantages of robotic surgery are loss of tactile feedback, large bulky robotic arms, limited variety of instrumentation, and cost. The daVinci® robotic system (Intuitive Surgical Corporation, Sunnyvale, CA) costs approximately $2000 for every 10 uses [33].

Data are very limited with regards to intra-operative surgical factors and complications of robotic versus open radical hysterectomies for the treatment of cervical cancer. However, studies have reported on the use of robotics for cervical cancer staging and have demonstrated the safety and effectiveness of this technique, reporting good outcomes for small number of patients. Recent comparison studies have also reported outcomes with robotic radical hysterectomies when compared to both laparotomy and laparoscopy [42].

In 2006, Sert and Abeler [43] published the first case report on robotic radical hysterectomy in cervical cancer. In September 2007, they described a small series [44] comparing 7 patients who underwent robotic radical hysterectomies with 8 patients who underwent radical hysterectomy by laparoscopy. Mean operative time for the robotic group was 241 min (range 160-445), and for the laparoscopy group was 300 min (range 225-375). The estimated blood loss in the robotic group was 71 mL and in the laparoscopic group was 160 mL (p 0.038). The length of hospitalization was 4 days in the robotic group and 8 days in the laparoscopy group (p 0.004). The mean docking time for the robotic surgery group was 25 minutes.

Magrina *et al.* [45] published a comparative series of patients undergoing robotic radical hysterectomy and patients undergoing laparoscopic radical hysterectomy or open abdominal radical hysterectomy. The robotic surgery group comprised 27 patients, 18 with cervical cancer and 9 with endometrial cancer. The laparoscopic surgery group comprised 31 patients, 18 with cervical cancer and 13 with endometrial cancer. The laparotomy group comprised 35 patients, 21 with cervical cancer and 14 with endometrial cancer. The mean operating time was significantly longer for the laparoscopic group compared to both the robotic and laparatomy groups ($p<0.001$). The mean estimated blood loss (133.1 mL for robotic surgery, 208.4 mL for laparoscopy, and 443.6 mL for laparotomy), and mean length of hospital stay (1.7 day for robotic surgery, 2.4 days for laparoscopy, and 3.6 days for laparotomy) were similar for the robotic and laparoscopy groups and significantly lower as compared to the laparotomy group ($p<0.05$). No conversions to laparotomy occurred in the robotic surgery group or the laparoscopic surgery group.

Most recently, Estape *et al.* [42] published the largest series comparing 32 patients undergoing robotic radical hysterectomy, 17 patients undergoing laparoscopic radical hysterectomy and 14 patients undergoing radical abdominal hysterectomy. Operative time for the robotic group was 2.4h and not significantly different from the laparoscopic group (2.2h) nor the laparotomy group (1.9h). The estimated blood loss for patients undergoing robotic surgery was significantly less than the laparotomy group, but not the laparoscopic group. The incidence of postoperative complications was lower in the robotic cohort (18.8%) as compared to the laparoscopic (23.5%), and laparotomy cohorts (28.6%). The robotic group had an average of 32.4 total nodes retrieved, as compared to 18.6 and 25.7 nodes retrieved in the laparoscopy and laparotomy cohorts, respectively; All differences were significant ($p<0.0001$ and $p<0.05$).

All these studies involved a relatively small number of patients, and it is crucial that more studies be conducted to further confirm the safety and feasibility of the robotic approach and ultimately prove the equivalency of this procedure to laparoscopy and laparotomy in terms of oncologic outcome and quality of life outcomes. Further cost-benefit analysis is also needed to determine whether the robot imposes a financial gain or burden to the healthcare system [46].

Residual Cervical Carcinoma

The incidence of residual cervical carcinoma is rare and the results of the treatment are equivalent to those of patients with an intact uterus [47].

The treatment of choice is radiotherapy. In a study of 145 cases [48] of residual cervical cancer, treated with radiotherapy, the authors found a higher complication rate than in patients with an intact uterus.

The surgery can be safely performed in patients with residual cervical cancer. This surgery consists of radical abdominal trachelectomy and pelvic lymphadenectomy for patients with tumors between Ia2 and IIa stages. Radical trachelectomy and laparoscopic pelvic lymphadenectomy have also been used for the treatment of residual cervical carcinoma. Liang *et al.* [49] reported six cases of residual carcinoma of the cervix or vaginal vault that were treated by means of this approach.

Radical Hysterectomy for Stage IIb

The standard treatment for stage IIb cervical cancer is concomitant radiotherapy and chemotherapy based on cisplatin. However, in some European and Asian countries, these patients are usually treated with radical hysterectomy. The advantages of this approach would be the lack of long-term complications caused by radio/chemotherapy in patients with no pathological risk factors for recurrence, such as positive lymph nodes, parametrial invasion and compromised surgical margins. In young patients, the ovarian function and vaginal elasticity can be preserved. Nevertheless, if positive lymph nodes are found, the ovaries can be transposed outside of the irradiation field. The main disadvantage of the primary radical surgery in these cases is the morbidity associated, thereby requiring postoperative radio/chemotherapy.

In a recent review [50], parametrial involvement was present in 21-55% of the patients who underwent radical hysterectomy for stage IIb cervical cancer, that is, approximately half of the patients were overstaged. These discrepancies in pathological clinical staging toward the standard originate from the inflammatory process, irregularities of the tumor, endometriosis and adhesions. The incidence of pelvic lymph node metastasis ranged from 35 to 45.8% and paraaortic lymph node metastasis from 4.5 to 7.2%.

The survival rate in five years for patients at stage IIb undergoing radical hysterectomy ranges from 55 to 77%. The most important prognostic factor is the number of positive lymph nodes. There are few studies broaching the complications of the surgical treatment in patients at stage IIb who have been treated with radical hysterectomy. Complications increase if patients receive adjuvant radiotherapy [50]. Complications are directly proportional to the dose of radiation. Intestinal obstruction, fistula and rectal bleeding occurred in 8.5% and 13.6% of the patients who received 5000cGy and 6000cGy, respectively [51].

Sentinel Lymph Node

Lymph node status remains the single most important independent prognostic factor in early stage cervical carcinoma. The application of the sentinel lymph node concept in oncology is recent, and it is currently part of the standard surgical management of cutaneous melanoma and breast cancer, and is a promising application in vulvar cancer treatment. Sentinel lymph node is the first regional lymph node to receive lymphatic drainage from primary tumor and represents the lymph node status. Lymphadenectomy, whether it is auxiliary, inguinal or pelvic, increases morbidity in cancer patients. The main morbidities caused by lymphadenectomy are lymphedema, seroma formation with the need for repeated punctures, lymphoceles, increased infection rate, longer period of hospitalization, injury to nerves and vascular system and, in a minor scale, lymphatic fistulas (Figure 4). The sentinel lymphadenectomy technique avoids morbidities caused by radical lymphadenectomy. Currently, in breast cancer and melanoma, it is the standard technique and technetium and/or patent blue are used to identify the sentinel lymph nodes (Figure 5).

In cases of carcinoma of the cervix, clinical application of sentinel lymphadenectomy is still being studied. However, results of early studies have shown that this technique appears to be promising. If this concept is validated, radical pelvic lymphadenectomy and resulting morbidities would be avoided.

In cervical carcinomas with less than 2cm of diameter the lymph node length ranges from 0 to 16% [52]. Therefore, many patients undergo unnecessary pelvic lymphadenectomy. Sentinel lymphadenectomy, either by laparotomy or by laparoscopy, if scientifically validated, will represent an advance in the surgical treatment of cervical cancer in early stages.

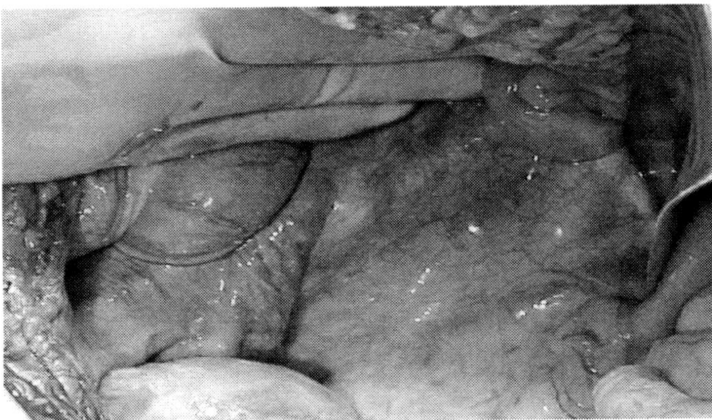

Figure 4. Patent blue used to identify the sentinel lymph node in external iliac chain.

The study of sentinel lymph nodes in cervical cancer was initially performed by Echt in 1999 [53].

The technique that has been used is the blue patent and/or technetium injection to the cervix. Generally, the injection is given to the four quadrants (3, 6, 9 and 12 hours) of the cervix.

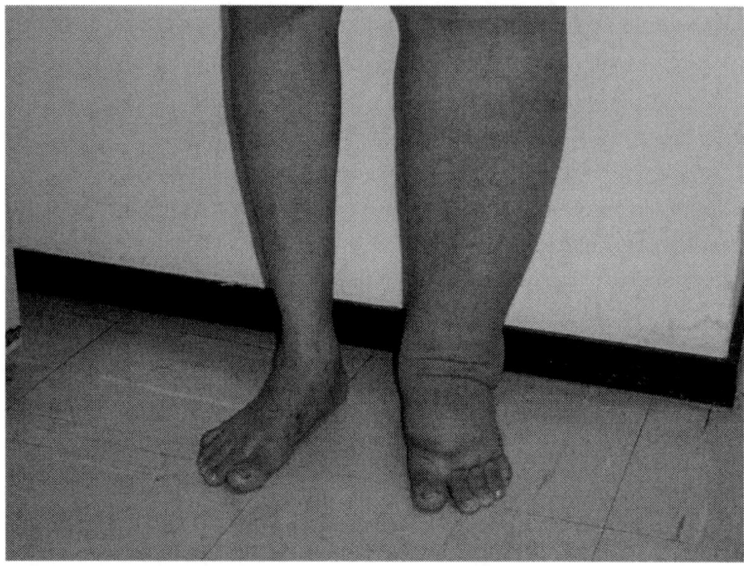

Figure 5. Lymphedema after radical hysterectomy and radiation therapy.

In a review with 669 published cases of various authors, the rate of identification of sentinel lymph nodes was 87.3% with a false negative rate of 3.3% and a negative predictive value of 99.2% [54].

The number of identified sentinel lymph nodes has varied widely, in which the external iliac chain is the site where they are most frequently found [55]. Complications caused by the procedure are secondary to the effects of patent blue. The most frequent event is decreased pulse oximetry, however, this occurs due to the interpretation of the pulse oximeter readings, causing no alteration in the arterial blood gas. Another effect is the bluish color of the skin and mucous membranes after the procedure using patent blue, but it is momentary and disappears within 48 hours [56]. The most feared complication is anaphylaxis, which must be readily treated.

The sentinel lymphadenectomy technique in cervical cancer is not standardized yet. Many studies are still required so that this technique can be routinely applied in patients with cervical cancer.

References

[1] Landoni F, Maneo A , Colombo A , Placa F, Milani R, Perego P, Favini G, Ferri L, Mangioni C. Randomized study of radicla surgery versus radiotherapy for stage Ib-IIa cervical cancer. *Lancet* 1997 ; 350:535-540.

[2] Ballon SC. The Wertheim hysterectomy. *Surg Gynecol Obstet* 1976 Jun;142(6):920-4.

[3] Whitney CW, Sause W, Bundy BN,Malfetano JH, Hannigan EV, Fowler WC Jr, Clarke-Pearson Dl, Liao SY. Randomized comparison of fluorouracil plus cisplatin versus hydroxyurea as an adjunct to radiation therapy in stage IIB-IVA carcinoma of the cervix with negative para-aortic lymph nodes: a Gynecologic Oncology Group and Southwest Oncology Group study. *J Clin Oncol* 1999;17 (5): 1339-348.

[4] Morris M, Eifel PJ, Lu J, Grigsby PW, Levenback C, Stevens RE, Rotman M, Gershenson DM, Mutch DG. Pelvic radiation with concurrent chemotherapy compared with pelvic and para-aortic radiation for high-risk cervical cancer. *N Engl J Med* 1999;340 (15): 1137-143.

[5] Rose PG, Bundy BN, Watkins EB, Thigpen JT, Deppe G, Maiman MA, Clarke-Peterson DL, Insalaco S. Concurrent cisplatin-based radiotherapy and chemotherapy for locally advanced cervical cancer. *N Engl J Med* 1999;340 (15): 1144-153.

[6] Keys HM, Bundy BN, Stehman FB, Muderspach LI, Chafe WE, Suggs CL 3rd, Walker JL, Gersell D. . Cisplatin, radiation, and adjuvant hysterectomy compared with radiation and adjuvant hysterectomy for bulky stage IB cervical carcinoma. *N Engl J Med* 1999;340 (15): 1154-161.

[7] Peters WA 3rd, Liu PY, Barrett RJ 2nd, Stock RJ, Monk BJ, Berek JS, Souhami L, Grigsby P, Gordon W Jr, Alberts DS. Concurrent chemotherapy and pelvic radiation therapy compared with pelvic radiation therapy alone as adjuvant therapy after radical surgery in high-risk early-stage cancer of the cervix. *J Clin Oncol* 2000;18 (8): 1606-613.

[8] Sedlis A, Bundy BN, Rotman MZ, Lentz SS, Muderspach LI, Zaino RJ. A randomized trial of pelvic radiation therapy versus no further therapy in selected patients with stage IB carcinoma of the cervix after radical hysterectomy and pelvic lymphadenectomy: A Gynecologic Oncology Group Study. *Gynecol Oncol* 1999 May;73(2):177-83.

[9] Landoni F, Maneo A, Cormio G, Perego P, Milani R, Caruso O, Mangioni C. Class II versus class III radical hysterectomy in stage IB-IIA cervical cancer: a prospective randomized study. *Gynecol Oncol* 2001 Jan ;80(1):3-12.

[10] Selman TJ, Luesley DM, Murphy DJ, Mann CH. Is radical hysterectomy for early stage cervical cancer an outdated operation?.*BJOG* 2005 Mar; 112(3):363-65.

[11] Steed H; Capstick V; Schepansky A; Honore L; Hiltz M; Faught W. Early cervical cancer and parametrial involvement: is it significant?. *Gynecol Oncl* 2006 Oct;103(1):53-7.

[12] Raspagliesi F, Ditto A, Fontanelli R, Zanaboni F, Solima E, Spatti G et al. Type II versus Type III nerve-sparing radical hysterectomy: comparison of lower urinary tract dysfunctions. *Gynecol Oncol* 2006; 102(2):256-62.

[13] Bloemers MC, Portelance L, Legler C, Renaud MC, Tan SL. Preservation of ovarian function by ovarian transposition prior to concurrent chemotherapy and pelvic radiation for cervical cancer. A case report and review of the literature. *Eur J Gynaecol Oncol* 2010;31(2):194-7.

[14] Sutton GP, Bundy BN, Delgado G, et al. Ovarian metastases in stage IB carcinoma of cervix: a Gynecology-Oncology Group of Study. *Am J Obstet Gynecol* 1992; 166:50-3.

[15] Yamamoto R, Okamoto K, Yukiharu T, et al. Study of risk factors for ovarian metastases in stage Ib-IIIb cervical carcinoma and analysis of ovarian function after a transposition. *Gynecol Oncol* 2001; 82:312-6.

[16] Vieira SC, Coelho FRG, Mourão Netto M. Risco de câncer de mama e endométrio em vigência de reposição hormonal na pós-menopausa: um problema para ginecologistas e cancerologistas. *Acta Oncol Bras* 2000; 20:32-7.

[17] Pete I, Bosze P. The fate of the retained ovaries following radical hysterectomy. *Eur J Gynaecol Oncol* 1998; 19:22-4.

[18] Morice P, Juncker L, Rey A, El-Hassan J, Haie-Meder C, Castaigne D. Ovarian transposition for patients with cervical carcinoma treated by radiosurgical combination. *Fertil Steril* 2000; 74:743-8.

[19] Ishii K, Aoki Y, Takakuwa K, Tanaka K. Ovarian function after radical hysterectomy with ovarian preservation for cervical cancer. *J Reprod Med* 2001; 46:347-52.

[20] Vieira SC, Silva AG, Vale, LRG, Lima MM. Preservação dos ovários em cirurgia radical para câncer do colo uterino. *Rev bra ginecol obstet* 2002 Nov-Dez;24(10):681-84.

[21] Shimada M, Kigawa J, Nishimura R, Yamaguchi S, Kuzuya K, Nakanishi T, Suzuki M, Kita T, Iwasaka T, Terakawa N. Ovarian metastasis in carcinoma of the uterine cervix. *Gynecol Oncol* 2006 May; 101(2):234-37.

[22] Parker WH, Broder MS, Chang E, Feskanich D, Farquhar C, Liu Z et al. Ovarian conservation at the time of hysterectomy and long-term health outcomes in the nurses' health study. *Obstet Gynecol* 2009;113(5):1027-37.

[23] Shepherd JH. Uterus-conserving surgery for invasive cervical cancer. *Best Pract Res Clin Obstet Gynaecol 2005* Aug;19(4):577-90.

[24] Gien LT, Covens A. Fertility-sparing options for early stage cervical cancer. *Gynecol Oncol* 2010;117:350-7.

[25] Hertel H; Köhler C; Grund D; Hillemanns P; Possover M; Michels W; Schneider A; Radical vaginal trachelectomy (RVT) combined with laparoscopic pelvic lymphadenectomy: prospective multicenter study of 100 patients with early cervical cancer. *Gynecol Oncol* 2006 Nov;103(2):506-11.

[26] Abu-Rustum NR; Sonoda Y; Black D; Levine DA; Chi DS; Barakat RR. Fertility-sparing radical abdominal trachelectomy for cervical carcinoma: technique and review of the literature. *Gynecol Oncol* 2006 Dec;103(3):807-13.

[27] Ungar L, Palfalvi L, Hogg R, Siklos P, Smith R, Del Priore G, et al. Abdominal radical trachelectomy: a fertility-preserving option for women with early cervical cancer. *Br J Obstet Gynecol* 2005;112:366–9.

[28] Canis M, Mage G, Wattiez A, Pouly JL, Manhes H, Bruhat MA. Does endoscopic surgery have a role in radical surgery of cancer of the cervix uteri?. *J Gynecol Obstet Biol Reprod* 1991;20(7):997.

[29] Spirtos NM, Schlaerth JB, Kimball RE, Leiphart VM, Ballon SC. Laparoscopic radical hysterectomy (type III) with aortic and pelvic lymphadenectomy. *Am J Obstet Gynecol* 1996 Jun;174(6):1763-67.

[30] Clinical Outcomes of Surgical Therapy Study Group. A comparison of laparoscopically assisted and open colectomy for colon cancer. *N Engl J Med* 2004 May;350(20):2050-59.

[31] Johnson N; Barlow D; Lethaby A; Tavender E; Curr E; Garry R. Surgical approach to hysterectomy for benign gynaecological disease. *Cochrane Database Syst Rev* 2005; 1:CD003677.

[32] Naik R, Jackson Ks, Lopes A, Cross P, Henry JA. Laparoscopic assisted radical vaginal hysterectomy versus radical abdominal hysterectomy – a randomised phase II trial: perioperative outcomes and surgicopathological measurements. *BJOG* 2010;117:746-51.

[33] Ramirez PT; Soliman PT; Schmeler KM; Reis R; Frumovitz M. Laparoscopic and robotic techniques for radical hysterectomy in patients with early-stage cervical cancer. *Gynecol Oncol* 2008; 110:S21-S24.

[34] Spirtos NM, Eisenkop SM, Schlaerth JB, Ballon SC. Laparoscopic radical hysterectomy (type III) with aortic and pelvic lymphadenectomy in patients with stage I cervical cancer: surgical morbidity and intermediate follow-up. *Am J Obstet Gynecol* 2002 Aug;187(2):340-48.

[35] Pomel C, Atallah D, Le Bouedec G, Rouzier R, Morice P, Castaigne D, Dauplat J. Laparoscopic radical hysterectomy for invasive cervical cancer: 8-year experience of a pilot study. *Gynecol Oncol* 2003;91(3):534-39.

[36] Ramirez PT, Slomovitz BM, Soliman PT, Coleman RL, Levenback C. Total laparoscopic radical hysterectomy and lymphadenectomy: the M. D. Anderson Cancer Center experience. *Gynecol Oncol* 2006 Aug;102(2):252-55.

[37] Belval CC; Barranger E; Dubernard G; Touboul E; Houry S; Daraï E. Peritoneal carcinomatosis after laparoscopic radical hysterectomy for early-stage cervical adenocarcinoma. *Gynecol Oncol* Sep 2006; 102(3):580-82.

[38] Martínez-Palones JM; Gil-Moreno A; Pérez-Benavente MA; Garcia-Giménez A; Xercavins J. Umbilical metastasis after laparoscopic retroperitoneal paraaortic lymphadenectomy for cervical cancer: a true port-site metastasis?. *Gynecol Oncol* 2005 Abr;97(1):292-95.

[39] Balli JE; Franklin ME; Almeida JA; Glass JL; Diaz JA; Reymond M. How to prevent port-site metastases in laparoscopic colorectal surgery.Surg Endosc 2000 Nov;14(11):1034-6.

[40] Lee CL, Wu KY, Huang KG, Lee PS, Yen CF. Long-term survival outcomes of laparoscopically assisted radical hysterectomy in treating early-stage cervical cancer. *Am J Obstet Gynecol* 2010; 202 [Epub ahead of print].

[41] Li G, Yan X, Shang H, et al. A comparison of laparoscopic radical hysterectomy and pelvic lymphadenectomy and laparotomy in the treatment of Ib-IIa cervical cancer: 8-year experience of a pilot study. *Gynecol Oncol* 2003;91:534-9.

[42] Estape R; Lambrou N; Diaz R; Estape E; Dunkin N; Rivera A. A case matched analysis of robotic radical hysterectomy with lymphadenectomy compared with laparoscopy and laparotomy. *Gynecol Oncol* 2009; 113:357-361.

[43] Sert BM; Abeler VM. Robotic-assisted laparoscopic radical hysterectomy (Piver type III) with pelvic node dissection – case report. *Eur J Gynaecol Oncol* 2006;27:531-3.

[44] Sert BM; Abeler VM. Robotic radical hysterectomy in early-stage cervical carcinoma patients, comparing results with total laparoscopic radical hysterectomy cases. The future is now? *Int J Med Robot* 2007;3:224-8.

[45] Magrina JF; Kho RM; Weaver AL; Montero RP; Magtibay PM. Robotic radical hysterectomy: Comparison with laparoscopy and laparotomy. *Gynecol Oncol* 2008;109:86-91.

[46] Ko EM; Muto MG; Berkowitz RS; Feltmate CM. Robotic versus open radical hysterectomy: A comparative study at a single institution. *Gynecol Oncol* 2008; 111:425-430.

[47] Deffieux X; Huel C; Cosson M; Leveque J; Bonnet K; Fernandez H. Hystérectomie sub-totale: données récentes et implications pratiques.*J Gynecol Obstet Biol Reprod* (Paris) 2006 Feb;35(1):10-5.

[48] Hellström AC, Sigurjonson T, Pettersson F. Carcinoma of the cervical stump. The radiumhemmet series 1959-1987. Treatment and prognosis.*Acta Obstet Gynecol Scand* 2001 Feb;80(2):152-7.

[49] Liang Z, Xu H, Chen Y, Li Y, Chang Q, Shi C. Laparoscopic radical trachelectomy or parametrectomy and pelvic and para-aortic lymphadenectomy for cervical or vaginal stump carcinoma: report of six cases. *Int J Gynecol Cancer* 2006 Jul-Aug;16(4):1713-6.

[50] Suprasert P, Srisomboon J, Kasamatsu T. Radical hysterectomy for stage IIB cervical cancer: a review. *Int J Gynecol Cancer* 2005 Nov-Dec; 15(6):995-1001.

[51] Matsuyama T, Inoue I, Tsukamoto N, Kashimura M, Kamura T, Saito T, Uchino H. Stage Ib, IIa, and IIb cervix cancer, postsurgical staging, and prognosis. *Cancer* 1984 Dec;54(12):3072-7.

[52] Delgado G, Bundy BN, Fowler WC, Stehman FB, Sevin B, Creasman WT, Major F, DiSaia P, Zaino R. A prospective surgical pathological study of stage I squamous carcinoma of the cervix: a Gynecologic Oncology Group Study. *Gynecol Oncol* 1989 Dec ;35(3):314-20.

[53] Echt ML; Finan MA; Hoffman MS; Kline RC; Roberts WS; Fiorica JV. Detection of sentinel lymph nodes with lymphazurin in cervical, uterine, and vulvar malignancies. *South Med J* 1999 Feb;92(2):204-8.

[54] Di Stefano AB; Acquaviva G; Garozzo G; Barbic M; Cvjeticanin B; Meglic L; Kobal B; Rakar S. Lymph node mapping and sentinel node detection in patients with cervical carcinoma: a 2-year experience. *Gynecol Oncol* 2005 Dec;99(3):671-9.

[55] Vieira SC, Zeferino LC, Silva BB, Santana JOB, Santos, LG, Carvalho TCB, Rocha MCBD. Estudo do linfonodo sentinela no câncer do colo uterino com azul patente . *Rev Assoc Med Bras* 2004 Jul-Set;50(3):302-304.

[56] Vieira SC, Silva BB, Zeferino LC. Diminuição da oximetria de pulso após injeção de corante azul patente para identificação do linfonodo sentinela em câncer do colo uterino. *Acta Oncol Bras* 2003 Out-Dez;23(3):511-513.

In: Brain Cancer, Tumor Targeting and Cervical Cancer
Editor: Elena K. Salvatti
ISBN: 978-1-61122-738-3

Chapter VII

Recent Advances in Glioma Gene Therapy

José Segovia[1,*] *and Adolfo López-Ornelas*[2]
[1]Departamento de Fisiología, Biofísica y Neurociencias and [2]Departamento de Farmacología, Centro de Investigación y de Estudios Avanzados del IPN, Mexico, 07300, D.F.

Abstract

Almost half of the primary tumors of the Central Nervous System (CNS) are gliomas, and the most malignant form of these tumors, glioblastoma multiforme (GBM), aggressively infiltrates into the brain parenchyma. Current treatments for GBM include surgical resection, chemotherapy and radiotherapy, but even after combined therapies, the prognosis remains very poor, with a median survival of roughly one year after diagnosis. Although the knowledge regarding the genetic basis, and the molecular signaling pathways involved in the origin and progression of the tumors has increased significantly in the last few years, allowing the generation of new chemotherapeutic agents that are used together with sophisticated surgical and radiation techniques, new therapeutic approaches are necessary for GBM. Gene therapy appears to be a relevant strategy for the treatment of gliomas, because it permits the delivery of therapeutic genes to tumors that rarely metastasize outside of the CNS. We have proposed the use of Growth arrest specific1 (Gas1) as an adjuvant for the treatment of gliomas, because of its capacity acting as a tumor suppressor. In this chapter, we will discuss the main characteristics of gliomas, the rationale for the application of gene therapy for their treatment, and the effects of the expression of Gas1 reducing the growth of these tumors.

* To whom correspondence should be addressed: Tel: (5255) 5747-3958; Fax: (5255) 5747-3754; E-mail: jsegovia@fisio.cinvestav.mx

Keywords: Glioblastoma, Gas1, gene therapy, tumor suppressor, apoptosis, cell arrest, GDNF

Introduction

According to data from the U.S.A. the incidence of primary brain and spinal cord tumors is 18.7/100.0000 inhabitants, and malignant tumors have an incidence of 7.2/100.000 inhabitants. The incidence of these tumors varies according to age, the incidence between 0 and 4 years is 5/100.000, between 15 and 45 years is 15/100.000, and increases for the population between 65 and 74 years to 56/100.000 [1].

Glioblastoma multiforme is the most common and deadly of the CNS primary tumors and has an incidence of 5-20 cases per 100,000 inhabitants. Glioblastoma represents 60 to 75% of malignant gliomas, anaplastic astrocytomas for 10 15%, and anaplastic oligodendrogliomas and anaplastic oligoastrocytomas 10% [1]. Malignant gliomas are invasive neoplasms which are derived from glial cells (astrocytes, oligodendrocytes or ependymal cells) and are histologically heterogeneous. The World Health Organization (WHO) classifies astrocytomas into four grades; grade I, pilocytic astrocytomas, are classically asymptomatic at the beginning, have a limited invasive capacity, and infrequently progress to anaplastic glioma. However, these neoplasms may have cellular pleomorphism and microvascular hyperplasia, despite their designation as grade I. Grade II, diffuse astrocytomas, have the capacity of expanding from the adjacent brain primary lesion and have a high rate of progression to anaplastic tumors, are well differentiated, with mild to moderate nuclear pleomorphism [2]. Grade III and IV tumors are malignant gliomas. Grade III, anaplastic astrocytomas, are characterized by increased cellularity, nuclear atypia, and proliferative activity. Grade IV, glioblastomas, have the same features that grade III tumors but also contain areas of microvascular proliferation, necrosis, or both [3]. Gliomas rarely metastasize but often generate satellite tumors in the brain parenchyma.

Glioma Risk Factors

The only exogenous environmental risk factor demonstrated as a cause inducing gliomas is the exposure to either therapeutic or high-dose radiation, even though high-dose chemotherapy for the treatment of cancer in sites other than brain has also been related to glioblastoma [4, 5]. Glioblastoma is slightly more frequent in men than in women [1].

There is an increased incidence of gliomas (including astrocytomas and glioblastomas) associated with various genetic disorders, including neurofibromatosis Von Hippel-Lindau, Li-Fraumeni and Turcot syndromes. On the other hand, the relationship between electromagnetic radiation and the use of mobiles phones with gliomas has not been supported by epidemiologic studies [6-8].

Genetic alterations have been observed and associated with the development of glioblastoma, for instance: extra copies of chromosome 7 were found, and the oncogene *egfr*, receptor for the Epidermal Growth Factor (EGF), was shown to be amplified in patients with gliomas [9]. Furthermore, karyotype and loss of heterozygosity studies defined the location of tumor suppressor loci on chromosomes 9, 10, and 17 [10]. plays an important role controlling genomic DNA damage, inducing cell cycle arrest to allow DNA repairs and can activate apoptosis to eradicate injured cells [11]. Thus, mutations of the *p53* tumor suppressor gene are recognized as the most frequent alterations on chromosome 17 in patients with glioblastoma.

Other tumor suppressor genes lost in gliomas are: the *p16* cell cycle inhibitor and the Phosphatase and Tensin Homolog *(pten)* on chromosomes 19 and 10, respectively. The *p16* gene can slow down cell cycle progression, whereas *pten* is a negative regulator of Phosphoinositide 3–kinase (PI3K) [12], which is the most important pathway to stimulate cellular proliferation in the presence of growth factors. The Isocitrate Dehydrogenase 1 (*idh1*) gene, and less frequently the *idh2* gene are mutated in one allele in glioblastoma [13]. The mutation leads to a single amino acid change (arginine 132) in the active site of IDH1. It was proposed that this might have an indirect oncogenic effect through the activation of the hypoxia-inducible factor pathway [14].

Clinical Presentation

Brain neoplasms present both generalized and focal neurologic symptoms. Generalized symptoms are related to increased intracranial pressure which, in turn, causes nausea and vomiting. Focal symptoms and signs reflect the intracranial location of the tumor. The duration and frequency of symptoms may change according with the type of intracranial glioma. A fast developing hemiparesis, for example, is typical of high-grade gliomas [15]. Headache is the most frequent symptom associated with brain tumor; it is typically diffuse and can indicate the hemisphere where the tumor is situated [16]. Secondly, seizures occur in 15 to 95 percent of the patients with brain tumors and are more frequent in low grade glioma. Classically, the seizures are focal, but may become generalized. Also, postictal hemiparesis or aphasia (Todd's phenomenon) can indicate the location of the lesion [15].

Diagnosis

Currently, Functional Magnetic Resonance Imaging (fMRI) and Magnetic Resonance Imaging (MRI) are the most common methods used for glioma diagnosis. Nevertheless, contrasted Computerized Tomography (head CT scan) still may suggest the presence of malignant gliomas showing rim-enhanced lesions. These imaging studies reveal a heterogeneously enhanced mass with surrounding edema. Usually, glioblastomas have central areas of necrosis and peri-tumoral edema, the later associated with anaplastic gliomas [17].

From a histological point of view the diagnosis is mainly based on high mitotic activity and nuclear atypia. The presence of areas with vascular proliferation, cellular atypia and necrosis in a glioma lesion is considered pathognomonic of glioblastoma [18].

The basis for the diagnosis of brain tumors is a clinical-pathological evaluation. However, gene expression studies in glioblastoma have established transcriptional profiles that reflect the tumor biology and may be used for tumor classification, and to predict patient outcome, and response to treatment [19].

Treatment

Treatment is tailored according with the tumor location, patient age and Karnofski grade (50 or less indicates a high risk of death within 6 months, mostly based on the degree of disability and secondary complications). In locations where it is possible, cytoreductive surgery is commonly followed by radiotherapy, which consists in the application of 60 Gy (twenty Gy over tumor and forty Gy holocranial). Frequently, a chemotherapy scheme can also be administered and consists of Temozolomide, an alkylating agent which mechanism of action is to inhibit DNA replication. After oral administration, Temozolamide crosses the brain-blood barrier, and reaches high concentrations in the CNS. The primary treatment with this drug is accompanied with radiotherapy and when followed by 6 monthly cycles of Temozolomide it provides an increased survival of patients with glioblastoma. On the other hand, the effect of Temozolomide is linked to the intracellular levels of MGMT (O-6-Methylguanine-DNA methyltransferase) which is a crucial protein that repairs DNA and protects tumor cells against alkylating drugs, thus, high levels of this protein have been associated with resistance to Temozolomide [20].

Other targets for the treatment of glioblastoma are proteins atypically activated during tumorigenesis. Different growth factor receptors, like Epidermal Growth Factor Receptor (EGFR), Vascular Endothelial Growth Factor Receptor (VEGFR) and Platelet-Derivated Growth Factor Receptor (PDGFR), are amplified and/or mutated and over-expressed in these tumors and these alterations result in increased cell survival, uncontrolled cell proliferation, angiogenesis and cell migration [21]. Currently, new drugs that target these receptors are been investigated.

Unfortunately, the use of new drugs for the treatment of glioblastoma has not had a considerable effect in the survival of patients with glioblastoma yet, who currently have an average survival of 15 months after diagnosis.

Gene Therapy in Glioblastoma

Because of the poor results obtained with current treatments, gene therapy emerges as a possible alternative to improve the survival of these patients. There are several clinical trials in progress and these enhance the possibility of finding adjuvant therapies against this type of tumors.

Gene therapy can be defined as the transfer of genetic material (transgene) into a somatic cell to induce a beneficial effect in the patient. The transfer is accomplished with the use of techniques involving gene delivery systems, either viral or non-viral systems [22]. The genetic information of a pathogenic cell may be amended by introducing a normal gene into the target organism to replace the function of the faulty gene, or in the case of cancer, the product of the transgene will eradicate the tumor cells.

Among the main different strategies employed in pre-clinical and clinical trials in gene therapy for glioma, we find the following:

- The use of cytokines (interleukins, tumor necrosis factors, colony stimulating factors and interferons) to boost the antitumor activity of immune cells [23].

- To express exogenous antigens (tumor vaccines), to increase the immune response [24].
- To express suicide genes or increasing the sensitivity of the tumor cells to specific drugs, for instance, transducing the gene of an enzyme (Thymidine Kinase) which selectively activates a prodrug (Aciclovir) [25].
- Using antisense therapy to block the expression of oncogenes [26].
- Expressing tumor suppressor genes, i.e. *p53* [27].
- Use of oncolytic adenoviruses to eliminate tumor cells [28].
- Transferring genes with antiangiogenic effect [29].
- Reducing the toxicity of chemotherapy, expressing drug resistance genes [30].

Another strategy of gene therapy in the field of cancer is to transfer genes involved in signaling pathways leading to cell death and/or genes that encode a molecule toxic to the tumor. A key aspect is to target the expression of the therapeutic transgenes exclusively to tumor cells to avoid affecting healthy cells, thus reducing potential harmful side-effects. Glioblastoma multiforme is an appropriate target for local gene therapy because it is constricted to an anatomical region and rarely presents metastases. This decreases the risk of adverse effects because the vectors can be administered in the vicinity or directly into the tumor [31].

Several viral vectors have been employed for gene therapy, and each has particular characteristics that can be considered either advantageous or as shortcomings, depending on the use and specific disease in which they will be used. For details of different viral vectors see Table 1.

Table 1. Advantages and shortcomings of viral vectors employed in gene therapy

Vector	Advantages	Shortcomings
Adenovirus	• High capacity to carry large genes. • Infect quiescent cells. • Easy to obtain in high titles. • Does not cause disease.	• Short transgene period of expression. • High immune response. • Infect proliferating cells.
Retrovirus	• Long transgene period of expression. • Low immune response. • Well studied biology.	• Integration into the genome.
Lentivirus	• Long time of expression. • Infect quiescent cells.	• Integration into the genome.
Adenovirus-associated virus	• Site-specific integration. • Stable expression. • Infect quiescent cells.	• Low transgene capacity.
Herpes simplex virus	• High capacity to carry large genes. • Efficient infecting neurons.	• Cytotoxicity of infected cells. • Short transgene period of expression.

Table 2. Clinical trials using suicide gene therapy

Phase	Vector	Transgene	MS	Reference
N/A	Adenovirus	HSV-TK	14.4	31
I	Adenovirus	HSV-TK	6.2	73
I	Retrovirus	HSV-TK	8	74
I	Retrovirus	HSV-TK	8.1	75
I	Retrovirus	HSV-TK	7.4	76
I	Adenovirus	HSV-TK	14	76
I	Adenovirus	HSV-TK	4	77
I	Adenovirus	HSV-TK	10.4	78
I	Retrovirus	HSV-TK	8.6	79
I	Adenovirus	HSV-TK	4	80
I/II	Retrovirus	HSV-TK	8.4	81
I/II	Retrovirus	HSV-TK	5	82
III	Retrovirus	HSV-TK	12	83

HSV-TK; herpes simplex virus-thymidine kinase. MS, mean survival.

Basically, three approaches have been used in clinical trials for gliomas: 1) suicide gene therapy 2) cytokine gene therapy and 3) oncolytic viral therapy. The most employed strategy has been suicide gene therapy (Table 2) and the more frequently transgene used in these trials has been the gene of Herpes Simplex Virus-Thymidine Kinase (HSV-TK). Adenovirus or retrovirus have been the gene transfer methods more employed, however, low transduction efficiency when using retrovirus, and limited effect in the site of application in case of adenovirus, have been observed.

Table 3. Clinical trials using cytokine gene therapy

Phase	Vector	Transgene	MS	Reference
I	Lipofection	IFN-β	NA	84
I	Retrovirus	IL-4	10	85
Pilot	Retrovirus	*B7-2, GM-CSF	NA	86
I	Retrovirus	Il-2, HSV-TK	NA	87

IFN; interferon. IL; interleukin. GM-CSF; granulocyte-macrophage colony stimulating factor. * T cell co-stimulatory molecules. MS, mean survival. NA; not available.

Table 4. Clinical trials using oncolytic virus

Phase	Vector	Transgene	MS	Reference
I	HSV	No transgene	9	88
I	HSV	No transgene	6	89
I	HSV	No transgene	6.5	90
I	HSV	No transgene	11.3	91

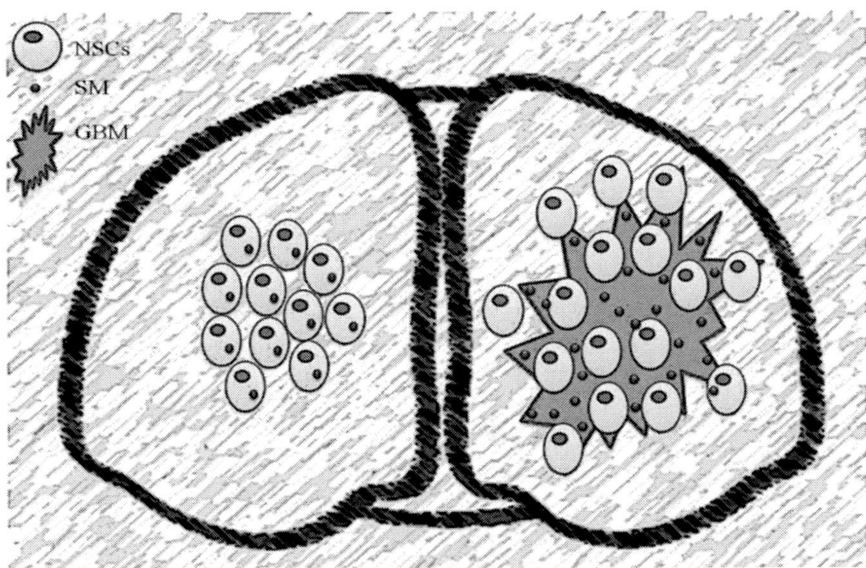

Figure 1. Schematic representation of the use of NSCs as a vehicle for transgenes with anti-tumoral effect. NSCs expressing a therapeutic transgene, coding for a secretable, soluble molecule (SM) can be implanted into the brain (or administered systemically) and migrate into a GBM, where they release the active molecule. NSCs, Neural Stem Cells; SM, secretable molecule; GBM, Glioblastoma.

For the cytokine gene therapy approach (Table 3) the following encoded proteins have been used: IL-2, IFN-β, IL-4 and GM-CSF (Granulocyte Macrophage-Colony Stimulating Factor) however, the brain edema caused by inflammatory responses generates CNS damage, regardless of the vector employed, either adenovirus or retrovirus. The oncolytic viral therapy (table 4) uses replication-competent viruses to infect and lyse the tumor cells. The main problem with this therapy is the limited effect in the site of administration.

Adenovirus and retrovirus have been the most commonly used vector for gene therapy in glioblastoma. Replication-deficient retroviruses infect only dividing cells and have been used in clinical trials, transducing suicide genes. However, low transduction efficiency has hindered the use of retroviral vectors, as well as the potential for insertional mutations [32, 33]. Adenoviral vectors transduce both dividing and quiescent cells, offer a high transgene expression, and appear to be safe for human use. The immune response, characteristic of this virus, may provide anti-tumor effects in the treatment of cancer [34–37]. In addition, adenoviruses can be engineered in different manners to achieve conditional replication or cell-type-specific expression [38, 39].

There are two major problems related with the efficacy of gene therapy strategies for the treatment of gliomas. The first one is that viral vectors infect a relatively small number of cells, this is, in part due, to poor diffusion of viral vectors within the brain parenchyma, that causes that only cells near to the site of administration become infected. This is also related, with the possibility that healthy cells can be infected and express a potentially harmful gene. The second problem is compounded by vectors that integrate the transgene into the genome of infected cells and may cause insertional mutations, as has occurred with retroviral vectors [25, 40-43].

An interesting approach to target gliomas is the use of Neural Stem Cells (NSCs) which are a population of cells with the ability to differentiate into any lineage of the CNS, such as neurons and glia. These cells are found in two regions of the adult human brain: the dentate gyrus and the subventricular zone [44, 45]. NSCs have the capacity to migrate towards brain tumors and this tropism provides a new system for the treatment of glioblastoma, by allowing the administration of anti-tumoral agents to the primary tumor as well as to satellite tumors [46, 47]. Several therapies based on the use of NSCs have been proposed, including the delivery of molecules to specific locations of the CNS, increasing the efficiency of their release, using these cells as vehicles for the gene products (figure 1). If the carrier protein is secreted, it might exert its action on cancer cells in autocrine and paracrine manners [48-50].

Significantly, the vector used for gene therapy in tumors of the CNS must achieve specific expression in target cells, thereby avoiding adverse effects. For this reason, we must find better and more specific vectors.

Gas1 as a Therapeutic Protein

A potentially interesting protein for the treatment of gliomas is Gas1, which is capable of arresting cell cycle and inducing apoptosis of glioma cells. The gene coding for *gas1* was originally isolated from mouse fibroblasts deprived of serum [51]. In humans, *gas1* is located on chromosome 9 and has only one exon with an open reading frame (ORF) of 1036 bp. The product of the *gas1* gene is 345 amino acid protein with a hydrophobic N-terminal domain, which functions as a signal peptide that guides Gas1 to the endoplasmic reticulum. There, the signal peptide is cleaved, the remaining of the protein glycosilated, and a Glycophosphatidylinositol (GPI) anchor is attached to the C-terminus, through posttranslational modifications, to the hydrophobic C-terminal domain. Finally, Gas1 is located extracellularly and the GPI anchor binds Gas1 to the lipid rafts on the outer layer of the cell membrane [52, 53].

Gas1 is expressed during development in the CNS and, in embryogenesis the expression of Gas1 has been associated with cell death through the development of the limbic system [54]. However, Gas1 induces cell proliferation in the cerebellum [55]. Therefore, the molecular function of Gas1 *in vivo* remains unclear. The contradictory effects of Gas1 in different cells suggest that Gas1 is a pleiotropic protein which functions depend on the cellular context.

Gas1 intracellular signaling pathways have not been completely elucidated, however, we have shown there is a significant homology between Gas1 and the GFRα (co-receptor for the Glial cell line-Derived Neurotrophic Factor) family of receptors, thus Gas1 inhibits the survival signaling evoked by GDNF. In this manner, Gas1 alters the pattern of autophosphorylaton of Ret, a receptor tyrosine kinase), thus inhibiting the activity of AKT. The inhibition of the GDNF intracellular signaling cascade then leads to cell arrest and the induction of apoptosis mediated by the activation of caspase-3 [56, 57].

There are also data indicating that Gas1 interacts with Shh (Sonic hedgehog), a protein involved with cell proliferation during embryogenesis. Early studies had proposed Gas1 as a possible antagonist of Shh but, more recent reports have shown that Shh and Gas1 have a synergistic effect during development, and that when this interaction is altered it is associated with craniofacial malformations, both in rodents and humans [58-63]. However, Gas1 induces cell arrest and apoptosis in primary cultures of human gliomas, in the absence of Shh [64].

Several studies have shown that the over-expression of Gas1 induces cell arrest and apoptosis in different tumor cells [65, 66]. The over-expression of Gas1 in glioma cell lines or primary cultures of human gliomas, using either retroviral or adenoviral vectors, both *in vitro* and employing *in vivo* models, induces cell cycle arrest and apoptosis [64, 67-69]. Interestingly, one of these studies analyzed the effect of the co-expression of Gas1 and p53, and demonstrated that Gas1 exerts its anti-tumoral effects independently of the presence of p53 [69]. This last fact is important because in the majority of tumors, up to 75% in the case of glioblastomas [70], genes involved in p53 signaling pathway are mutated or inactivated, thus the capacity of Gas1 as an anti-tumoral agent even in the absence of p53 indicates that Gas1 could have a significant effect in most gliomas. In these studies, the vectors employed the *gfa2* promoter, a glial specific promoter, to drive the expression of *gas1*, thus achieving specific expression in cells of glial origin compared with other studies that used strong heterologous viral promoters, like that form Cytomegalovirus (CMV) [71, 72].

It has been shown that the anti-tumoral effect of Gas1 resides on its amino terminus, particularly in the domain comprised between amino acids 182 to 234. Furthermore, medium collected from fibroblasts treated with Phospholipase C (PLC), an enzyme that cleaves the phospho-glycerol link of the GPI that anchors Gas1, was applied to proliferating cells and it was observed that it induced a decrease in the proliferation of cells of independent cultures. Also, the expression of Gas1 without its GPI anchor retained its inhibitory activity on cell proliferation [52, 53]. These studies indicate that the activity of Gas1 resides in the N-terminal, and the GPI anchor is not required for its function. Based on this information, we recently generated lentiviral vectors that express a truncated, secretable, form of Gas1, which transcription is induced by Tetracycline. We demonstrated the release of the mutated protein, as well as its effect inducing glioma cell death [92]. We believe that this form of Gas1 will increase its therapeutic range, because of both the autocrine and paracrine mechanisms of action, that eradicate both cells producing Gas1, and neighboring, non-infected cells.

Conclusion

There is as a need for more effective treatments for gliomas. Gene therapy offers an interesting approach to treat gliomas, and although no clear beneficial clinical effects have been reported yet, current data indicate it is safe for the patients. There are several technical aspects that must be overcome in order for this strategy to be successful. One is related with a more efficient delivery of the transgenes into the tumor, meaning affecting more tumor cells and satellite tumors; this can be achieved by the use of secreted soluble molecules, as well as employing NSCs as vehicles to target the tumors. We consider that both techniques are promising approaches to overcome the previously mentioned hurdles. We also discussed the experimental, pre-clinical, results obtained with Gas1. Moreover, we presented data showing that this protein efficiently eradicates tumors of glial origin, and can also function as a soluble secreted molecule, thus enhancing its therapeutic range, by eliminating not only the cells that express the transgene, but neighboring non-infected cells. Thus, we consider that Gas1 is a potentially relevant adjuvant agent for the treatment of gliomas.

Acknowledgments

This work was partially supported by Conacyt grant 54756 (JS). We thank N. Zarco and R. Navarro for their careful reading and comments to the manuscript.

References

[1] CBTRUS 2010 statistical report: primary brain tumors in the United States, 2004-2006. Central Brain Tumor Registry of the United States, 2000-2004. (Accessed July 2, 2010, http://www.cbtrus.org).

[2] Sanai N, Alvarez-Buylla A, Berger MS. 2005. Neural stem cells and the origin of gliomas. *N Engl J Med.* 25(8): 811-22.

[3] Wen PY, Kesari S. 2008. Malignant gliomas in adults. *N Engl J Med.* 31(5): 492-507.

[4] Edick MJ et al. 2005. Lymphoid gene expression as a predictor of risk of secondary brain tumors. *Genes Chromosomes Cancer*. 42: 107–116.

[5] Ohgaki H and Kleihues P. 2005. Epidemiology and etiology of gliomas. *Acta Neuropathol (Berl).* 109: 93–108.

[6] Ohgaki H. 2009. Epidemiology of brain tumors. Methods Mol Biol. 472:323-42.

[7] Kyritsis AP, Bondy ML, Rao JS, Sioka C. 2010. Inherited predisposition to glioma. *Neuro Oncol.* 12(1):104-13.

[8] Inskip PD, Tarone RE, Hatch EE, Wilcosky TC, Shapiro WR, Selker RG, et al. 2001. *N Engl J Med. Cellular-telephone use and brain tumors*. 11(2):79-86

[9] Libermann TA, Nusbaum HR, Razon N, et al. 1985. Amplification, enhanced expression and possible rearrangement of EGF receptor gene in primary human brain tumours of glial origin. *Nature.* 313:144-147.

[10] James CD, Carlbom E, Nordenskjold M, Collins VP, Cavenee WK. 1989. Mitotic recombination of chromosome 17 in astrocytomas. *Proc Natl Acad Sci U S A.* 86:2858-2862.

[11] Yin S, Van Meir EG. p53 pathway alterations in brain tumors. In: Van Meir EG, ed. *CNS Cancer: Models, Markers, Prognostic Factors, Targets and Therapeutic Approaches.* 1st ed. New York: Humana Press (Springer). 2009:283-314.

[12] Stokoe D, Furnari FB. The PTEN/PI3 kinase pathway in human glioma. In: Van Meir EG, ed. *CNS Cancer: Models, Markers, Prognostic Factors, Targets and Therapeutic Approaches.* 1st ed. New York: Humana Press (Springer). 2009:315-357.

[13] Parsons DW, Jones S, Zhang X, et al. 2008. An integrated genomic analysis of human glioblastoma multiforme. *Science.* 321:1807-1812.

[14] Zhao S, Lin Y, Xu W, et al. 2009. Glioma-derived mutations in IDH1 dominantly inhibit IDH1 catalytic activity and induce HIF- 1alpha. *Science.* 324:261-265.

[15] DeAngelis LM. 2001. Brain tumors. *N Engl J Med.* 11: 114-23.

[16] Forsyth PA, Posner JB. 1993. Headaches in patients with brain tumors: a study of 111 patients. *Neurology.* 43: 1678-83.

[17] Cha S. 2006. Update on brain tumor imaging: from anatomy to physiology. *AJNR Am J Neuroradio.* 27:475-87.

[18] Brat DJ, Van Meir EG. 2004. Vaso-occlusive and prothrombotic mechanisms associated with tumor hypoxia, necrosis, and accelerated growth in glioblastoma. *Lab Invest.* 84:397-405.

[19] Sulman EP, Guerrero M, Aldape K. Transcription profiling of brain tumors: tumor biology and treatment stratification. In: Van Meir EG, ed. *CNS Cancer: Models, Markers, Prognostic Factors, Targets and Therapeutic Approaches.* 1st ed.NewYork: Humana Press (Springer). 2009:529-552.

[20] Baer JC, Freeman AA, Newlands ES, Watson AJ, Rafferty JA and Margison GP: 1993. Depletion of O6-alkylguanine-DNA alkyltransferase correlates with potentiation of temozolomide and CCNU toxicity in human tumour cells. *Br J Cancer.* 67: 1299-1302.

[21] Minniti G, Muni R, Lanzetta G, Marchetti P, Enrici RM. 2009. Chemotherapy for glioblastoma: current treatment and future perspectives for cytotoxic and targeted agents.; *Anticancer Res.* 29(12):5171-84.

[22] Anson, D. S. 2004. The use of retroviral vectors for gene therapy-what are the risks? A review of retroviral pathogenesis and its relevance to retroviral vector-mediated gene delivery. *Genet Vaccines Ther.* 2(1): 9.

[23] Butowski N. 2010. Immunostimulants for malignant gliomas. *Neurosurg Clin N Am.* 21(1):53-65.

[24] Van Gool S, Maes W, Ardon H, Verschuere T, Van Cauter S, De Vleeschouwer S. 2009. Dendritic cell therapy of high-grade gliomas. *Brain Pathol.* 19(4):694-712.

[25] Natsume A, Yoshida J. 2008. Gene therapy for high-grade glioma: current approaches and future directions. *Cell Adh Migr.* 2(3):186-91.

[26] Trojan J, Cloix JF, Ardourel MY, Chatel M, Anthony DD. 2007. Insulin-like growth factor type I biology and targeting in malignant gliomas. *Neuroscience.* 30(3):795-811.

[27] Hamel W, Westphal M. 2003. Gene therapy of gliomas. *Acta Neurochir Suppl.* 88:125-35.

[28] Pulkkanen KJ, Yla-Herttuala S. 2005. Gene therapy for malignant glioma: current clinical status. *Mol Ther.* 12(4):585-98.

[29] Van Meir EG, Hadjipanayis CG, Norden AD, Shu HK, Wen PY, Olson JJ. 2010. Exciting new advances in neuro-oncology: the avenue to a cure for malignant glioma. *CA Cancer J Clin.* 60(3):166-93.

[30] Lu C, Shervington A. 2008. Chemoresistance in gliomas. *Mol Cell Biochem.* 312(1-2):71-80.

[31] Immonen, A., et al. 2004. AdvHSV-tk gene therapy with intravenous ganciclovir improves survival in human malignant glioma: a randomised, controlled study. *Mol. Ther.* 10: 967 – 972.

[32] Vile, R. G., and Russell, S. J. 1995. Retroviruses as vectors. Br. Med. Bull. 51: 12 -30.

[33] Rainov, N. G., and Ren, H. 2003. Clinical trials with retrovirus mediated gene therapy—What have we learned? *J. Neurooncol.* 65: 227 – 236.

[34] Kay, M. A., Glorioso, J. C., and Naldini, L. 2001. Viral vectors for gene therapy: the art of turning infectious agents into vehicles of therapeutics. *Nat. Med.* 7: 33 – 40.

[35] Sandmair, A. M., et al. 2000. Thymidine kinase gene therapy for human malignant glioma, using replication-deficient retroviruses or adenoviruses. *Hum. Gene Ther.* 11: 2197 – 2205.

[36] Danthinne, X., and Imperiale, M. J. 2000. Production of first generation adenovirus vectors: a review. *Gene Ther.* 7: 1707 – 1714.

[37] Romano, G., Michell, P., Pacilio, C., and Giordano, A. (2000). Latest developments in gene transfer technology: achievements, perspectives, and controversies over therapeutic applications. *Stem Cells.* 18: 19 – 39.

[38] Alemany, R., Balague, C., and Curiel, D. T. 2000. Replicative adenoviruses for cancer therapy. *Nat. Biotechnol.* 18: 723 – 727.

[39] Kanerva, A., and Hemminki, A. 2004. Modified adenoviruses for cancer gene therapy. *Int. J. Cancer.* 110: 475 – 480.

[40] Kim SM, Lim JY, Park SI, Jeong CH, Oh JH, Jeong M, et al. 2008. Gene therapy using TRAIL-secreting human umbilical cord blood-derived mesenchymal stem cells against intracranial glioma. *Cancer Res.* 68: 9614-23.

[41] Hacein-Bey-Abina S, Garrigue A, Wang GP, et al. 2008. Insertional oncogenesis in 4 patients after retrovirusmediated gene therapy of SCID-X1. *J Clin Invest.* 118:3132-3142.

[42] Hacein-Bey-Abina S, von Kalle C, Schmidt M, et al. 2003. A serious adverse event after successful gene therapy for X-linked severe combined immunodeficiency. *N Engl J Med.* 348:255-256.

[43] Howe SJ, Mansour MR, Schwarzwaelder K, et al. 2008. Insertional mutagenesis combined with acquired somatic mutations causes leukemogenesis following gene therapy of SCID-X1 patients. *J Clin Invest.* 118:3143-3150.

[44] Landgren H, Curtis MA. 2010. Locating and labeling neural stem cells in the brain. *J Cell Physiol.* Jul 23. [Epub ahead of print]

[45] Hirabayashi Y, Gotoh Y. 2010 Epigenetic control of neural precursor cell fate during development. *Nat Rev Neurosci.* 11(6):377-88.

[46] Uhl, M., M. Weiler, et al. 2005. Migratory neural stem cells for improved thymidine kinase-based gene therapy of malignant gliomas. *Biochem Biophys Res Commun.* 328(1): 125-9.

[47] Muller, F. J., E. Y. Snyder, et al. 2006. Gene therapy: can neural stem cells deliver? *Nat Rev Neurosci.* 7(1): 75-84.

[48] Kim MH, Billiar TR, Seol DW. 2004. The secretable form of trimeric TRAIL, a potent inducer of apoptosis. *Biochem Biophys Res Commun. 321*: 930-5.

[49] Hingtgen S, Ren X, Terwilliger E, Classon M, Weissleder R, Shah K. 2008. Targeting multiple pathways in gliomas with stem cell and viral delivered S-TRAIL and Temozolomide. *Mol Cancer Ther*. 7: 3575-85.

[50] Menon LG, Kelly K, Yang HW, Kim SK, Black PM, Carroll RS. 2009. Human bone marrow-derived mesenchymal stromal cells expressing S-TRAIL as a cellular delivery vehicle for human glioma therapy. *Stem Cells.* 27: 2320-30.

[51] Schneider C, King RM, Philipson L. 1988. Genes specifically expressed at growth arrest of mammalian cells. *Cell.* 9;(6):787-93.

[52] Ruaro, M. E., M. Stebel, et al. 2000. Analysis of the domain requirement in Gas1 growth suppressing activity. *FEBS Lett.* 481(2): 159-63.

[53] Stebel, M., P. Vatta, et al. 2000. The growth suppressing gas1 product is a GPI-linked protein. *FEBS Lett.* 481(2): 152-8.

[54] Lee, C. S. and C. M. Fan 2001. Embryonic expression patterns of the mouse and chick Gas1 genes. *Mech Dev.* 101(1-2): 293-7.

[55] Liu, Y., N. R. May, et al. 2001. Growth arrest specific gene 1 is a positive growth regulator for the cerebellum. *Dev Biol.* 236(1): 30-45.

[56] Schueler-Furman, O., E. Glick, et al. 2006. Is GAS1 a co-receptor for the GDNF family of ligands? *Trends Pharmacol Sci.* 27(2): 72-7

[57] López-Ramirez, M. A., G. Dominguez-Monzon, et al. 2008. Gas1 reduces Ret tyrosine 1062 phosphorylation and alters GDNF-mediated intracellular signaling. *Int J Dev Neurosci.* 26(5): 497-503.

[58] Lee CS, Buttitta L, Fan CM. 2001. Evidence that the WNT-inducible growth arrest-specific gene 1 encodes an antagonist of sonic hedgehog signaling in the somite. *Proc Natl Acad Sci U S A.* 25(20):11347-52.

[59] Allen, B. L., T. Tenzen, et al. 2007. The Hedgehog-binding proteins Gas1 and Cdo cooperate to positively regulate Shh signaling during mouse development. *Genes Dev.* 21(10): 1244-57.

[60] Martinelli, D. C. and C. M. Fan. 2007. Gas1 extends the range of Hedgehog action by facilitating its signaling. *Genes Dev.* 21(10): 1231-43.

[61] Martinelli, D. C. and C. M. Fan. 2007. The role of Gas1 in embryonic development and its implications for human disease. *Cell Cycle.* 6(21): 2650-5.

[62] Seppala, M., M. J. Depew, et al. 2007. Gas1 is a modifier for holoprosencephaly and genetically interacts with sonic hedgehog. *J Clin Invest.* 117(6): 1575-84.

[63] Ribeiro LA, Quiezi RG, Nascimento A, Bertolacini CP, Richieri-Costa A. 2010. Holoprosencephaly and holoprosencephaly-like phenotype and GAS1 DNA sequence changes: Report of four Brazilian patients. *Am J Med Genet A.* 152A (7):1688-94.

[64] Domínguez-Monzón G, Benítez JA, Vergara P, Lorenzana R, Segovia J. 2009. Gas1 inhibits cell proliferation and induces apoptosis of human primary gliomas in the absence of Shh. *Int J Dev Neurosci.* 27: 305-13.

[65] Evdokiou A, Cowled PA. 1998. Tumor-suppressive activity of the growth arrest-specific gene *GAS1* in human tumor cell lines. *Int J Cancer.* 75:568-77.

[66] Gobeil S, Zhu X, Doillon CJ, Green MR. 2008. A genome-wide shRNA screen identifies GAS1 as a novel melanoma metastasis suppressor gene. *Genes Dev.* 22: 2932-40.

[67] Zamorano A, Lamas M, Vergara P, Naranjo JR, Segovia J. 2003. Transcriptionally mediated gene targeting of Gas1 to glioma cells elicits growth arrest and apoptosis. *J Neurosci Res.* 71: 256-63.

[68] Zamorano A, Mellström B, Vergara P, Naranjo JR, Segovia J. 2004. Glial-specific retrovirally mediated gas1 gene expression induces glioma cell apoptosis and inhibits tumor growth in vivo. *Neurobiol Dis.* 5: 483-91.

[69] Benítez JA, Arregui L, Vergara P, Segovia J. 2007. Targeted-simultaneous expression of Gas1 and p53 using a bicistronic adenoviral vector in gliomas. *Cancer Gene Ther.* 14: 836-46.

[70] Facoetti A, Ranza E, Nano R. 2008. Proliferation and programmed cell death: role of p53 protein in high and low grade astrocytoma. *Anticancer Res.* 28(1A):15-9.

[71] Shirsat NV, Shaikh SA. 2003. Overexpression of the immediate early gene fra-1 inhibits proliferation, induces apoptosis, and reduces tumourigenicity of c6 glioma cells. *Exp Cell Res.* 15;(1):91-100.

[72] Biglari A, Bataille D, Naumann U, Weller M, Zirger J, Castro MG, Lowenstein PR. 2004. Effects of ectopic decorin in modulating intracranial glioma progression in vivo, in a rat syngeneic model. *Cancer Gene Ther.* 11(11):721-32.

[73] Chiocca EA, Abbed KM, Tatter S, Louis DN, Hochberg FH, Barker F, Kracher J, Grossman SA, Fisher JD, Carson K, Rosenblum M, Mikkelsen T, Olson J, Markert J, Rosenfeld S, Nabors LB, Brem S, Phuphanich S, Freeman S, Kaplan R, Zwiebel J. 2004. A phase I open-label, dose-escalation, multi-institutional trial of injection with an E1B-Attenuated adenovirus, ONYX-015, into the peritumoral region of recurrent malignant gliomas, in the adjuvant setting. *Mol Ther.* Nov; 10(5):958-66.

[74] Harsh GR, Deisboeck TS, Louis DN, Hilton J, Colvin M, Silver JS, Qureshi NH, Kracher J, Finkelstein D, Chiocca EA, Hochberg FH. 2000. Thymidine kinase activation of ganciclovir in recurrent malignant gliomas: a gene-marking and neuropathological study. *J Neurosurg.* 92(5):804-11.

[75] Ram Z, Culver KW, Oshiro EM, Viola JJ, DeVroom HL, Otto E, Long Z, Chiang Y, McGarrity GJ, Muul LM, Katz D, Blaese RM, Oldfield EH. 1997. Therapy of malignant brain tumors by intratumoral implantation of retroviral vector-producing cells. *Nat Med.* 3(12):1354-61

[76] Sandmair AM, Loimas S, Puranen P, Immonen A, Kossila M, Puranen M, Hurskainen H, Tyynelä K, Turunen M, Vanninen R, Lehtolainen P, Paljärvi L, Johansson R, Vapalahti M, Ylä-Herttuala S. 2000. Thymidine kinase gene therapy for human malignant glioma, using replication-deficient retroviruses or adenoviruses. *Hum Gene Ther.* 1(16):2197-205.

[77] Trask TW, Trask RP, Aguilar-Cordova E, Shine HD, Wyde PR, Goodman JC, Hamilton WJ, Rojas-Martinez A, Chen SH, Woo SL, Grossman RG. 2000. Phase I study of adenoviral delivery of the HSV-tk gene and ganciclovir administration in patients with current malignant brain tumors. *Mol Ther.* 1(2):195-203.

[78] Germano IM, Fable J, Gultekin SH, Silvers A. 2003. Adenovirus/herpes simplex-thymidine kinase/ganciclovir complex: preliminary results of a phase I trial in patients with recurrent malignant gliomas. *J Neurooncol.* 65(3):279-89.

[79] Shand N, Weber F, Mariani L, Bernstein M, Gianella-Borradori A, Long Z, Sorensen AG, Barbier N. 1999. A phase 1-2 clinical trial of gene therapy for recurrent glioblastoma multiforme by tumor transduction with the herpes simplex thymidine kinase gene followed by ganciclovir. GLI328 European-Canadian Study Group. *Hum Gene Ther.* 20(14):2325-35.

[80] Smitt PS, Driesse M, Wolbers J, Kros M, Avezaat C. 2003. Treatment of relapsed malignant glioma with an adenoviral vector containing the herpes simplex thymidine kinase gene followed by ganciclovir. *Mol Ther.* 7(6):851-8.

[81] Prados MD, McDermott M, Chang SM, Wilson CB, Fick J, Culver KW, Van Gilder J, Keles GE, Spence A, Berger M. 2003. Treatment of progressive or recurrent glioblastoma multiforme in adults with herpes simplex virus thymidine kinase gene vector-producer cells followed by intravenous ganciclovir administration: a phase I/II multi-institutional trial. *J Neurooncol.* 65(3):269-78.

[82] Izquierdo M, Martín V, de Felipe P, Izquierdo JM, Pérez-Higueras A, Cortés ML, Paz JF, Isla A, Blázquez MG. 1996. Human malignant brain tumor response to herpes simplex thymidine kinase (HSVtk)/ganciclovir gene therapy. *Gene Ther.* 3(6):491-5.

[83] Rainov NG. 2000. A phase III clinical evaluation of herpes simplex virus type 1 thymidine kinase and ganciclovir gene therapy as an adjuvant to surgical resection and radiation in adults with previously untreated glioblastoma multiforme. *Hum Gene Ther.* 20(17):2389-401.

[84] Wakabayashi T, Natsume A, Hashizume Y, Fujii M, Mizuno M, Yoshida J. 2008. A phase I clinical trial of interferon-beta gene therapy for high-grade glioma: novel findings from gene expression profiling and autopsy. *J Gene Med.* 10(4):329-39.

[85] Okada H, Lieberman FS, Edington HD, Witham TF, Wargo MJ, Cai Q, Elder EH, Whiteside TL, Schold SC Jr, Pollack IF. 2003. Autologous glioma cell vaccine admixed with interleukin-4 gene transfected fibroblasts in the treatment of recurrent glioblastoma: preliminary observations in a patient with a favorable response to therapy. *J Neurooncol.* 64(1-2):13-20.

[86] Parney IF, Chang LJ, Farr-Jones MA, Hao C, Smylie M, Petruk KC. 2006. Technical hurdles in a pilot clinical trial of combined B7-2 and GM-CSF immunogene therapy for glioblastomas and melanomas. *J Neurooncol.* 78(1):71-80.

[87] Barzon L, Pacenti M, Franchin E, Colombo F, Palù G. 2009. HSV-TK/IL-2 gene therapy for glioblastoma multiforme. *Methods Mol Biol.* 542:529-49.

[88] Rampling R, Cruickshank G, Papanastassiou V, Nicoll J, Hadley D, Brennan D, Petty R, MacLean A, Harland J, McKie E, Mabbs R, Brown M. 2000. Toxicity evaluation of replication-competent herpes simplex virus (ICP 34.5 null mutant 1716) in patients with recurrent malignant glioma. *Gene Ther.* 7(10):859-66.

[89] Markert JM, Medlock MD, Rabkin SD, Gillespie GY, Todo T, Hunter WD, Palmer CA, Feigenbaum F, Tornatore C, Tufaro F, Martuza RL. 2000. Conditionally replicating herpes simplex virus mutant, G207 for the treatment of malignant glioma: results of a phase I trial. *Gene Ther.* 7(10):867-74.

[90] Papanastassiou V, Rampling R, Fraser M, Petty R, Hadley D, Nicoll J, Harland J, Mabbs R, Brown M. 2002. The potential for efficacy of the modified (ICP 34.5(-)) herpes simplex virus HSV1716 following intratumoural injection into human malignant glioma: a proof of principle study. *Gene Ther.* 9(6):398-406.

[91] Harrow S, Papanastassiou V, Harland J, Mabbs R, Petty R, Fraser M, Hadley D, Patterson J, Brown SM, Rampling R. 2004. HSV1716 injection into the brain adjacent to tumour following surgical resection of high-grade glioma: safety data and long-term survival. *Gene Ther.* 11(22):1648-58.

[92] López-Ornelas A, Mejía-Castillo T, Vergara P and Segovia J. 2010. Lentiviral transfer of an inducible transgene expressing a soluble form of Gas1 causes glioma cell arrest, apoptosis and inhibits tumor growth. Cancer Gene Ther. doi:10.1038/cgt.2010.54

In: Brain Cancer, Tumor Targeting and Cervical Cancer
Editor: Elena K. Salvatti
ISBN: 978-1-61122-738-3

Chapter VIII

Congenital Tumors of the Central Nervous System

Melinda Czeh[1,2,*] ***and Angela M. Kaindl***[1,2]
[1]Department of Pediatric Neurology, Charité – Universitätsmedizin Berlin, Campus Virchow-Klinikum, Augustenburger Platz 1, 13353 Berlin, Germany
[2]Institute of Neuroanatomy and Cell Biology, Charité – Universitätsmedizin Berlin, Campus Mitte, Charitéplatz 1, 10117 Berlin, Germany

Abstract

Congenital tumors of the central nervous system are rare but have a poor prognosis due to their size and location as well as the contraindication of radiotherapy at this age. Various classifications have been described based on the age of the patient at clinical manifestation. The short latency between tumor onset and clinical signs suggests a genetic predisposition, as has been described for some hereditary tumor syndromes that manifest in infancy (astrocytoma in tuberous sclerosis, choroid plexus tumor in Li Fraumeni syndrome, medulloblastoma in Gorlin syndrome, atypical rhabdoid/teratoid tumor in rhabdoid tumor predisposition syndrome). Here, we describe the clinical signs, pre- and postnatal diagnostic paradigms, histopathological and genetic characterization of the most frequent congenital tumors of the central nervous system: teratomas, astrocytomas, choroid plexus papillomas, primitive neuroectodermal tumors, medulloblastomas, atypical rhabdoid/terotoid tumors, craniopharyngiomas. Treatment options and pitfalls will be discussed.

Keywords: CNS tumors, brain, teratoma, astrocytoma, choroid plexus papilloma, PNET, ATRT, medulloblastoma, craniopharygioma, lipoma

* Corresponding author: Melinda Czeh MD, PhD, Department of Pediatric Neurology, Charité – Universitätsmedizin Berlin, Campus Virchow Klinikum, Augustenburger Platz 1, 13353 Berlin, Germany. Tel/ Fax: +49 30 450 566112/ 920. Email: melinda.czeh@charite.de.

Table 1. Familial syndromes associated with CNS tumors

Syndrome	CNS Manifestations	OMIM	Gene	Chromosome
Neurofibromatosis type 1 (AD)	Optic pathway glioma (pilocytic astrocytoma)	#162200	*NF1*	17q11.2
Tuberous sclerosis (AD)	Subependymal giant cell astrocytoma	#191100 #613254	*TSC1* *TSC2*	9q34 16p13.3
Li Fraumeni (AD)	Choroid plexus papilloma Medulloblastoma Astrocytoma	#151623 +604373	*TP53* *CHEK2*	17p13.1 22q12.1
Rhabdoid tumor predisposition syndrome (AD)	ATRT	#609322	*INI1*	22q11
Gorlin syndrome (AD)	Medulloblastoma	*601309	*PTCH1*	9q22.3
Von Hippel-Lindau syndrome (AD)	Hemangioma of cerebellum, spinal cord	#193300	*VHL*	3p26-p25

Abbreviations: AD, autosomal dominant; NF1, neurofibromin 1; TSC, tuberous sclerosis complex; TP53, tumor protein p53; CHEK2, checkpoint kinase 2; ATRT, atypical teratoid/rhabdoid tumor; INI1, integrase interactor 1; PTCH1, drosophila homolog of patched 1; VHL, von Hippel-Lindau.

Introduction

Congenital tumors of the central nervous system (CNS) are rare, but have a poor prognosis. The true prevalence of fetal tumors is difficult to determine since it varies widely among institutions [1,2]. Furthermore, the fetal prevalence is surely much higher, given the undetected cases by *in utero* deaths, stillbirth and terminated pregnancies. The prevalence for all congenital tumors lies within the range of 1,7-13,3 per 100 000 live births [2,3]. CNS tumors account for about 10% of all congenital tumors ranking behind neuroblastomas, extracranial teratomas and soft-tissue tumors [2]. They represent approximately 0,5-1,9% of all pediatric brain tumors [2,3]. Still, their size and location as well as the contraindication of radiotherapy at this age lead to high mortality rates of approximately 5-20% of deaths in the fetal and neonatal period [1].

Definition

There are various definitions of congenital CNS tumors: While a tumor that produces symptoms at birth is doubtlessly congenital, the classification of tumors that are diagnosed at later age is under discussion. Arnstein *et al.* defined CNS tumors as congenital if they were diagnosed within the first 2 months of life [4]. Solitare and Krigman, on the other hand, defined three groups of congenital CNS tumors: (i) *definitely congenital* tumors producing symptoms at birth, (ii) *probable congenital* tumors producing symptoms within the first week of life and (iii) *possible congenital* tumors producing symptoms within the first months of life [5]. In subsequent studies the definition of the third group was often modified, limiting the age of patients to be included to 2 [6,7] or 3 months [3]. Congenital tumors of the CNS differ from those observed in older childen and adolescents with regard to prevalence, clinical signs, location, biological behavior, radiological signs, histological charactestics, response to therapy and prognosis [1,8]. The most common congenital tumors of the CNS are teratomas, accounting for approximately half of all reported cases [2]. They are followed in frequency by astrocytomas, choroid plexus papillomas, primitive neuroectodermal tumors (PNET), atypical teratoid/rhabdoid tumors (ATRT), medulloblastomas, craniopharyngiomas and lipomas [2,3]. In the following, we discuss common clinical presentation *in utero* or in the neonatal age and review current knowledge on the most common congenital CNS tumors. Various hereditary syndromes are associated with an increased risk of developing CNS tumors (see Table 1) [9,10,11,12,13,14,15,16,17,18].

Clinical Symptoms, Prenatal Diagnosis and Complications

Congenital tumors are often diagnosed incidentally by routine ultrasound scanning during pregnancy [8]. In case of detection of a congenital brain tumor by ultrasound scanning, high-resolution magnetic resonance (MR) imaging can further advance the precisely diagnosis [8]. While the tumor is growing, patients are still *in utero* or have just been born. Therefore, clinical signs differ from those in older children. Congenital tumors are most frequently detected in the third trimester of pregnancy [3]. Polyhydramnion and large-for-gestational age uterus are common signs noticed prenatally in approximately one third of the cases [1,3]. A common initial symptom observed in the fetus or newborn is macrocephaly (in about 60% of symptomatic cases), caused either directly by the tumor or secondarily through a hydrocephalus [1,3,6]. The hydrocephalus can be the result of a compression of the ventricular system by the tumor, by a hemorrhage or by hypersecretion of cerebrospinal fluid (CSF) [1], and it may be more life threatening than the tumor itself [8]. According to the current opinion, delivery should be induced if the fetus is affected severely by a hydrocephalus and has reached a gestation age of more than 34 weeks. If the gestation age is below 34 weeks and the fetal condition is critical, fetal lung maturation should be induced in preparation of delivery as soon as possible. Dexamethason, used to accelerate fetal lung maturation, also reduces the cerebral edema produced by the tumor [8] but may have negative effects on brain development. Prior to delivery induction, repeated cephalocenteses can decrease fetal intracranial pressure [8]. Since the pressure by the mother´s abdominal wall during uterine contraction is transmitted to the intracranial cavity of the fetus, raising the intracranial pressure, fetuses with brain tumors and/or hydrocephalus should be delivered by cesarean section [8]. The signs and symptoms of *increased* intracranial pressure in *infants* differ from those of older children because of the presence of open *fontanelles*. In infants, increased intracranial pressure through rapidly growing CNS tumors is often associated with separation of cranial sutures and bulging fontanelles; in progressed stages vomiting and lethargy may appear [1]. Other neurological symptoms and signs such as seizures, hemiparesis and cranial nerve abnormalities are uncommon in newborns with congenital tumors.

An accurate prenatal diagnosis is crucial for an appropriate planning of the further treatment possibilities and/or, if treatment is not possible, for counseling parents regarding outcome and the possibility of a termination of pregnancy. The most important prenatal differential diagnosis of a brain tumor is an intracranial hemorrhage, which can also present as a disorganized intracranial mass with an accompanying hydrocephalus. This differential diagnosis is even more complicated as congenital brain tumors are often accompanied by intracranial hemorrhage [3]. Thus an underlying brain tumor should always be considered in case of intracranial hemorrhage [2,3]. Such hemorrhages have been reported in 3 to 18% of patients with congenital brain tumors [1,6,7]. Regardless of the tumor type, the incidence of hemorrhage in newborn with congenital tumors was much higher than in older child and adult, probably due to the larger size of the tumor in the neonatal period [6] and to the rapidly changing forces during delivery [1]. Large tumors can cause stillbirth and dystocia and require caesarean section (especially by intracranial teratoma, glioblastomas, PNET) [1]. Moreover, large benign masses with massive blood perfusion may result in cardiovascular compromise and inhibit normal organ development [3]. Based of the high mitotic potential of embryonic cells, fetal tumors grow very fast and reach enormous sizes, regardless whether they are benign or malignant. It is possible that a large mass develops within weeks following a normal ultrasound screen [2].

Imaging studies (ultrasound, MR) are very helpful to differentiate early on between curable tumors such as choroid plexus tumors and fatal ones such as teratoma [1]. The majority of fetal and neonatal intracranial tumors are located supratentorially in contrast to the intracranial tumors in older children and adolescent in whom they are most commonly infratentorial [1]. Almost half of the perinatal CNS tumors are teratomas, a finding that is uncommon in older patients [1,2]. Associated congenital anomalies are present in 14-20% of all cases with CNS tumors (particularly with teratomas), most common as cleft lip or cleft palate [2,3]. Lipomas have a strong association with agenesis of corpus callosum [2].

The prognosis of congenital CNS tumors is universally poor because of their size and location [1,3], even though they are histologically benign [2]. The outcome is dependent on the size, location, surgical resectability, histologic type and condition of the newborn [19]. Infants with choroid plexus papillomas, low-grade astrocytomas, lipomas and ganglioglioma have much better prognosis than fetuses with other congenital brain tumors [2,19]. In contrast, teratomas, primitive neuroectodermal tumors, craniopharyngioma and glioblastomas are often fatal [2]. Treatment of congenital brain tumors is predominantly surgical, but because of the large tumor masses and the condition of the infant, complete surgical resection of the tumor is often not possible. Unresectable tumors may be treated with chemotherapy. Irradiation is generally contraindicated because of the detrimental effects on brain development [3]. Since the presence of large tumor mass and the treatment possibilities disrupts normal neuronal development, patients often suffer from psychomotor deficits despite the removal of CNS tumor and successful therapy [2]. Cognitive deficits are reported in 40-100% of long-term survivors [19,20]. Two important studies performed with an interval of 9 years reflect the improved survival due to better treatment possibilities, but also the remaining poor prognosis: Davis and Raisanen reported a 11% survival rate in a cohort of 225 patients in 1993 (56% of patients with choroid plexus tumors, 10% of those with astrocytomas, 12% of those with medulloblastomas and 14% of those with meningeoma) [21]. In contrast, Isaacs *et al* reported a 28% survival of 250 patients in 2002 (73% of patients with choroid plexus tumors, 34% of those with astrocytoma, 36% of those with meningeoma and 5% of those with medulloblastoma) [19].

Teratoma

Teratomas are the most common congenital CNS tumors, reported as frequent as one third to one half of all cases, in contrast to older children (9% of all brain tumors in children younger than 2 years and 2% in children younger than 15 years) [8]. Teratomas are generally large, complex, mixed cystic and solid tumor masses with or without foci of calcification [2,7]. These tumors are composed of all three germ cell layers: the dominant ectodermal layer (neural tissue), the mesodermal layer (fat, bone, cartilage and muscle) and the endodermal layer (respiratory and gastrointestinal tissue). Teratomas can be mature or immature. Mature teratomas consist exclusively of fully differentiated tissue, while immature teratomas contain incompletely differentiated tissues. Immature teratomas can also contain malignant tissues such as embryonal carcinomas or yolk sac tumor [3]. The immaturity of the congenital teratoma may be more reflective of the immaturity of the fetus and not necessarily reflect the biological behavior of the tumor. The prognosis is influenced more by the location and size of the congenital teratoma than its histology [2]. In some cases, a malformed fetus can be found in a teratoma, and this is considered as a *fetus-in-fetu* if a vertebral column is present [8]. Teratoma can be localized in the sacrococcygeal region, in the head and neck region or in the intracranial cavity. The sacrococcygeal region is the most common site for teratomas, 70-80% of all teratomas originate from this region [2]. Sacrococcygeal teratomas are classified according to the American Acedemy of Pediatrics Surgery Section Survey into four categories depending on the location (externally versus internally) of tumor mass. Type I of sacrococcygeal teratoma contains predominantly external (sacrococcygeal) component with only a minimal internal (presacral) component. Type II tumors are external with a significant intrapelvic extension. Type III tumors are visible externally but the dominant mass is pelvic and extends into the abdomen. Type IV tumors are presacral with no external presentation [2,22]. This classification has an important relevance, since sacrococcygeal tumors with an undiagnosed internal part can undergo malignant transformation [2,22]. Malignant transformation of a sacrococcygeal teratoma occurs in 5-10% of cases diagnosed in the first 2 months and up to 90% diagnosed at age of 2-4 months [2,3]. The most important prognostic indicator is the size of solid component of the tumor mass [2]. They can be associated with anomalies, most frequent with renal dysplasia, hydronephrosis, urethral atresia, undescended testes, rectal atresia and stenosis and hip dislocation [2,23,24,25]. The highest incidence of urological complications (vesicoureteral reflux, ureteral and urethral obstruction and neurogenic bladder) was seen in patients with type IV sacrococcygeal teratoma [23]. In the head and neck region, teratomas can develop in the thyrocervical area, the palate, the nasopharynx and the orbita. Intracranial teratomas often arise in the midline, predominantly from the pineal region and the hypothalamic area and involve the third ventricle [2]. They also can originate from the suprasellar region, the cerebral hemispheres, the ventricles and the posterior fossa [1,8]. Such tumors can be smaller and 'merely' produce a hydrocephalus or presenting as an incidental postmortem finding in a stillborn fetus. However, they can also be large with the tumor mass replacing intracranial tissues and extending into the orbita and the neck [1,3]. In large tumor masses, the origin often cannot be determined anymore, especially if, as in at least one third of such cases, the brain is replaced by the tumor [1]. Prognosis of infants with congenital brain tumors is very poor, and patients with congenital teratomas have the lowest survival rates of all infants with perinatal brain tumors [1]. Most infants die before or shortly after birth. According to a review by Isaacs [1], the survival rate of patients with intracranial teratoma is less than 10%. The prognosis worsens with increasing size of the tumor and with decreasing age of the fetus at time of diagnosis [1]. The mortality rate for infants with congenital sacrococcygeal teratoma is as high as 50% [2].

Astrocytoma

Astrocytomas are the second most frequent congenital tumors of the CNS, accounting for 18-47% of all congenital tumors [3]. They are neuroglial tumors composed of astrocytes, varying from the well-differentiated low-grade pilocytic astrocytomas to the poorly differentiated glioblastoma multiforme [2,3]. Similar to other congenital CNS tumors, astrocytomas diagnosed perinataly differ from those found in older children regarding their location, clinical manifestation and histological features. Cerebellar pilocytic astrocytoma have not been found in fetuses, although typically found in pediatric population [2,3]. Astrocytomas originate from cerebral hemispheres, optic nerves, thalamus-hypothalamus, mesencephalon and pons [1]. Astrocytomas from the cerebral hemisphere often involve more than one lobe, and displace the lateral ventricle [1]. The low-grade pilocytic astrocytomas arise most common in the optic nerve and hypothalamic/chiasmatic region [1]. There is a high association (up to 50%) between optic pathway pilocytic astrocytoma and neurofibromatosis type I (Table 1) and is prognostically unfavorable [1,9]. The frequent combination of mental deterioration, EEG abnormalities and seizures in these patients indicate the presence of more diffuse pathological changes in the brain than those caused by the tumor [9]. Previous studies shown, that there is no correlation between the extent of surgical resection, irradiation and the length of survival of patients by optic astrocytoma. According to the low grade of the tumor, the long period of dormancy and the spontaneous regression, conservative treatment should be the best option in these cases [1,9]. Certain forms of low-grade astrocytomas, like the subependymal giant cell astrocytomas (SEGA) associated with tuberous sclerosis (Table 1) arise most frequently in the first and second decades of life, are not a common diagnosis in the perinatal period [3]. Tuberous sclerosis (TS) is a familial autosomal dominant disease. It is commonly accepted that mutations in the two genes *TSC1* (tuberous sclerosis complex) and *TSC2* are responsible for the disease, at least in a large proportion of patients (Table 1) [10]. SEGA arise almost exclusively around the Monroi foramina. Although they are benign in nature and grow slowly, they can obstruct cerebrospinal fluid pathways leading to hydrocephalus [26,27,28]. Newborns with tuberous sclerosis die more often of their cardiac complications (rhabdomyoma) than of the intracranial manifestation [1]. Anaplastic astrocytomas of intermediate-grade malignancy (WHO grade II-III) are diffuse infiltrating lesions that are predominantly located in the cerebral hemispheres [3]. Newborns with anaplastic astrocytomas have a better prognosis than those with glioblastoma multiforme [3]. Some long-term survivors, treated only with surgical resection without adjuvant chemotherapy and radiotherapy, have been reported [3,29,30]. Similarly, congenital glioblastomas are very rare and have a poor prognosis; the latter is, however, better than in adult patients [3]. They derive from the cerebral hemisphere and basal nuclei [1]. They grow very rapidly, resulting in a large unilateral tumor mass with a possible shift of midline structures, obstructive hydrocephalus and hemorrhage [1]. Glioblastomas are immunoreactive with GFAP and unreactive with synaptophysin. The overall survival for newborns with congenital astrocytomas is around 30%. According to the review of 250 cases with brain tumors by Isaacs *et al*, 37% of the patients with congenital histologically benign astrocytomas survived. In contrast, in the group of malignant astrocytomas, only 14% of patients survived [1].

Choroid Plexus Papilloma

Choroid plexus papilloma (CPP) are more common in the newborn than in older children. They account for 5-20% of all perinatal CNS tumors, however, only 0,5-3% of intracranial tumors in children [2,3]. More than 80% of the congenital CPPs are localized in the lateral ventricle in contrast to those detected in older children and in adults, in whom they are often localized in the fourth ventricle [3]. Since CPPs often overproduce CSF, they lead to communicating hydrocephalus and raise intracranial pressure [2,3]. They were frequently found in Li-Fraumeni syndrome (a cancer-predisposing syndrome by mutation of p53) (Table 1) [3,11]. CPPs are WHO grade I tumors composed of mature epithelial cells and characterized by highly vascularization. Radiological signs are well-defined, lobular, homogenous intraventricular masses, which can be differentiated from other tumors that are usually heterogenous [2,3]. It is important to differentiate choroid plexus papilloma from the much more common choroid plexus cyst [2] and from the less frequent but much more malignant choroid plexus carcinoma [2]. Most choroid plexus carcinomas occur in the lateral ventricles, similar to the benign chorid plexus papilloma. However, unfortunately they disseminate throughout the CSF early on so that metastases may already be present at the time of diagnosis [19]. Choroid plexus papillomas have one of the best survival rates of all neonatal brain tumors. If they are fully resectable they have a good prognosis [7]. Isaacs *et al* reported about 73% survival rate of the 33 patients with these tumors. Moreover, three of these patients had carcinomas [19]. Choroid plexus papillomas spread only very rarely into the CSF [3,31]. The survival rate of choroid plexus carcinomas are improving with therapy combining surgery, chemotherapy and radiation, allowing children with a subtotal resection to have prolonged disease-free intervals, as reported by Duffner *et al.* Respective patients had a 3-year survival rate of around 70% [19,32].

Embryonal Tumors

The group of embryonal tumors encompasses primitive neuroectodermal tumors (PNET), atypical teratoid/rhabdoid tumors (ATRT) and medulloblastomas. All of these tumors are highly malignant (WHO grade IV) and generally have a bad prognosis through metastasis in the CSF and invasion of the meninges and surrounding cerebral tissue and spinal cord [1,3]. Some of them metastasize not only in the CNS, but also into distant organs (lungs, liver, bone marrow and lymph nodes) [33,34].

PNET are more common in the first year of life (up to 27% of CNS tumors in children) [35] than in the newborn (3,4-13,2% of fetal CNS tumors) [2,3]. They originate from the cerebral hemisphere, brainstem, suprasellar region, pineal region and spinal cord. Upon imaging, they are heterogeneous tumor masses and are often associated with hemorrhage and hydrocephalus [36]. They contain highly malignant small tumor cells, which are undifferentiated or poorly differentiated neuroepithelial cells [3]. Based on the aggressiveness of the tumor, the prognosis is very bad. According to the review by Isaacs et al. 2 out of 7 patients with PNET survived following isolated surgery, and one further patient who also had received chemotherapy survived [1]. In further studies, few survivors have been reported who received chemotherapy in addition to surgery [1,37,38].

ATRT of the CNS are rare, highly malignant embryonal brain tumor associated with poor prognosis [39]. They represent 1,3% of the CNS tumors in the general pediatric population and 6,7% of those younger than 2 years [39,40]. ATRT can be found in several organs including the CNS (also lungs, liver, kidney, skin). ATRTs of the CNS are more commonly found in the infratentorial compartment, in the posterior fossa, predominantly in the cerebellum/ cerebellopontine angle [3,39]. Based on the aggressiveness of ATRTs, metastases via leptomeningeal dissemination into the CSF is common and can be detected in about 14-20% of patients at the time of diagnosis [3,39]. Tumors invading the cerebellopontine angle may be associated with VI and VII cranial nerve palsies [39]. On imaging, ATRTs are heterogeneous masses with hemorrhagic and necrotic area and marginal cysts. They can rapidly enlarge [3]. It is important to distinguished ATRT from other embryonal tumors (PNET, medulloblastomas) because the diagnosis is predictive for the prognosis. They can be characterized by the presence of rhabdoid cells, which contain an eccentric round nucleus with a prominent nucleolus, an eosinophilic cytoplasm and pale intracytoplasmic rhabdoid inclusions. Morphological features of these rhabdoid cells, however, vary: some are small and spindle shaped with an ovoid nucleus, while others are large [19,41]. ATRTs of the CNS are characterized by monosomy 22 or a partial deletion of chromosome band 22q11.2, which contains the tumor suppressor gene *hSNF5/INI1* [3,42]. The latter has been detected in about 76% of both CNS and extra-CNS ATRTs [3]. Germline mutation in patients with a familiar syndrome, the rhabdoid tumor predisposition syndrome, can cause ATRT in several organs (Table 1) [3,14,15]. Because of the poor prognosis, multimodality therapy (resection followed by local or craniospinal irradiation and adjuvant chemotherapy) is commonly applied [39]. All of the treatment possibilities have their limitations. Total resection is often difficult due to the size of the tumor, the surrounding infiltration and location. However, it is important, that long-term survivors have generally undergone gross total excision [9,43]. Based on the poor prognosis and frequent local recurrences, radiation therapy is an important part of the therapy, despite its side effects. 75% of patients who survived longer than 18 months had at least local radiotherapy [39,44]. Based on brain invasion, recurrence and spreading into the CNS, prognosis is still poor, although survival of the patients improved during the last years. Isaacs *et al.* reported an overall survival of 6-11 months in 2002 [19], which is similar to the one reported by Rorke *et al.* in 1996 [41]. However, based on a multimodality therapy with complete resection in about 70% of the cases, chemotherapy (intrathecal) and radiation therapy, Hilden reported about 33% median event free survival of 42 children at 42 months from diagnosis [43]. A recent study from the Boston Children´s Hospital, reported about a 50% 2-year progression-free survival rate and an about 70% overall survival rate using an intensive multimodality therapy approach [39,45]

Medulloblastoma have been separated from supratentorial PNET in the WHO classification of CNS tumors due to their differing biological behavior and treatment responses [3]. Similar to PNET, they account for up to 25% of CNS tumors in children.

However, they are not common in the newborn [3,46,47]. The most common clinical signs of congenital medulloblastomas are macrocephaly and hydrocephalus [1]. Medulloblastomas originate from the vermis of the cerebellum and grow into the fourth ventricle and cerebellar hemispheres [1], spreading in the CSF and metastasizing in 5-18% of cases through invasion of blood vessels into organs outside of the CNS (liver, lung, bone marrow, lymph nodes) [1,34,46]. Congenital anomalies in newborns with medulloblastomas have been described and were most frequent omphalocele, cleft palate, imperforate anus, myelomeningocele and cerebellar agenesis [1]. From the five histological variants of medulloblastomas (classic, desmoplastic, extensive nodularity, large cell and anaplastic), congenital medulloblastomas with extensive nodularity (MBEN) is the most common one [3,16]. This form has a much better prognosis than medulloblastomas in general [16]. MBEN is often present in patients with Gorlin syndrome (naevoid basal cell carcinoma syndrome) and is the exclusively form of medulloblastoma in these patients [3]. Gorlin syndrome is an autosomal-dominant syndrome caused by mutations in the *PTCH1* (drosophila homolog of patched 1) gene on chromosome 9 and is associated with a predisposition to various tumors including basal cell carcinomas, bone cysts, fibroma, rhabdomyosarcoma and MBEN [3,16,17]. The prognosis is poor with a 3-year progression-free survival of about 20% and a 5-year progression-free survival was around 30%, as resported by the Children´s Cancer Group in 1994 and in the Baby POG I (Pediatric Oncology Group) in 2009, respectively [48,49].

Craniopharyngioma

Craniopharyngioma are also a rarity in the perinatal period although they represent 5-10% of pediatric tumors [3]. They are benign tumors (WHO grade I) and arise from the Rathke pouch, an ectodermal diverticulum at the roof of the mouth [2]. They are large, suprasellar, heterogeneous tumors with large cystic masses, sometimes calcifications and thus difficult to distinguish from teratoma [2,3]. Fetuses and neonates with large craniopharyngioma have a poor prognosis with only few survivors [3]. If complete surgical resection is possible, patients can be cured [50]. However, this is often a high-risk procedure in the neonate. Life-threatening electrolyte disturbances caused by pituitary insufficiency is an important problem. Irradiation is contraindicated because of the age. According to a review by Isaacs *et al.* [19] the survival of patients with congenital craniopharyngioma was 23% in 2002 [19], despite the improvement of survival rates through modern neurosurgical techniques.

Lipoma

The prevalence of lipoma is often underestimated because they are small (average diameter 1,6 cm) and often asymptomatic. However, in prenatal ultrasound scanning they represent approximately 10% of cases [2]. They often originate in the midline and are associated with an agenesis of the corpus callosum [2]. Lipoma have the best prognosis of all intracranial tumors [2].

Other Rare Congenital Tumors of the CNS

Desmoplastic infantile tumors (desmoplastic infantile astrocytomas, desmoplastic infantile gangliogliomas) are found almost exclusively in infants below 6-12 months of age. They contain a desmoplastic stroma with a neuroepithelial population and are benign tumors (WHO grade I) with a good prognosis except for seizure activity that may start in some patients after surgery [3]. Although cutaneous hemangiomas are the most frequent tumors in infants, their intracranial manifestation is extremely rare, occurring often in patients with PHACE syndrome (posterior fossa malformation, facial hemangiomas, arterial anomalies, cardiac/aortic anomalies, eye abnormalities and sterna defects) and with von Hippel-Lindau syndrome [3,18]. Anaplastic ependymomas, also rare tumors, can be located supratentorially, reach a large size and have a poor prognosis [3]. Further rare tumors include congenital meningeoma, choriocarcinomas, gangliogliomas and anaplastic oligodendrogliomas [3].

Outlook

The congenital CNS tumors described above are rare but account for a large part of infant mortality due to their poor prognosis. Therapy is limited not only due to the size and location of these tumors, but also based on the limited therapeutic approaches that can be applied at this age. Future studies will need to focus on new, age-matched therapies that are not detrimental to the fragile, developing brain.

Acknowledgments

Our research is supported by the German Research Foundation (DFG; SFB665 and IB of the BMBF) and the Sonnenfeld Stiftung.

References

[1] Isaacs H: I. Perinatal brain tumors: a review of 250 cases. *Pediatr Neurol.* 2002, 27:249-261.

[2] Woodward PJ, Sohaey R, Kennedy A, Koeller KK: From the Archives of the AFIP: a comprehensive review of fetal tumors with pathologic correlation. *Radiographics.* 2005, 25:215-242.

[3] Severino M, Schwartz ES, Thurnher MM, Rydland J, Nikas I, Rossi A: Congenital tumors of the central nervous system. *Neuroradiology* 2010, 52: 531-548.

[4] Arnstein LH, Boldrey E, Naffziger HC: A case report and survey of brain tumors during the neonatal period. *J Neurosurg.* 1950, 8:315-319.

[5] Solitare GB, Krigman MR: Congenital intracranial neoplasm. A case report and review of the literature. *J Neuropath Exp Neurol.* 1964, 23: 280-292.

[6] Wakai S, Arai T, Nagai M: Congenital Brain Tumors. *Surg Neurol.* 1984, 21:597-609.

[7] Buetow PC, Smirniotopoulos JG, Done S: Congenital brain tumors: A review of 45 cases. *AJR Am J Roentgenol.*1990, 155:587-593.

[8] Cavalheiro S, Moron AF, Hisaba W, Dastoli P, Silva NS: Fetal brain tumors. *Childs Nerv Syst.* 2003, 19: 529-536.

[9] Borit A, Richardson EP: The biological and clinical behaviour of pilocytic astrocytomas of the optic pathways. *Brain.* 1982, 105:161-187.

[10] Jiang T, Jia G, Ma Z, Luo S, Zhang Y: The diagnosis and treatment of subependymal giant cell astrocytoma combined with tuberous sclerosis. *Childs Nerv Syst.* 2010

[11] Yuasa H, Tokito S, Tokunaga M: Primary carcinoma of the choroid plexus in Li-Fraumeni syndrome: case report. *Neurosurgery.* 1993. 32: 131-133.

[12] Simon M, Ludwig M, Fimmers R, Mahlberg R, Müller-Erkwoh A, Köster G, Schramm J: Variant of the CHEK2 gene as a prognostic marker in glioblastoma multiforme. *Neurosurgery.* 2006, 59:1078-1085.

[13] Rieber J, Remke M, Hartmann C, Korshunov A, Burkhardt B, Sturm D, Mechtersheimer G, Wittmann A, Greil J, Blattmann C, Witt O, Behnisch W, Halatsch ME, Orakcioglu B, von Deimling A, Lichter P, Kulozik A, Pfister S: Novel oncogene amplifications in tumors from a family with Li-Fraumeni syndrome. *Genes Chromosomes Cancer.* 2009, 48:558-568.

[14] Sévenet N, Sheridan E, Amram D, Schneider P, Handgretinger R, Delattre O: Constitutional mutations of the hSNF5/INI1 gene predispose to a variety of cancers. *Am J Hum Genet.* 1999, 65:1342-1348.

[15] Taylor MD, Gokgoz N, Andrulis IL, Mainprize TG, Drake JM, Rutka JT: Familial posterior fossa brain tumors of infancy secondary to germline mutation of the hSNF5 gene. *Am J Hum Genet.* 2000, 66:1403-1406.

[16] Garre ML, Cama A, Bagnasco F, Morana G, Giangaspero F, Brisigotti M, Gambini C, Forni M, Rossi A, Haupt R, Nozza P, Barra S, Piatelli G, Viglizzo G, Capra V, Bruno W, Pastorino L, Massimino M, Tumolo M, Fidani P, Dallorso S, Schuhmacher RF, Milanaccio C, Pietsch T: Medulloblastoma variants: age-dependent occurence and relation to Gorlin Syndrome-a new clinical perspective. *Clin Cancer Res.* 2009, 15:2463-2471.

[17] Lo Muzio L: Nevoid basal cell carcinoma syndrome (Gorlin syndrome). *Orphanet J Rare Dis.* 2008, 3:32.

[18] Kim JJ, Rini BI, Hansel DE. Von Hippel Lindau syndrome. *Adv Exp Med Biol.* 2010, 68: 228-249

[19] Isaacs H: II. Perinatal brain tumors: a review of 250 cases. *Pediatr Neurol.* 2002, 27:333-342.

[20] Glauser TA, Packer RJ. Cognitive deficits in long-term survivors of childhood brain tumors. *Childs Nerv Syst.* 1991, 7:2-12.

[21] Raisenen JM, Davis RL: Congenital brain tumors. *Pathology* (Phil) 1993 2:103-116.

[22] Altman RP, Randolph JG, Lilly JR: Sacrococcygeal teratoma: American Academy of Pediatrics Surgical Section Survey – 1973. *J Pediatr Surg.* 1974, 9: 389-398.

[23] Reinberg Y, Long R, Manivel JC, Resnick J, Simonton S, Gonzalez R: Urological aspects of sacrococcygeal teratoma in children. *J Urol.* 1993, 150: 948-949.

[24] Milam DF, Cartwright PC, Snow BW: Urological manifestations of sacrococcygeal teratoma. *J Urol.* 1993, 149: 574-576.

[25] Hedrick HL, Flake AW, Crombleholme TM, Howell LJ, Johnson MP, Wilson RD, Adzick NS: Sacrococcygeal teratoma: prenatal assessment, fetal intervention, and outcome. *J Pediatr Surg.* 2004, 39:430-438.

[26] Berhouma M: Management of subependymal giant cell tumors in tuberous sclerosis complex: the neurosurgeon´s perspective. *World J Pediatr.* 2010, 6:103-110.

[27] Hussain N, Curran A, Pilling D, Malluci CL, Ladusans EJ, Alfirevic Z, Pizer B: Congenital subependymal giant cell astrocytoma diagnosed on fetal MRI. *Arch Dis Child.* 2006, 91:520.

[28] Raju GP, Urion DK, Sahin M: Neonatal subependymal giant cell astrocytoma: new case and review of literature. *Pediatr Neurol.* 2007, 36:128-131.

[29] Roosen N, Deckert M, Nicola N, Wechsler W, Schober R, von Voss H, Mayer P, Werner C: Congenital anaplastic astrocytoma with favorable prognosis. Case report. *J Neurosurg.* 1988, 69:604-609.

[30] Connolly B, Blaser SI, Humphreys RP, Becker L: Long-term survival of an infant with anaplastic astrocytoma. *Pediatr Neurosurg.* 1997, 26:97-102.

[31] Jinhu Y, Jianping D, Jun M, Hui S, Yepeng F: Metastasis of a histologically benign choroid plexus papilloma: case report and review of the literature. *J Neurooncol.* 2007, 83: 47-52.

[32] Duffner PK, Kun LE, Burger PC, Horowitz ME, Cohen ME, Sanford RA, Krischer JP, Mulhern RK, James HE, Rekate HL et al: Postoperative chemotherapy and delayed radiation in infants and very young children with choroid plexus carcinomas. The Pediatric Oncology Group. *Pediatr Neurosurg.* 1995, 22:189-196.

[33] Mitchell CS, Wood BP, Shimada H: Neonatal disseminated primitive neuroectodermal tumor. *AJR Am J Roentgenol.* 1994, 162:1160.

[34] McComb JG, Davis RL, Isaacs H: Extraneural metastatic medulloblastoma during childhood. *Neurosurgery.* 1981, 9:548-551.

[35] Isaacs H: Fetal brain tumors: a review of 154 cases. *Am J Perinatol.* 2009, 26: 453-466.

[36] Yamada T, Takeuchi K, Masuda Y, Moriyama T, Kitazawa S, Maruo T: Prenatal imaging of congenital cerebral primitive neuroectodermal tumor. *Fetal Diagn Ther.* 2003, 18:137-139.

[37] Haddad SF, Menezes AH, Bell WE, Godersky JC, Afifi AK, Bale JF: Brain tumors occurring before 1 year of age: a retrospective review of 22 cases in an 11-year period (1977-1987). *Neurosurgery.* 1991, 29:8-13.

[38] Ashwal S, Hinshaw DB, Bedros A: CNS primitive neuroectodermal tumors of childhood. *Med Pediatr Oncol.* 1984, 12:180-188.

[39] Biswas A, Goyal S, Puri T, Das P, Sarkar C, Julka PK, Bakhshi S, Rath GK: Atypical teratoid rhabdoid tumor of the brain: case series and review of literature. *Childs Nerv Syst.* 2009, 25: 1495-1500.

[40] Rickert CH, Paulus W: Epidemiology of central nervous system tumors in childhood and adolescence based on the new WHO classification. *Childs Nerv Syst.* 2001 17:503-511.

[41] Rorke LB, Packer RJ, Biegel JA: Central nervous system atypical teratoid/rhabdoid tumors of infancy and childhood: definition of an entity. *J Neurosurg.* 1996, 85:56-65.

[42] Biegel JA, Kalpana G, Knudsen ES, Packer RJ, Roberts CW, Thiele CJ, Weissman B, Smith M: The Role of INI1 and the SWI/SNF complex in the development of rhabdoid

tumors: meeting summary from the workshop on childhood atypical teratoid/rhabdoid tumors. *Cancer Res.* 2002, 62: 323-328.

[43] Hilden JM, Meerbaum S, Burger P, Finlay J, Janss A, Scheithauer BW, Walter AW, Rorke LB, Biegel JA: Central nervous system atypical teratoid/rhabdoid tumor: results of therapy in children enrolled in a registry. *J Clin Oncol.* 2004, 22: 2877-2884.

[44] Packer RJ, Biegel JA, Blaney S, Finlay J, Geyer JR, Heidemann R, Hilden J, Janss AJ, Kun L, Vezina G, Rorke LB, Smith M: Atypical teratoid/rhabdoid tumor of the central nervous system: report on workshop. *J Pediatr Hematol Oncol.* 2002 24:337-342.

[45] Chi SN, Zimmermann MA, Yao X, Cohen KJ, Burger P, Biegel JA, Rorke-Adams LB, Fisher MJ, Janss A, Mazewski C, Goldman S, Manley PE, Bowers DC, Bendel A, Rubin J, Turner CD, Marcus KJ, Goumnerova L, Ullrich NJ, Kieran MW: Intensive multimodality treatment for children with newly diagnosed CNS atypical teratoid rhabdoid tumor. *J Clin Oncol.* 2009, 27: 385-389.

[46] Kayama T, Yoshimoto T, Shimizu H, Sakurai Y: Neonatal medulloblastoma. *J Neurooncol.* 1993, 15:157-163.

[47] Ermis B, Aydemir C, Taspinar O, Cagavi F, Bahadir B, Ozdemir H: Congenital medulloblastoma. *Arch Dis Child Fetal Neonatal Ed.* 2006, 91:F373.

[48] Geyer JR, Zeltzer PM, Boyett JM, Rorke LB, Stanley P, Albright AL, Wisoff JH, Milstein JM, Allen JC, Finlay JL, Ayers GD, Shurin SB, Stevens K, Bleyer WA. Survival of infants with primitive neuroectodermal tumors or malignant ependymomas of the CNS treated with eight drugs in 1 day. A report from the Children´s Cancer Group. *J Clin Oncol.*1994.12:1607-1615

[49] Duffner PK, Horowitz ME, Krischer JP, Burger PC, Cohen ME, Sanford RA, Friedman HS, Kun LE and Pediatric Oncology Group (POG): The treatment of malignant brain tumors in infants and very young children: an update of the Pediatric Oncology Group experience. *Neuro Oncol.* 1999, 1:152-161.

[50] Müller-Scholden J, Lehrnbecher T, Müller HL, Bensch J, Hengen RH, Sörensen N, von Stockhausen HB: Radical surgery in a neonate with craniopharyngioma. *Pediatr Neurosurg.* 2000, 33:265-269.

In: Brain Cancer, Tumor Targeting and Cervical Cancer
Editor: Elena K. Salvatti
ISBN: 978-1-61122-738-3

Chapter IX

Management of Recurrent Cervical Carcinoma

C. Iavazzo and G. Vorgias
Department of Gynecology, Metaxa Memorial Hospital,
Cancer Hospital, Piraeus, Greece

Abstract

Background:

Cervical carcinoma recurrence is defined as a de novo tumor development, at least three months after the radical excision of the disease or after radiotherapy treatment. One third of cervical carcinomas are going to recur after initial treatment.

Method:

We present a review on the field of recurrent cervical carcinoma regarding the diagnosis, treatment options and prognosis of such an entity.

Results:

The risk factors for recurrence include: stage, tumor size, histopathologic characteristics. 80% of recurrences occur in the first two years after treatment for the initial disease. 2/3 of the local recurrences after surgical treatment occur in the pelvis and are central. Classification of patients is based on the Shingelton and Hatch criteria:

a. Adherence to the pelvic side wall
b. Tumor < or > of 3 cm

c. Time interval between the primary treatment and the relapse.

The indication for pelvic exenteration is non-metastatic persistent or recurrent cervical malignancy after prior pelvic irradiation. The disease must be confined to the central pelvis in order to be completely respectable. The so-called 'triad of trouble' (including ureteral obstruction, neural sheath involvement and venous or lymphatic compromise of the iliac vessels) is used for the determination of respectability (possible when only one of them is present). Radiotherapy is used in pelvic recurrences after surgery and include a combination of external radiotherapy and brachytherapy plus chemotherapy with platinum-based regimens. The possibility of treatment of patients with recurrences outside the pelvis is rather small and for this reason such patients are candidates for palliative treatment.

Conclusion:

Treatment options in patients with persistent or locally recurrent cervical cancer are limited. Less than 20% of such patients are going to survive. Patient selection is crucial. Such patients should be treated in an organized oncologic centre under the continued supervision of surgeon, radiotherapist, internist, and psychiatrist.

Introduction

Epidemiological Characteristics

Disease recurrence of cervical cancer is defined as a de novo tumor development, three months after the radical excision of the disease or after radiotherapy treatment. One third of cervical carcinomas are going to recur after initial treatment. Moreover, 55-65% of the recurrences occur in the first postoperative year while 80% of recurrences occur in the first two years after treatment for the initial disease. For this reason, close follow-up is mandatory. Nevertheless, preventive examination of such patients identify early only 1/3 of cases with recurrence.

Risk Factors

The risk factors of the recurrence include: stage, tumor size, histopathologic characteristics, mode of initial treatment, age and performance status of the patient. More specifically:

a. The risk of recurrence increases with tumor's stage. In patients with early disease who underwent radical hysterectomy +/- radiotherapy, the recurrence rates are 10% and 20% for stage IB and IIA, respectively. Patients who received only radiotherapy show pelvic or distant metastases in 9.6% and 17.5% in stage IB, respectively, compared to 41% and 42% in stage III, respectively.

b. The influence of tumor size in the survival rates does not depend on the treatment option. The tumor size has a negative correlation with the survival rates in patients with early stage disease (stage IB-IIA) compared to patients with late disease (stage IIB-III). So, tumors with diameter > 4 cm (stage IB2) have worse prognosis than smaller tumors.

c. Regarding the histopathological characteristics that influence the recurrence rates, it is known that the recurrence rates differ in the different histological types. The majority of patients with recurrences (80-90%) are of squamous cell carcinoma. However, it has been referred that adenocarcinomas, clear-cell carcinomas and micro-cell carcinomas, as well as low grade tumors have increased risk of recurrences. More specifically, local recurrence rate for primary squamous cell tumor was found to be 53.6%, adenocarcinoma - 6.3% and poorly-differentiated cell carcinoma - 4.9%. Moreover, in patients who underwent radical hysterectomy, the presence of lymph node metastasis and the positive surgical margins are independent risk factors (recurrence in 30%-80% of cases). The depth of invasion of the cervical stroma, as well as invasion of vascular spaces increase the recurrence risk in patients with negative lymph nodes and disease-free surgical margins. In a recent study, Munagala et al showed that significant prognostic factors for recurrence include, stage (P=0.019), pelvic lymph node metastasis (P=0.013), parametrial involvement (P=0.025), and tumor size (P=0.034).

d. In a recent Cochrane Library review, it was shown that there is no significant difference between high dose rate and low dose rate intracavitary brachytherapy when considering recurrence for women with cervical carcinoma.

e. Age and performance status of the patient are two other factors that influence the risk of recurrence. Younger patients in good general condition have usually a better prognosis. This could be explained by the fact that elderly patients or patients with co morbidities usually undergo less radical operation or radiotherapy.

Recurrence Location

2/3 of the local recurrences after surgical treatment occur in the pelvis and are central. Less common locations are in the pelvic wall, upper abdomen or other organs (e.g. lungs). On the other hand, after radiotherapy the recurrence in the early stage disease is found usually outside of the pelvis (76%). Cases with high risk characteristics (advanced stage, positive lymph nodes, unfavorable histopathology) recur in the pelvic walls or outside the pelvis. However, it should be mentioned that distant metastases can always be found simultaneously with the central recurrence. The central pelvic recurrences present with vaginal bleeding or pathological vaginal discharge. The characteristic symptoms include low extremities edema, pelvic pain, weight loss or ureter obstruction. In rare cases the chest X-ray reveals lung metastases.

Clinical Evaluation of the Patient

The diagnosis of the recurrence is mainly based on the clinical findings. Pap smear might be useful in the central pelvic recurrences. However, the definite diagnosis is made by biopsy of the recurrence. In cases of pelvic wall recurrences, U/S or CT-guided FNA or core biopsies might be useful. If all such methods fail then exploratory laparotomy is proposed. Patient selection is crucial as treatment options in patients with persistent or locally recurrent cervical cancer are limited. For this reason, history of the patient should be evaluated. Distant metastases must be excluded using C/T scan, MRI, PET C/T scan and/or bone scanning, intravenous pyelogram, rectoscopy and cystoscopy. MRI is an excellent modality for depicting invasive cervical cancer: it can provide objective measurement of tumor size and provides a high negative predictive value for parametrial invasion and stage IVA disease. MRI and positron emission tomography (PET)/computed tomography (CT) play key roles in identifying recurrent disease. PET/CT is also useful in detecting nodal and distant metastases and in radiotherapy planning. For example, according to Chang et al preoperative [(18)F]FDG PET/CT maximum standardized uptake value predicts recurrence of uterine cervical cancer. Moreover, nodal involvement detected by [(18)F]fluorodeoxyglucose (FDG-PET) in cervical cancer relates to clinical stage and stratifies patient recurrence and survival outcomes.

Classification

Based on the selected information, the classification of patients is made according to Shingelton and Hatch criteria which include:

a. Adherence to the pelvic side wall
b. Tumor < or > of 3 cm
c. Time interval between the primary treatment and the relapse

According to these criteria, patients are classified as:

a. Low risk patients: no adherence to the pelvic side wall, tumor < 3cm, time interval > 1 year

b. Medium risk patients: other combination of the three factors

c. High risk patients: adherence to the pelvic side wall, tumor > 3cm, time interval < 1 year

Treatment

Treatment of cervical cancer recurrences is difficult. Such patients should be treated in an organized oncologic centre under the continued supervision of surgeon, radiotherapist, internist, and psychiatrist. Overall, less than 20% of such patients are going to survive. The possibility of treatment of patients with recurrences outside the pelvis is rather small and for this reason such patients are candidates for palliative treatment.

In pelvic recurrences where no previous radiotherapy has been used, the optional treatment is radiotherapy +/- chemotherapy. On the other hand if previous radiotherapy has been used, surgical treatment is proposed. Then, pelvic exenteration is the option. Pelvic exenteration depends on the location, tumor size, general condition of the patient and the equipment or the experience of the oncologic centre.

Pelvic exenteration

Total pelvic exenteration includes removal of the genital organs, bladder and rectum and is used during an explorative laparotomy since Brunschwig's first surgical evisceration in 1948. So, Brunschwig of Memorial Sloan Kettering Center, New York was the first to describe in 1948 pelvic exenteration as a treatment option of advanced or recurrent cervical tumors. With improving surgical technology and increasing surgical experience, exenteration is a logical extension of current laparoscopic practice. In 2003, Pomel et al. achieved to proceed the first laparoscopic total pelvic exenteration and then on some case series of this method were presented in the literature. The morbidity and mortality rates initially ranged between 50-70%. However, nowadays, the 5-year survival rates could reach 50%.

Indications

The disease must be confined to the central pelvis in order to be completely respectable. The so-called 'triad of trouble' (including ureteral obstruction, neural sheath involvement and venous or lymphatic compromise of the iliac vessels) is used for the determination of respectability (possible when only one of them is present). The indication of pelvic exenteration is non-metastatic persistent or recurrent cervical malignancy after prior pelvic irradiation. More specifically, the method is indicated in recurrent or persistent cervical squamous cell carcinoma, recurrent adenocarcinoma of the cervix isolated to the central pelvis, and severely damaged by irradiation pelvic tissues (although it is thought to be a controversial indication).

Contraindications

On the other hand, the method is contraindicated in:

1. Ureteral obstruction in the intravenous pyelogram.
2. Sciatic nerve distribution of pain.
3. Low-extremity edema due to compromise of the iliac vessels.
4. Distant metastases (outside the pelvis, lymph node metastases).
5. Recurrent adenocarcinoma of the cervix because of its hematologic or lymphatic routes of metastasis.
6. Obesity.
7. Age>70 years.
8. Co-morbidities (diabetes mellitus, infections, cardiopulmonary, renal status).

Description of the Operation

Pelvic exenteration is classified according to the Mayo Clinic Criteria proposed by JF Magrina as following:

1. Total pelvic exenteration is an operation which includes removal of the vagina, the uterus, the fallopian tubes, the ovaries, the bladder, and the rectosigmoid.
2. Anterior pelvic exenteration is defined as the removal of the genital organs plus the bladder and low part of ureters.
3. Posterior pelvic exenteration is defined as the removal of the genital organs plus the rectosigmoid.
4. Extended exenteration is defined as total pelvic exenteration plus removal of pelvic wall muscles, bones, small bowel.

The reconstruction phase of the procedure includes: bowel reconstruction, urinary reconstruction, pelvic floor reconstruction and vaginal reconstruction.

Laparotomic Versus Laparoscopic Pelvic Exenteration

During our Medline search, we found three recent retrospective studies of pelvic exenteration with the classical method. The operative time was 7-7.75 hours and the mean intraoperative blood loss was 2,500 ml. Early and late complications are noticed such as: anastomosis leak, bleeding, wound infection, fistulas, acute renal failure, deep vein thrombosis, pelvic abscess, bowel obstruction (early), and UTIs, ureteral stenosis, fistulas, pelvic abscess, bowel obstruction (late). The total number of patients was 218. The median disease free-survival was 32 months. The 5-year survival rates were from 48 up to 54%. (table 1)

On the other hand, we found 6 relevant studies which included 25 patients who underwent laparoscopic pelvic exenteration until now. The mean operating time ranged from 3 up to 9 hours but there was a variety of anterior, posterior or total pelvic exenteration. From this, of course we can understand that it is a very difficult operation for the surgical team. The perioperative blood loss is limited with the laparoscopic method (ranging from 370 up to 500 cc). Finally, postoperative, the mean hospital stay ranged from 3 up to 27 days with results similar to the laparotomic method. Minor complications of the method such as intraoperative internal iliac artery injury, perineal or wound abscesses, and major complications including ureteric leak or subacute intestinal obstruction are mentioned. The follow-up time reached 16 months with no differences in the overall survival rate. (table 2)

There is only a limited number of studies regarding the use of laparoscopic pelvic exenteration as a new option in the treatment of cervical cancer and therefore additional multi-center studies are necessary to understand the indications and optimize the results of the method.

Table 1: Classical pelvic exenteration

Author/Year	Number of patients	Mean time of procedure	Blood loss	Complications	Follow-up
Goldberg et al/2006	95			14% ureteral anastomosis leaks, 17% wound complications, 4% parastomal leak hernias, 36% UTIs/pyelonephritis, 11% gastrointestinal fistulas, pouch incontinence, 9% small bowel obstruction, 7% thromboembolic complications	48% 5-year survival
Berek et al/2005	75	7-7.75 hours	2.5 lt	UTIs, wound infection, intestinal fistula, intestinal obstruction	54% 5-year survival
Sharma et al/2005	48			27% early complications: anastomotic leak, bleeding, fistulas, acute renal failure, deep vein thrombosis, pelvic abscess, bowel obstruction 75% late postoperative complications: UTIs, ureteral stenosis, fistulas, pelvic abscess, bowel obstruction	Median survival: 35 months Median disease-free survival: 32 months

Table 2: Laparoscopic pelvic exenteration

Author/Year	Type of article	Number of	Surgical method	Mean time of procedure	Blood loss	Complications	Hospital stay	Follow-up
Ferron et al./2006	Case series	7	Laparoscopy-assisted pelvic exenteration: 2 patients total 3 patients anterior 2 patients posterior	6.5 hours	less than 500ml	4 patients minor complications	Mean hospital stay: 27 days	Mean follow-up: 14 months 2 patients free of disease 1 patient local recurrence 4 patients died (3 were
Uzan et al./2006	Case series	5	Laparoscopic pelvic exenteration: 2 patients total 1 patient posterior	4.5-9 hours	370ml			3 patients died (3 were metastatic) 2 patients alive for 11 and 15 months
Puntambekar et al./2006	Case series	12	Laparoscopic anterior pelvic exenteration	Median operative time: 180 min	100-500 ml	1 patient: internal iliac artery injury 3 patients postoperative complications 2 patients subacute intestinal obstruction 1 patient ureteric leak	Median hospital stay: 3 days	Median follow-up: 15 months
Ferron et al./2006	Case series	5	Laparoscopy-assisted pelvic exenteration: 1 patient total 2 patients anterior 2 patients posterior	6 hours (range 4.5-9 hours)	less than 500ml	2 patients minor complications (perineal and abdominal wound abscess)	27 days (13-33days)	Follow-up time: 3-16 months 3 patients free of disease 2 patients groin metastasis 1 patient died after 8 months
Lin et al./2004	Case report	1	Laparoscopic total pelvic exenteration	6 hours		Wound infection 50 days postoperatively		Free of disease 1 year after
Pomel et al./2003	Case report	1	Laparoscopic total pelvic exenteration	9 hours		Uneventful postoperative course		

Furthermore, randomized control trials of pelvic exenteration (laparotomy versus laparoscopy) may be useful for optimizing the method. However, till now, laparoscopic pelvic exenteration seems to be an optimal method of pelvic exenteration with similar results with the classical method. Laparoscopic pelvic exenteration seems to have minimal intraoperative blood loss or complications, less postoperative complications, shorter hospital stay, and better cosmetic result. On the other hand, further time in the follow-up period of these patients is needed in order to have the ability to find out the late postoperative complications and the 5-year survival of them. Gynecologists and especially laparoscopists should never forget that such a technically challenging procedure need to be used with caution as special laparoscopic experience is necessary.

The Da Vinci Surgical System Experience.

Robot-assisted laparoscopy is feasible after concurrent chemoradiation and brachytherapy in cases of recurrent cervical cancer. Lim et al in 2009 presented the first robotic assisted total pelvic exenteration followed by Davis et al and Lambaudie et al in 2010 who used the method in other five patients. This new surgical approach reduces hospital stay, and seems to result in less severe complications than conventional laparotomy without modifying the oncological outcome. However, urinary diversion is usually made extracorporeally by a transrectal laparotomy.

Prognosis

Postoperatively, the method is characterized by high morbidity and mortality rates. Often, there is need of admission to the intensive care units.

30-50% of patients face postoperative complications (problems from the reconstruction of the urinary or gastrointestinal tract, bleeding, infection, thromboembolism, ureteral stricture, urinary leakage, intestinal obstruction or fistulas). Recently, in a retrospective study, it has been shown that 70.2% of patients had at least one major complication, including ileus (8.5%), intestinal-fistula (29.8%), ureteral anastomotic insufficiency (6.4%), abscess (6.4%), and cardiothrombotic events (23.4%).

Surgical mortality after pelvic exenteration ranges between 5% and 8%. Total pelvic exenteration is accompanied with considerable morbidity, but good local control and acceptable overall survival. In the recent retrospective study, at a median follow-up of 7 months (range: 1-42), 46.8% patients died and 46.8% experienced a relapse. All the high risk patients died of operative complications and/or persistent cancer within 18 months of exenteration.

Almost one out of two patients will suffer complications of some kind, and one out of three will have a severe complication with pelvic exenteration.

However, the cure rates range from 20% up to 60% of those patients. The possibility of treatment success is increased if the disease is central with small tumor size and is presented after a long disease free interval.

Role of Radiotherapy

Most regional recurrences after definitive radiotherapy for cervical cancer include a component of marginal failure, usually immediately superior to the radiation field. These recurrences suggest a deficiency in target volume. Recurrences also occur in-field, suggesting a deficiency in dose. Developments in pretreatment staging, field delineation, dose escalation, and post-treatment surveillance may help to improve outcome in these patients.

Radiotherapy is used in pelvic recurrences after surgery and include a combination of external beam radiotherapy and brachytherapy plus chemotherapy with platinum-based regimens.Vaginal recurrences with low depth of invasion which present 5 years after the radiotherapy could be treated only with brachytherapy.

In radiotherapy naïve patients, pelvic irradiation of 40-50 Gy to the tumor is indicated. ^{198}Au or ^{192}Ir implant for recurrent cervical cancer are often used. Doses of 20-40 Gy in 3 to 5 fractions over 3 to 5 weeks with LDR and HDR techniques have been used.

Furthermore, it has been shown that concurrent chemoradiation achieves significantly better outcome than radiation alone in patients with recurrences after primary radical hysterectomy.

Prognosis

Thomas et al. noted a median survival of only 7 months for 242 patients with recurrent cervical cancer who were primarily treated with irradiation. Re-irradiation with interstitial or intracavitary methods could be curative in patients with small volume of central recurrences with long disease free interval.

CORT and LEER Techniques

a. Hockel and Knapstein described the combined operative and radiotherapeutic treatment which combines pelvic exenteration plus pelvic wall radiotherapy. CORT method (Combined operative and radiotherapeutic treatment) is used as a treatment option of post-irradiation recurrences infiltrating the pelvic sidewalls. After the surgical procedure, a single interstitial implant is placed encompassing the entire area of potential microscopic residual disease with a 2 cm margin. By using the CORT procedure, pelvic side-wall recurrences of gynecologic malignancies arising in a previously irradiated pelvis may be locally controlled. The first treatment results of the newly designed combined operative- and radiochemotherapy concept led to an improvement of the prognosis of patients with recurrences of cervical cancer at the pelvic wall.

b. Laterally extended endopelvic resection (LEER) as well as pre-exenterative chemotherapy are promising novel therapeutic modalities for recurrent cervical cancer with pelvic extension in an irradiated area. LEER is characterized by the resection of the viscera en bloc with the side wall muscles and major vessels in the lesser pelvis of the female. For this reason, a vascular implant is usually used to preserve limb hematosis. Until recently, surgical treatment of locally advanced and recurrent cervical carcinoma was performed only with a central disease location. The required operation, pelvic exenteration, was contraindicated for tumors fixed to the pelvic wall. The laterally extended endopelvic resection (LEER) now allows the exstirpation of a subset of mesenteric pelvic side-wall tumors with clear margins and opens up a 50% chance of long-term survival for the affected patients.

Role of Chemotherapy - Palliative Treatment

The recurrences after exenteration are cured with chemotherapy and the 1-year mortality rates reach 90%.

If radiotherapy or pelvic exenteration could not be used or in cases with distant metastases then platinum based chemotherapy is proposed with poor results. Generally, radiation therapy in previously irradiated patients is considered palliative. For this reason quality of life as well as the survival are parameters that should also be considered. Cis-platinum based chemotherapy has been proven beneficial. Cisplatin is an effective radiosensitizer, but its single agent activity in recurrent cervical cancer, especially after prior cisplatin exposure, is disappointing, with a response rate of only 13%. Carboplatin-paclitaxel is active and well tolerated in patients with recurrent cervical cancer after definitive radiotherapy. This combination should be considered as an alternative regimen to cisplatin-paclitaxel in this patient population.

Oxaliplatin has preclinical activity in cisplatin-resistant tumors and may have synergic activity when combined with paclitaxel. The combination of paclitaxel and oxaliplatin is an effective and safe regimen in patients with recurrent cervical cancer including the majority of those previously exposed to cisplatin. Moreover, paclitaxel-ifosfamid-cis-platin or topotecan-cis-platin combinations could be used as active regimens with acceptable toxicity in relapsed cervical cancer.

Role of Imatinib, Bevacizumab and other Proposed Techniques

Imatinib mesylate inhibits platelet-derived growth factor receptor (PDGFR), and there are evidences that the PDGFR participates in development and progression of cervical cancer. However lack of activity in cervical cancer recurrences was found in a recent study.Angiogenesis is central to cervical cancer development and progression. The dominant role of angiogenesis in cervical cancer seems to be directly related to HPV inhibition of p53 and stabilization of HIF-1 alpha, both of which increase VEGF. Bevacizumab binding and subsequent inactivation of VEGF seem to shrink cervical tumors and delay progression without considerable toxicity, and are therefore being studied in a Gynecologic Oncology Group (GOG) phase III trial. Other intracellular tyrosine kinase inhibitors (TKIs) of angiogenesis such as pazopanib are also encouraging, especially in lieu of their oral administration.

It is also proposed recently that complete remission of relapsed uterine cervical cancers by using weekly administration of bevacizumab and paclitaxel/ carboplatin could be succeeded.

Following publication of five pivotal randomized trials of concurrent chemo-irradiation for patients with locally advanced cervical cancer (ie, International Federation of Gynecology and Oncology (FIGO] stages IB2-IVA) in 1999 and 2000, the National Cancer Institute issued a Clinical Alert advising that concurrent chemotherapy (typically, single-agent cisplatin) be incorporated into the treatment program of women scheduled to receive definitive pelvic radiotherapy. Although the adoption of this new standard has improved overall survival and decreased the recurrence rate by 50%, for those patients who do relapse, the prognosis is very poor and, ultimately, therapy in this setting is palliative in nature. The Gynecologic Oncology Group (GOG) has now completed eight randomized trials for metastatic and recurrent cervical cancer, all of which have studied cisplatin-based regimens. The eighth trial (protocol 204) compared four cisplatin-based doublets containing paclitaxel, topotecan, vinorelbine, or gemcitabine. Because the vast majority of patients are now expected to receive cisplatin "up front" as part of primary therapy with pelvic radiation, there are concerns for the development of drug-resistant clones in recurrences both inside and outside of the radiation field. The GOG has recently reported the results from a phase II trial evaluating the anti-vascular agent, bevacizumab, in women who were eligible for second-line or third-line therapy for metastatic and/or recurrent disease (protocol 227C).

Further Discussion

Radiotherapy and/or pelvic exenteration represent the treatment of vaginal recurrence, but the prognosis remains unsatisfactory and with long-term complications. Benedetti-Panici et al investigated the possible role of vaginectomy for isolated vaginal relapse in cervical cancer. Twenty-nine patients with isolated vaginal relapse from cervical cancer were enrolled. After a median follow-up of 57.5 months (range, 8-100 months), 55% of patients were disease-free. The 5-year overall survival and progression-free survival rates were 70.5% and 59.4%, respectively. For this reason, in carefully selected patients, vaginectomy may be considered a therapeutic option for isolated vaginal relapse in cervical cancer. Older patients with long disease-free interval and small recurrences benefit the most from this bladder-sparing surgical technique.

In a recent study the chance of cure and the associated morbidity following pelvic exenteration was determined. More specifically, initial therapy prior to diagnosis of persistent or locally recurrent disease included radiation therapy in 51%. Anterior exenteration was performed in 86% and total exenteration in 14%. Half of the women underwent additional procedures. A continent urinary diversion was constructed in 16 and an ileal conduit in 27 patients. Early postoperative complications were generally minor and only 2 patients required surgical intervention. Four intestinal fistulas were successfully treated conservatively. Late complications were mainly tumor-related. Complication rates associated with the urinary diversion were low and there was no difference in complications between continent and incontinent diversions. The overall disease-specific 5-year survival rate after exenteration was 36.5%. Survival correlated significantly with surgical margin status.

Counselling

Women facing an exenterative procedure must be counseled carefully about the risks and long-term concerns related to the procedure. Each should undergo a comprehensive evaluation to make sure there is no evidence of unresectable or metastatic disease that would make her an unsuitable candidate for exenteration. Central and pelvic wall recurrences of cervical cancer do not exhibit pronounced biologic differences. Patients with large (> or = 5 cm) recurrences have a poor prognosis in spite of extended radical treatment irrespective of tumor location. Efforts should be made to detect isolated pelvic relapses at a smaller tumor size to enhance the chance for long term survival after local control by exenteration and CORT.

Conclusion

Treatment for cervical cancer is very successful, especially in early stages. However, most patients presenting in advanced stages of disease will experience recurrence. The prognosis of recurrent disease is very poor and treatment options are limited. The diagnosis of recurrence may be apparent or difficult, but determining the extent of disease is always complex. Routine follow-up of asymptomatic patients has other objectives and is not a reliable way to detect recurrences. Symptomatic patients require extensive investigation to detect the extent of the disease. For patients with central pelvic recurrences, exenteration offers the prospect of survival in more than one-third of cases. Newer developments include laterally extended endopelvic resection that may become an option for patients with more extensive pelvic recurrence. For patients with recurrences of cervical cancer, the roles of second-time radiotherapy or postradiation chemotherapy are very limited. Palliative treatment is important for all patients with untreatable disease. Pain relief forms a central part of palliative care.

References

[1] Monk BJ, Lopez LM, Zarba JJ, Oaknin A, Tarpin C, Termrungruanglert W, Alber JA, Ding J, Stutts MW, Pandite LN. Phase II, Open-Label Study of Pazopanib or Lapatinib Monotherapy Compared With Pazopanib Plus Lapatinib Combination Therapy in Patients with Advanced and Recurrent Cervical Cancer. J Clin Oncol. 2010 Jul 6.

[2] Paton F, Paulden M, Saramago P, Manca A, Misso K, Palmer S, Eastwood A. Topotecan for the treatment of recurrent and stage IVB carcinoma of the cervix. Health Technol Assess. 2010 May;14 Suppl 1:55-62.

[3] Davis MA, Adams S, Eun D, Lee D, Randall TC Robotic-assisted laparoscopic exenteration in recurrent cervical cancer Robotics improved the surgical experience for 2 women with recurrent cervical cancer. Am J Obstet Gynecol. 2010 Jun;202(6):663.e1.

[4] Fotopoulou C, Neumann U, Kraetschell R, Schefold JC, Weidemann H, Lichtenegger W, Sehouli J. Long-term clinical outcome of pelvic exenteration in patients with advanced gynecological malignancies. J Surg Oncol. 2010 May 1;101(6):507-12.

[5] Lambaudie E, Narducci F, Leblanc E, Bannier M, Houvenaeghel G. Robotically-assisted laparoscopic anterior pelvic exenteration for recurrent cervical cancer: report of three first cases. Gynecol Oncol. 2010 Mar;116(3):582-3. Epub 2009 Nov 27.

[6] Mäenpää JU, Kangasniemi K, Luukkaala T. Pelvic exenteration for gynecological malignancies: an analysis of 15 cases operated on at a single institution. Acta Obstet Gynecol Scand. 2010;89(2):279-83.

[7] Jurado M, Alcázar JL, Martinez-Monge R. Resectability rates of previously irradiated recurrent cervical cancer (PIRCC) treated with pelvic exenteration: is still the clinical involvement of the pelvis wall a real contraindication? a twenty-year experience. Gynecol Oncol. 2010 Jan;116(1):38-43.

[8] Lim PC. Robotic assisted total pelvic exenteration: a case report. Gynecol Oncol. 2009 Nov;115(2):310-1.

[9] Marnitz S, Dowdy S, Lanowska M, Schneider A, Podratz K, Köhler C. Exenterations 60 years after first description: results of a survey among US and German Gynecologic Oncology Centers. Int J Gynecol Cancer. 2009 Jul;19(5):974-7.

[10] Ferenschild FT, Vermaas M, Verhoef C, Ansink AC, Kirkels WJ, Eggermont AM, de Wilt JH. Total pelvic exenteration for primary and recurrent malignancies. World J Surg. 2009 Jul;33(7):1502-8.

[11] Iavazzo C, Vorgias G, Akrivos T. Laparoscopic pelvic exenteration: a new option in the surgical treatment of locally advanced and recurrent cervical carcinoma. Bratisl Lek Listy. 2008;109(10):467-9.

[12] Jover R, Lourido D, Gonzalez C, Rojo A, Gorospe L, Alfonso JM. Role of PET/CT in the evaluation of cervical cancer. Gynecol Oncol. 2008 Sep;110(3 Suppl 2):S55-9.

[13] Davis MA, Adams S, Eun D, Lee D, Randall TC. Robotic-assisted laparoscopic exenteration in recurrent cervical cancer Robotics improved the surgical experience for 2 women with recurrent cervical cancer. Am J Obstet Gynecol. 2010 Jun;202(6):663.e1.

[14] Schulz-Wendtland R, Krämer S, Säbel M, Heller F, Keilholz L, Jäger W, Sauer R, Lang N, Bautz W. [Pelvic wall recurrence of cervix carcinomas. Combined surgical-radio-chemotherapeutic procedure (CORCT)] Strahlenther Onkol. 1998 May;174(5):279-83.

[15] Tewari KS, Monk BJ. The rationale for the use of non-platinum chemotherapy Clin Adv Hematol Oncol. 2010 Feb;8(2):108-15.

[16] Brunschwig A, Pierce VK. Partial and complete pelvic exenteration; a progress report based upon the first 100 operations. Cancer 1950(Nov);3(6):972– 4.

[17] Pomel C, Rouzier R, Pocard M, Thoury A, Sideris L, Morice P, et al. Laparoscopic total pelvic exenteration for cervical cancer relapse. Gynecol Oncol 2003 (Dec);91(3):616– 8.

[18] Ferron G, Querleu D, Martel P, Letourneur B, Soulie M. Laparoscopy-assisted vaginal pelvic exenteration. Gynecol Oncol. 2006 Mar;100(3):551-5. Epub 2005 Oct 24.

[19] Ferron G, Querleu D, Martel P, Chopin N, Soulie M. [Laparoscopy-assisted vaginal pelvic exenteration.] Gynecol Obstet Fertil. 2006 Nov 27; [Epub ahead of print]

[20] Uzan C, Rouzier R, Castaigne D, Pomel C. [Laparoscopic pelvic exenteration for cervical cancer relapse: preliminary study] J Gynecol Obstet Biol Reprod (Paris). 2006 Apr;35(2):136-45.

[21] S Puntambekar, RJ Kudchadkar and AM Gurjar et al., Laparoscopic pelvic exenteration for advanced pelvic cancers: a review of 16 cases. Gynecol Oncol 102 (2006),pp.513-6

[22] Lin MY, Fan EW, Chiu AW, Tian YF, Wu MP, Liao AC. Laparoscopyassisted transvaginal total exenteration for locally advanced cervical cancer with bladder invasion after radiotherapy. J Endourol 2004 (Nov); 18(9):867–70.

[23] Goldberg GL, Sukumvanich P, Einstein MH, Smith HO, Anderson PS, Fields AL. Total pelvic exenteration: the Albert Einstein College of Medicine/Montefiore Medical Center Experience (1987 to 2003). Gynecol Oncol. 2006 May;101(2):261-8.

[24] JS Berek, C Howe, LD Lagasse and NF Hacker, Pelvic exenteration for recurrent gynecologic malignancy: survival and morbidity analysis of the 45-year experience at UCLA, Gynecol Oncol 99 (2005), pp. 153–159

[25] Sharma S, Odunsi K, Driscoll D, Lele S. Pelvic exenterations for gynecological malignancies: twenty-year experience at Roswell Park Cancer Institute. Int J Gynecol Cancer. 2005 May-Jun; 15(3):475-82

[26] Magrina JF. Types of pelvic exenterations: a reappraisal. Gynecol Oncol 1990; 37(3):363-6

[27] Höckel M, Knapstein PG. The combined operative and radiotherapeutic treatment (CORT) of recurrent tumors infiltrating the pelvic wall: first experience with 18 patients. Gynecol Oncol 1992;46(1):20-8

[28] Candelaria M, Arias-Bonfill D, Chávez-Blanco A, Chanona J, Cantú D, Pérez C, Dueñas-González A. Lack in efficacy for imatinib mesylate as second-line treatment of recurrent or metastatic cervical cancer Int J Gynecol Cancer. 2009 Dec;19(9):1632-7.

[29] Takano M, Kikuchi Y, Kita T, Goto T, Yoshikawa T, Kato M, Watanabe A, Sasaki N, Miyamoto M, Inoue H, Ohbayashi M. Complete remission of metastatic and relapsed uterine cervical cancers using weekly administration of bevacizumab and paclitaxel/carboplatin. Onkologie. 2009 Oct; 32(10):595-7.

[30] Tewari KS, Monk BJ. Recent achievements and future developments in advanced and recurrent cervical cancer: trials of the Gynecologic Oncology Group. Semin Oncol. 2009 Apr;36(2):170-80.

[31] Monk BJ, Sill MW, Burger RA, Gray HJ, Buekers TE, Roman LD. Phase II trial of bevacizumab in the treatment of persistent or recurrent squamous cell carcinoma of the cervix: a gynecologic oncology group study. J Clin Oncol. 2009 Mar 1;27(7):1069-74.

[32] Munagala R, Rai SN, Ganesharajah S, Bala N, Gupta RC. Clinicopathological, but not socio-demographic factors affect the prognosis in cervical carcinoma. Oncol Rep. 2010 Aug;24(2):511-20.

[33] Benedetti Panici P, Manci N, Bellati F, Di Donato V, Marchetti C, De Falco C, Di Tucci C, Angioli R. Vaginectomy: a minimally invasive treatment for cervical cancer vaginal recurrence. Int J Gynecol Cancer. 2009 Dec; 19(9):1625-31.

[34] Wang X, Liu R, Ma B, Yang K, Tian J, Jiang L, Bai ZG, Hao XY, Wang J, Li J, Sun SL, Yin H. High dose rate versus low dose rate intracavity brachytherapy for locally advanced uterine cervix cancer. Cochrane Database Syst Rev. 2010 Jul 7; 7:CD007563.

[35] Mabuchi S, Isohashi F, Yoshioka Y, Temma K, Takeda T, Yamamoto T, Enomoto T, Morishige K, Inoue T, Kimura T. Prognostic factors for survival in patients with recurrent cervical cancer previously treated with radiotherapy. Int J Gynecol Cancer. 2010 Jul;20(5):834-40.

[36] Höckel M. Laterally extended endopelvic resection. Novel surgical treatment of locally recurrent cervical carcinoma involving the pelvic side wall. Gynecol Oncol. 2003 Nov;91(2):369-77.

[37] Chou HH, Chang HP, Lai CH, Ng KK, Hsueh S, Wu TI, Chen MY, Yen TC, Hong JH, Chang TC. (18)F-FDG PET in stage IB/IIB cervical adenocarcinoma/adenosquamous carcinoma. Eur J Nucl Med Mol Imaging. 2010 Apr;37(4):728-35.

[38] Holtz DO, Dunton CJ. Chemotherapy: cisplatin combinations in cervical cancer-which is best? Nat Rev Clin Oncol. 2010 Feb;7(2):74-5.

In: Brain Cancer, Tumor Targeting and Cervical Cancer
Editor: Elena K. Salvatti
ISBN: 978-1-61122-738-3

Chapter X

Real-Time Tumor Targeting in External Beam Radiotherapy

Marco Riboldi[1], Matteo Seregni[1], Andrea Pella[1], Ahmad Esmaili Torshabi[2], Roberto Orecchia[3,4] and Guido Baroni[1,2]

[1]Department of Bioengineering, Politecnico di Milano, Milano, Italy
[2]Bioengineering Unit, Centro Nazionale di Adroterapia Oncologica, Pavia, Italy
[3]Radiotherapy Division, European Institute of Oncology, Milano, Italy
[3]Department of Science & Biomedical Technologies, Università di Milano, Milano, Italy

Abstract

Tumor targeting in external beam radiotherapy is a crucial issue, as the accuracy that can be achieved is directly related to local tumor control and side effects. This applies both to conventional radiation therapy, that uses photons and electrons, and to particle therapy, where protons and ions provide higher geometrical selectivity and enhanced biological effects. Current strategies prescribe the massive use of image guidance technologies to maximize the targeting accuracy (Image Guided RadioTherapy, IGRT).

When IGRT is applied to moving tumors, such as in the lung or liver, image guidance becomes challenging, as motion leads to increased uncertainty. Furthermore, the need of continuously monitoring tumor motion to achieve adequate tumor targeting accuracy limits the applicability of IGRT technologies due to imaging dose constraints. Therefore, minimally invasive technologies for continuous monitoring, such as optical tracking and surface detection, have been proposed for real-time tumor targeting.

In this chapter we focus on real-time tumor targeting of moving tumors, investigating the applicability of using external surrogates combined with periodic imaging to model tumor motion. We first present the theoretical background of computing targeting errors with moving targets. Then we review the state of the art technologies and quantify the actual targeting accuracy based on published data and retrospective clinical studies. Finally we compare different strategies to model tumor motion in a patient database, discussing the use of different correlation models, the need of model updating, and the best trade off between targeting accuracy and imaging dose.

Introduction

Tumor targeting in external beam radiotherapy entails the confinement of the prescribed radiation dose to the target volume, thus reducing secondary damage [35]. This is needed to provide maximal dose deposition in the tumor with contemporary sparing of healthy structures. The achievement of optimal tumor targeting requires mainly:

- Accurate treatment planning, aiming at the definition of the target volume from the anatomical and functional point of view;
- Continuous and precise information concerning target position during treatment, i.e. when radiation dose is actually being delivered.

The former requirement is typically accomplished through specific imaging protocols, relying on current technologies for high definition volumetric imaging, such as Computer Tomography (CT), Position Emission Tomography (PET) and Magnetic Resonance Imaging (MRI). Image fusion is applied to combine anatomical and functional information for treatment planning, providing high resolution images for target definition [1]. Current imaging technologies also allow time-resolved acquisition protocols to be used for those sites when target motion is an issue, such as in the lung or liver [3]. For example, four-dimensional CT (4D CT) is commonly used to image anatomy variations as a function of patient breathing [13].

On the other hand, target localization during treatment requires the use of imaging technologies inside of the treatment room, to image the target or specific target surrogates when the patient is in position for treatment. The exploitation of in-room imaging to maximize the targeting accuracy is conventionally referred to as Image Guided RadioTherapy (IGRT) [1]. When IGRT is applied to moving tumors image guidance becomes challenging, as motion leads to increased uncertainty [10]. Furthermore, the need of continuously monitoring tumor motion to achieve adequate targeting accuracy limits the applicability of IGRT technologies due to imaging dose constraints [22]. Specifically, when ionizing radiation is applied in IGRT techniques, the imaging dose must be traded off with the required accuracy according to the ALARA principle (As Low As Reasonably Achievable) [22].

The interest in real-time tumor targeting based on IGRT technologies is growing momentum in external beam radiotherapy. This applies both to conventional therapy, that uses photons and electrons, and to particle therapy, where protons and ions provide higher geometrical selectivity and enhanced biological effects [12]. Current techniques ensure high selectivity in dose delivery, when combined with high definition treatment planning images for the individuation of anatomical and biological target structures [17]. Advanced centers are exploring the applicability of beam scanning, that requires accurate tumor tracking and effective motion compensation [4,16,43]. This is needed to make the most of the increased dose conformation and biological effect that can be achieved through dose painting techniques [30].

Accurate knowledge of real-time target location is challenging due to breathing motion variability during consecutive breathing cycles [26,33]. Furthermore, time prediction must be embedded to account for the time delay (latency) in reacting to motion compensation information. The effective latency is mainly a function of dose delivery technologies: under this perspective particle therapy features numerous advantages, as charged particles can be effectively and promptly steered by magnetic forces.

Different strategies may be applied to extract real-time target position information over time, ranging from continuous imaging [37,38] to the use of external surrogates [7,18,33]. The selection of the most appropriate approach is essentially dependent on the available technologies for in-room imaging and on the limits in imaging dose [10]. Therefore, minimally invasive technologies for continuous monitoring, such as optical tracking, are put forward to represent a viable solution [40]. Such technology must be combined with prediction methods based on the correlation between external surrogates and internal tumor motion.

Different studies have shown that the accuracy in real-time tumor targeting is mostly dependent on the uncertainties in external/internal correlation, whereas the prediction error is mainly a function of the system latency [7,18,33]. In this chapter we will focus on external/internal correlation models for real-time tumor targeting as the main source of uncertainty. We will first describe the basic theory for the quantification of targeting errors [29]. Then, we will consider state of the art commercial solutions and investigate the major error components in detail, relying on retrospective clinical data analysis. Finally, we will discuss the latest research in the development of patient-specific correlation models, aiming at enhanced accuracy in real-time tumor targeting [40].

Tumor Targeting Error Theory

Historically, the development of IGRT techniques has an analogy with the application of intra-operative imaging in image guided surgery [19]. Both fields share the need of accurate targetry to make the most of image guidance in increasing the accuracy. As far as image guided surgery in concerned, such a concept has been crystallized in the formalism of Target Registration Error (TRE) [14,15,25,46]. Fundamentally, TRE is the residual error in reaching the target in a surgical procedure. Other useful examples can be drawn from radiofrequency ablation [2] and interstitial brachytherapy [23,45], where the accuracy of needle insertion is a crucial issue. Similarly, the TRE concept can be extended to IGRT in terms of tumor localization uncertainties, i.e. in terms of relative motion of the tumor ("target") with respect to the incident beam ("surgical tool").

We present here the targeting error formalism applied to IGRT treatment of moving targets [29]. Such a formalism is suited to represent a general quantitative index of the targeting accuracy of any IGRT based technique. This requires the extension to a 4D motion space, where targeting of a moving tumor can be conveniently described. Using a time-weighted histogram as basic data representation, both intra- and inter-fractional variations can be mapped to distribution of target positions relative to the treatment beam, in order to define the actual targeting error [29].

The definition of targeting errors for moving tumors requires the availability of the following information (Figure 1):

- Reference tumor trajectory, representing the expected tumor motion path over the course of treatment; this can be defined on the basis of the treatment planning 4DCT
- 4D motion data acquired during treatment, by means of appropriate IGRT technologies, providing a quantitative description of target positional variations over time; ideally, such description needs to include both inter and intra-fractional changes.

Both datasets are defined according to the implemented delivery technique, such as ITV (Internal Target Volume) irradiation, respiratory gating and tumor tracking. Each motion mitigation strategy implies a different definition of reference data, as the reference position can either be a static volume (ITV, gating) or a trajectory over time (tracking). Measurements are combined to describe the variations of actual target position during treatment from the expected tumor trajectory, defined as a 3D vector field as a function of time (4D dataset). Accurate targeting error measurement requires the available information to be a meaningful representation of the statistical fluctuations in tumor position [6,8,42]. This can be done separately for inter- and intra-fractional variations or accounting for both factors in a comprehensive motion Probability Density Function (*mPDF*).

The described framework compares the different approaches to the compensation of inter- and intra-fractional motion in terms of the statistical distribution of residual uncertainties in tumor targeting. The cumulative *mPDF* function can be estimated in a twofold approach, depending on the specific modalities used to gather quantitative information. When inter and intra-fractional variations are measured separately, a first order estimate of the motion probability density function *mPDF* can be obtained by convolving the corresponding inter and intra-fractional contributions. Alternatively, when imaging of ground truth tumor trajectory is available throughout the treatment, the motion probability function *mPDF* can be quantified through direct measurement. Specifically, *mPDF(X)*, i.e. the probability of targeting errors to be lower or equal than a given value *X*, can be formalized as follows:

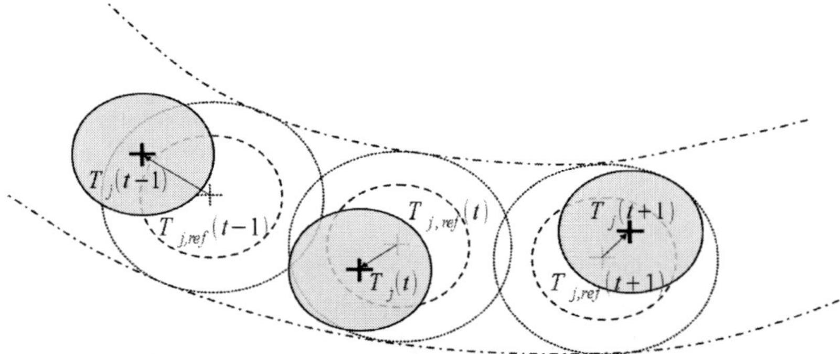

Figure 1. Definition of targeting errors with moving targets. Dotted + symbols represent the reference tumor trajectory as defined from the treatment planning 4D CT. The vector distance between current positions acquired during treatment and reference tumor trajectory define the distrubution of targeting errors in 4D, as reported in equation (1).

$$mPDF(X) = P\left[\left\{\left\|T_j(t) - T_{j,ref}(t)\right\| \leq X\right\}, j = 1,\dots,N_{fx}; t \in TI_j\right] \quad (1)$$

where TI_j is the time interval when breathing motion at fraction j was sampled, $T_j(t)$ and $T_{j,ref}(t)$ are the instantaneous and reference tumor positions at time t, N_{fx} is the total number of fractions, and symbol $\|\ \|$ denotes the Euclidean norm.

State of the Art Analysis

We analyzed a database of 339 fractions in 86 patients, who received hypofractionated stereotactic body radiotherapy with a robotic linear accelerator (CyberKnife®, Accuray Inc., Sunnyvale, CA) between July 2005 and June 2007. The patient database was made available by the Georgetown University Medical Center (Washington, DC). Such database includes patients treated with real-time compensation of tumor motion by means of the Synchrony® respiratory tracking module, as available in the Cyberknife® system [7]. The system provides tumor tracking relying on an external/internal correlation model between the motion of external infrared markers and of clips implanted near the tumor. The model is built at the beginning of each irradiation session using synchronized external/internal markers motion as training dataset. The initial model is updated as needed over the course of treatment with intermittent X-ray imaging. Infrared tracking is utilized to monitor patient breathing (at ~25 Hz) on the basis of 3 external markers placed on a vest. Implanted clips are visualized through orthogonal X-ray imaging and localized with dynamic tracking. The external/internal correlation model is exploited to adjust the position of the robotic linear accelerator according to the predicted tumor motion. The Synchrony® module models instantaneous tumor motion as a function of the external marker coordinates (Modeler) and predicts tumor motion in the near future (Predictor) to account for the system lag τ (~115 ms) [33]. As the treatment proceeds, relevant data, including tumor motion model/prediction, the coordinates of external markers and intermittent clip localization, are logged and stored in ASCII format. These latter were analyzed to compute the actual targeting error, by separating the contribution due to the external/internal correlation model from the need for predicting in the future to account for the system lag (Figure 2).

Non parametric statistical analysis (Wilcoxon test) was applied to compare the patient breathing motion in the time instants when implanted clips were imaged with the whole treatment session. This was obtained by relying on the external markers as breathing surrogates, in order to check whether intermittent imaging provides a reliable description of breathing motion over the course of treatment (Figure 3). Breathing motion in the time instants when intermittent imaging was acquired was not statistically different than the one measured over the whole fraction in 89% of the cases (at 99% statistical confidence), thus indicating that intermittent imaging was able to capture breathing motion reliably in most of the fractions.

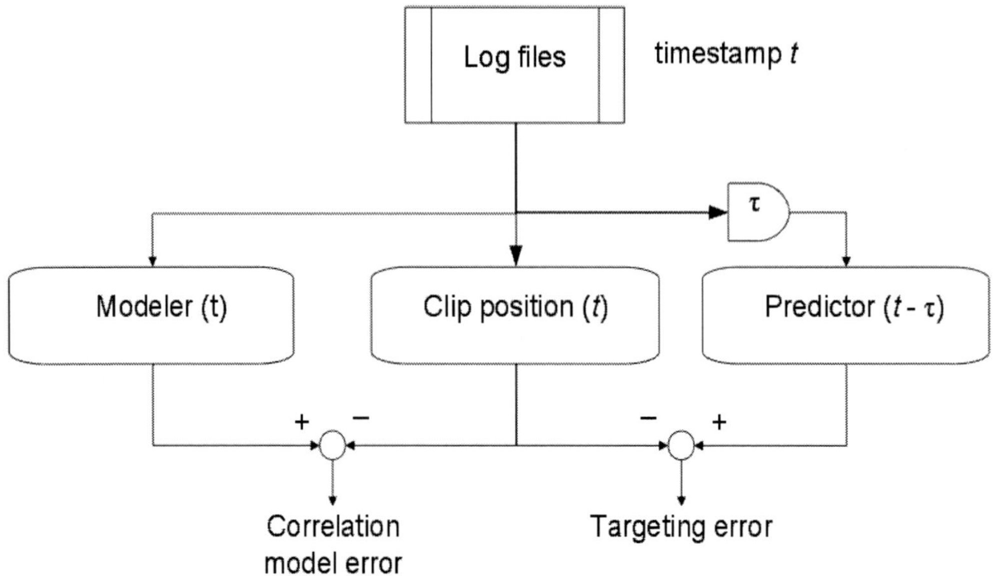

Figure 2. Flowchart of the performed retrospective clinical data analysis.

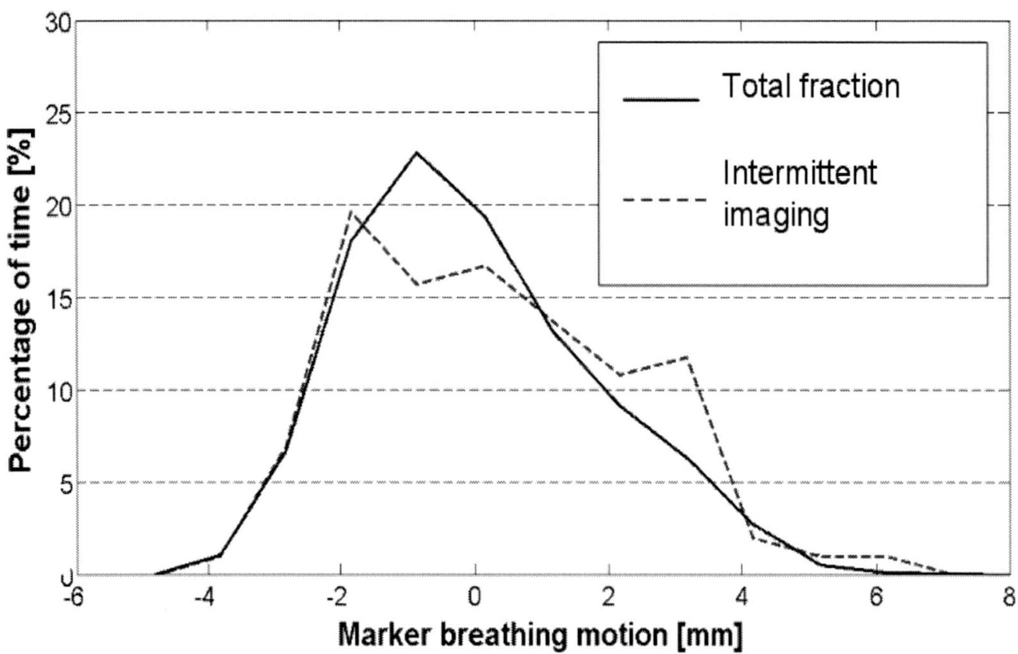

Figure 3. Distribution of breathing motion measured on the external markers, where the total fraction is compared with the instants in time when intermittent imaging was acquired.

Clinical data show that the mean error contribution due to the need for prediction is approximately 0.5 mm (Figure 4). The overall average accuracy in tumor tracking, expressed as the 95% confidence interval of targeting errors, is 2.85 mm for the Modeler and 3.43 mm for the Predictor (Figure 4). Hence, the overall accuracy when the time lag needed for prediction is accounted for is approximately 3.4 mm, as given by the Predictor.

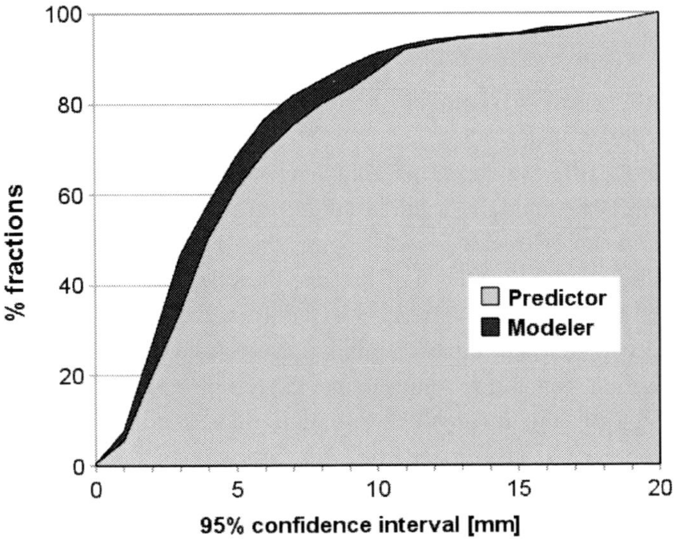

Figure 4. Cumulative distrubution of Modeler vs. Predictor targeting errors.

We proved that the targeting error increases as a function of the tumor range of motion and treatment site (Spearman rank correlation analysis, p-value < 0.01). Larger effects were measured in the superior-inferior direction, with a 3 mm error increase per 1 cm range of motion variation. Also, a positive correlation between the standard deviation value of imaging intervals and targeting errors was found (Spearman rank correlation, p-value $< 10^{-6}$). Therefore, unevenly spaced model updating was proved to significantly reduce the accuracy in tumor tracking (Figure 5).

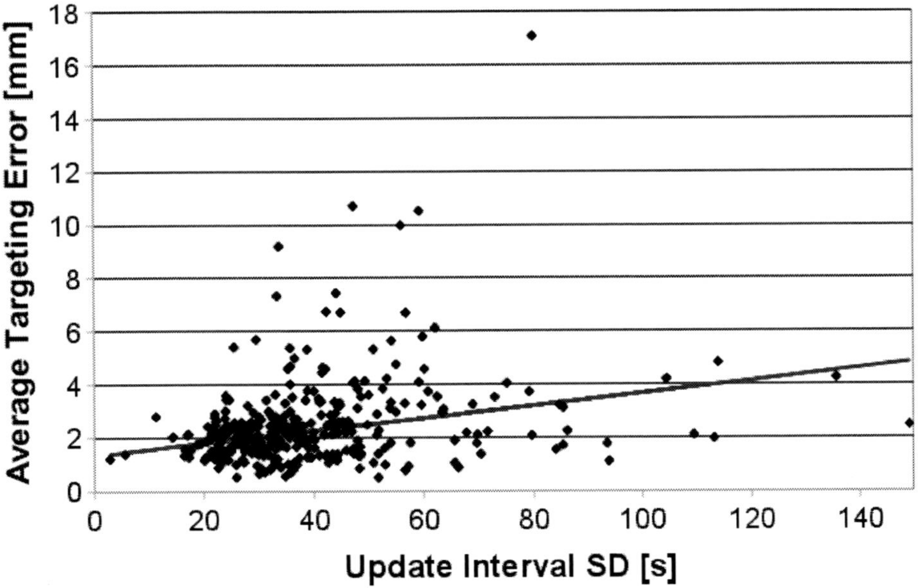

Figure 5. Targeting error dependence as a function of the variability in updating time intervals (SD). The depicted linear fit highlights the increasing trend.

Innovative Strategies

Recent efforts concerning respiratory motion modeling have addressed the problem of external/internal correlation and developed several competitive solutions [20,26-28,32,44]. The various methods differ in terms of complexity and range from linear correlation to adaptive non-deterministic models. Limited studies are available to assess the accuracy of different strategies in tumor tracking over a long time period, that is typically required for Cyberknife® treatments [7,18,33]. We present here a comparison between different approaches to infer tumor motion based on external surrogates. The implemented models were computed relying on the same training dataset used by Synchrony® and used to compute tumor motion for future timestamps. Specific model updating strategies were implemented relying on the available correlation data acquired during treatment, thus mimicking the Synchrony® update scheme. The following strategies for external/internal correlation were considered, in order to include models with an increasing level of complexity:

1. linear/quadratic correlation with implicit modeling of inhale/exhale variations [32,33]. Such model is meant to represent a basic mathematical description of the regression curves that correlate the motion of external surrogates with internal tumor motion. The main advantage of this approach is the reduced computational cost, allowing near real-time model training and updating. We will refer to this strategy as breathing *state model* in the following. The 3D movement of external markers was transformed into a mono-dimensional signal, by projecting the three-dimensional coordinates in the principal component space and considering the first principal component (PCA). The state model was implemented as a linear/quadratic correlation between external marker motion $m(t)$ and internal tumor motion $\boldsymbol{T(t)}$. Hysteresis is modeled through a time lag parameter τ, i.e. by considering $\boldsymbol{T(t)}$ to be correlated with both $m(t)$ and $m(t - \tau)$. The selection of the proper time lag was limited by the sparse imaging data included in the database. We assumed τ to be coincident with the time passing between two subsequent imaging acquisitions. Least-square minimization of tracking errors was applied to compute the matrix. After each of the imaging data was acquired, the model was updated using a first-in-first-out (FIFO) approach, with a data buffer limited to 25 imaging points. On-line monitoring of 3D tracking errors was also implemented: when errors exceeded the 5 mm threshold, the PCA analysis was updated and the state model was rebuilt as a function of the 25 most recent imaging points. At each timestamp t, the selection of linear vs. quadratic correlation was made on-line according to $m(t)$: for values outside the 80% of the total range resulting from PCA analysis, the linear model was applied to limit extrapolation errors [33]. Separate models were developed independently for the three external markers that are available in the database: final $\boldsymbol{T(t)}$ position was computed as the average position of the three models.
2. artificial neural networks (ANNs) to correlate multiple external variables through non linear transfer functions [31,34,39]. Such approach features specific generalization capabilities, as neural networks are able to integrate multiple factors without the need of a priori information, and to estimate the influence of input

parameters on the end result. In order to estimate tumor position, the outputs of three different ANNs were averaged, assuming as inputs for each network the coordinates of one of the external markers. The network architecture was designed as a single layer perceptron. The training phase in supervised learning requires that the input pattern is simulated and compared with the desired output signal. The number of example shown to the network during training has to be representative of the signal to be interpreted. Out of range peaks were then removed from training data, in order to maximize the consistency in the definition of network parameters. Since in clinical activities the number of X-ray acquisitions should be limited as much as possible, training data were interpolated to increment the examples used in training. The perceptron transfer function was an equivalent of a hyperbolic tangent sigmoid: this function is non linear and continuous, reasonably fine to manage natural events (typically non linear). As this function is derivable, it was possible to use a common back propagation algorithm during the training phase, intended here as the minimization of the mean square error between simulated and desired outputs. Given the simple architecture, it was possible to re-train at every incoming X-ray image without affecting the computational speed significantly. The re-training process relied on the same data interpolation procedure as done during the first training. Moreover, the re-training strategy was implemented according to different error-based approaches. If the maximum error measured on the incoming X-ray image was <0.3 mm, the network was just re-trained with the 30% of the most recent data; if the maximum error was not negligible (>0.3 mm), a new network was built based on the 90% of the most recent imaging points for training.

3. fuzzy logic approach based on iterative data clustering and improved membership functions, that allow intermediate values to be defined between conventional evaluations like true/false, yes/no, high/low [47-51]. Fuzzy logic offers several unique features for the implementation of correlation model. Fuzzy modeling is inherently robust since it does not require precise and noise-filtered inputs. The output is a smooth function despite a wide range of input variations. In such a model, user-defined parameters can be modified easily to improve the model performance. Fuzzy models are not limited to a few inputs and one or two outputs, nor is it necessary to measure or compute the rate-of-change of the input dataset. Due to the rule-based operation, any data point that provides some indication on internal or external motion can be included into the model. Fuzzy models can be applied on nonlinear systems that would be difficult or impossible to model mathematically [48,50]. Therefore, due to the intrinsic variability of breathing motion, a fuzzy environment may be optimal for tumor motion estimation [20]. In our proposed fuzzy correlation model, subtractive and Fuzzy C-Means (FCM) clustering algorithms [5,9,11] were considered in initial data grouping and membership function generation. First of all, the training dataset must be grouped into *n* clusters and this step is preformed using subtractive clustering with a specific influence range on training dataset. Small influence range yields many small clusters in the dataset and specifying large influence range results in few large clusters. In the implemented model the value of such influence range was set to one third of the width of the training data space, which acts as an optimal value. Then, based on FCM analyze properties, each data point in the dataset is allocated to every cluster with a certain

degree according to its distance from the cluster centers. In the other word, each data point that lies close to the center of a cluster will have a high probability of belonging to that cluster (membership grade), and another data point that lies far away from the center of a cluster will have a lower grade. The structure of the fuzzy inference system used for our predication model is based on Sugeno (or Takagi-Sugeno-Kang) model [41]. This model is computationally more efficient and gives a faster response, which is critical in situations where quick decisions should be taken. Same as ANN and state models, the fuzzy logic model can also be updated using X-ray imaging points during the treatment. In this model, updating step is performed by adding the last imaging data to the previous data and then re-building the fuzzy model using all the gathered imaging data.

Two patient groups were extracted from the database presented in the previous paragraph: 10 worst cases and 10 control cases randomly selected among the population. The worst cases were defined as those fractions exhibiting the largest average tracking errors, as stored in the log files [18]. Comparable results were highlighted for all strategies on the control cases, as confirmed by statistical non parametric tests (Friedman). The cumulative probability distribution functions (PDFs) of the 3D targeting errors are depicted in Figure 6 for worst cases. Each PDF takes into account all the imaging points acquired during treatment for model validation and update.

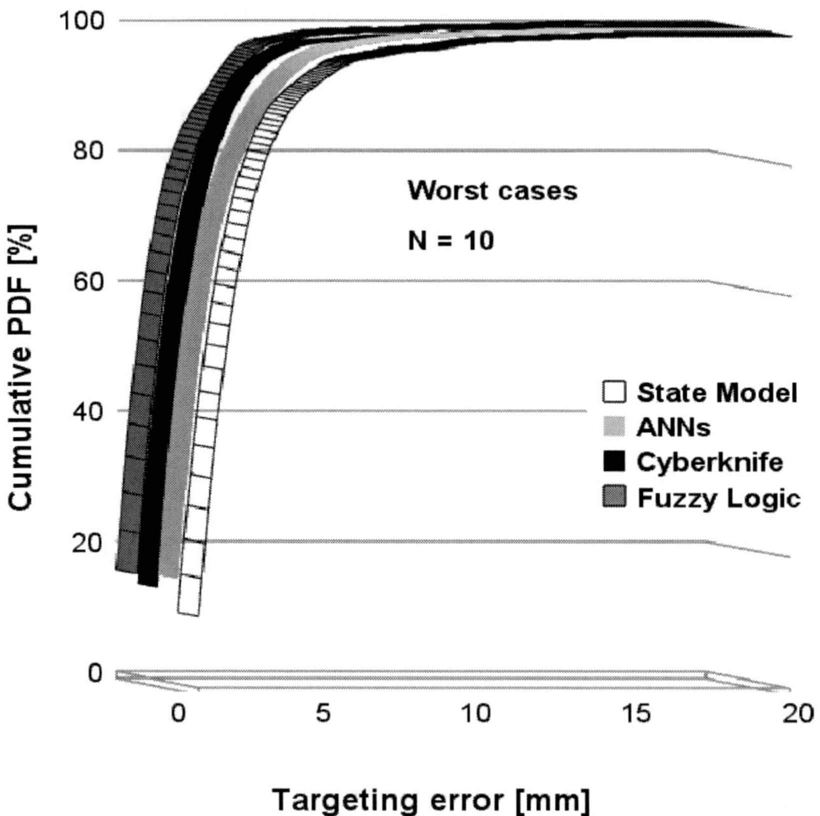

Figure 6. Cumulative PDF of targeting errors for the implemented correlation models.

Non parametric statistical analysis (Friedman test) applied to worst cases shows a significant difference among the implemented strategies (p-value$<1.5\cdot10^{-10}$). A post-hoc comparison (Dunn test) indicates that the state model has the worst performance, whereas the output of Cyberknife Synchrony®, of the ANN-based and fuzzy logic algorithms are comparable at 99% confidence. It can be noted from Figure 6 that the implemented alternative strategies ensure a reduction in the percentage of larger errors. For example, the error reduction with respect to Synchrony® at the 95% confidence level is 0.6% for the state model, 8.7% for ANNs and 10.8% for fuzzy logic.

The reported results show that all models have comparable performance in control cases, whereas fuzzy logic has the best performance when worst cases are considered. In particular, the fuzzy logic algorithm is able to partly recover tracking errors above 6 mm (Figure 6), resulting in improved accuracy for larger discrepancies. When considering the variability of results between different patients, the state model and ANN-based strategies exhibited a better stability. On the other hand, fuzzy logic seemed to be more sensitive to the available database for training. Though the overall performance was superior for fuzzy logic, one case where tracking errors are larger was exhibited.

Conclusion

Real-time tumor targeting in external beam radiotherapy is strictly dependent on the ability to correlate tumor motion with external surrogates. This is needed to optimize the therapeutic approach by minimizing unnecessary imaging dose to the patient. The reported analysis, where the state of the art for tumor targeting in high precision radiotherapy is analyzed, shows that adequate accuracy can be achieved. Caveats apply when the range of motion is large and the model updating scheme is irregular, suggesting that complex models are required to predict tumor motion over long time periods. An effective improvement with respect to current state of the art models can be obtained using a fuzzy logic approach to track target motion. Future studies will investigate the sensitivity of the implemented correlation models to the input database, in order to define optimal strategies.

Acknowledgments

The authors acknowledge Dr. Sonja Dieterich for providing access to the clinical database. The research leading to these results has received funding from the European Community's Seventh Framework Programme ([FP7/2007-2013] under grant agreement n° 215840-2).

References

[1] Balter, JM; Kessler, ML. Imaging and alignment for image-guided radiation therapy. *J. Clin. Oncol.*, 2007 25,931-937.

[2] Banovac, F; Tang, J; Xu, S; Lindisch, D; Chung, HY; Levy, EB; Chang, T; McCullough, MF; Yaniv, Z; Wood, BJ; Cleary, K. 2005 Precision targeting of liver lesions using a novel electromagnetic navigation device in physiologic phantom and swine. *Med. Phys.,* 2005 32, 2698-2705.

[3] Belderbos, J; Sonke, JJ. State-of-the-art lung cancer radiation therapy. *Expert Rev. Anticancer Ther.*, 2009 9,1353-1363.

[4] Bert, C; Saito, N; Schmidt, A; Chaudhri, N; Schardt, D; Rietzel, E. Target motion tracking with a scanned particle beam. *Med. Phys.*, 2007 34,4768-4771.

[5] Bezdek, JC. Pattern Recognition with Fuzzy Objective Function Algoritms. New York: Plenum Press; 1981.

[6] Bortfeld, T; Jokivarsi, K; Goitein, M; Kung, J; Jiang, SB. Effects of intra-fraction motion on IMRT dose delivery: statistical analysis and simulation. *Phys. Med. Biol.*, 2002 47, 2203-2220.

[7] Brown, WT; Wu, X; Fayad, F; Fowler, JF; Amendola, BE; García, S; Han, H; de la Zerda, A; Bossart, E; Huang, Z; Schwade, JG. CyberKnife radiosurgery for stage I lung cancer: results at 36 months. *Clin. Lung Cancer.,* 2007 8, 488-492.

[8] Chan, TC; Bortfeld, T; Tsitsiklis, JN A robust approach to IMRT optimization. *Phys. Med. Biol.,* 2006 51, 2567-2583.

[9] Chiu, S. Fuzzy Model Identification Based on Cluster Estimation. *Journal of Intelligent & Fuzzy Systems*, 1994 2, 267-278.

[10] Dieterich, S; Cleary, K; D'Souza, W; Murphy, M; Wong, KH; Keall, P. Locating and targeting moving tumors with radiation beams. *Med. Phys.*, 2008 35, 5684-5694.

[11] Dunn, JC. A Fuzzy Relative of the ISODATA Process and Its Use in Detecting Compact Well-Separated Clusters. *Journal of Cybernetics*, 1973 3, 32-57.

[12] Durante, M; Loeffler, JS. Charged particles in radiation oncology. *Nat. Rev. Clin. Oncol.*, 2010 7, 37-43.

[13] Evans, PM. Anatomical imaging for radiotherapy. *Phys. Med. Biol.*, 2008 53, R151-R191.

[14] Fitzpatrick, JM; West, JB. The distribution of target registration error in rigid-body point-based registration. *IEEE Trans. Med. Imaging,* 2001 20, 917-927.

[15] Fitzpatrick, JM; West, JB; Maurer, CR Jr. Predicting error in rigid-body point-based registration. *IEEE Trans. Med. Imaging,* 1998 17, 694-702.

[16] Grözinger, SO; Bert, C; Haberer, T; Kraft, G; Rietzel, E. Motion compensation with a scanned ion beam: a technical feasibility study. *Radiat. Oncol.,* 2008 14, 3-34.

[17] Guha, C; Alfieri, A; Blaufox, MD; Kalnicki, S. Tumor biology-guided radiotherapy treatment planning: gross tumor volume versus functional tumor volume. *Semin. Nucl. Med.*, 2008 38, 105-113.

[18] Hoogeman, M; Prevost, JB; Nuyttens, J; Poll, J; Levendag, P; Heumen, B. Clinical accuracy of the respiratory tumor tracking system of the cyberknife: assessment by analysis of log files. *Radiat. Oncol.,* 2009 74, 297-303.

[19] Jolesz, FA. RSNA Eugene P. Pendergrass New Horizons Lecture. Image-guided procedures and the operating room of the future. *Radiology,* 1996 204, 601-612.

[20] Kakar, M; Nyström, H; Aarup, LR; Nøttrup, TJ; Olsen, DR. Respiratory motion prediction by using the adaptive neuro fuzzy inference system (ANFIS). *Phys. Med. Biol.,* 2005 50, 4721-4728.

[21] Kang, SJ; Woo, CH; Hwang, HS; Woo, KB. Evolutionary design of fuzzy rule base for nonlinear system modeling and control. *IEEE Trans. Fuzzy Syst.,* 2000 8, 37-45.

[22] Keall, PJ; Mageras, GS; Balter, JM; Emery, RS; Forster, KM; Jiang, SB; Kapatoes, JM; Low, DA; Murphy, MJ; Murray, BR; Ramsey, CR; van Herk, MB; Vedam, SS; Wong, JW; Yorke, E. The Management of Respiratory Motion in Radiation Oncology report of AAPM task group 76. *Med. Phys.,* 2006 33, 3874-3900.

[23] Krempien, R; Hassfeld, S; Kozak, J; Tuemmler, HP; Dauber, S; Treiber, M; Debus, J; Harms, W. Frameless image guidance improves accuracy in three-dimensional interstitial brachytherapy needle placement *Int. J. Radiat. Oncol. Biol. Phys.,* 2004 60, 1645-1651.

[24] Lin, CK; Wang, SD. Fuzzy system identification using an adaptive learning rule with terminal attractors. *J. Fuzzy Sets Syst.,* 1999 101, 343-352.

[25] Maurer, CR Jr; Aboutanos, GB; Dawant, BM; Maciunas, RJ; Fitzpatrick, JM. 1996 Registration of 3-D images using weighted geometrical features. *IEEE Trans. Med. Imag.,* 1996 15, 836-849.

[26] Murphy, MJ; Dieterich, S. Comparative performance of linear and nonlinear neural networks to predict irregular breathing. *Phys. Med. Biol.,* 2006 51, 5903–5914.

[27] Ramrath, L; Schlaefer, A; Ernst, F; Dieterich, S; Schweikard, A. Prediction of respiratory motion with a multi-frequency based Extended Kalman Filter. Proceedings of the 21st International Conference and Exhibition on Computer Assisted Radiology and Surgery (CARS'07), Vol. 21, CARS, Berlin, Germany (2007).

[28] Riaz, N; Shanker, P; Gudmundsson, O; Wiersrma, R; Mao, W; Widrow, B; Xing, L. Predicting respiratory tumor motion with Multi-dimensional Adaptive Filters and Support Vector Regression. *Phys Med Biol.,* 2009 54, 5735-5718.

[29] Riboldi, M; Sharp, GC; Baroni, G; Chen, GT. Four-dimensional targeting error analysis in image-guided radiotherapy. *Phys. Med. Biol.*, 2009 54, 5995-6008.

[30] Rietzel, E; Bert, C. Respiratory motion management in particle therapy. *Med. Phys.*, 2010 37, 449-460.

[31] Robert, C; Gaudy, JF; Limoge, A. Electroencephalogram processing using neural networks. *Clinical Neurophysiology,* 2002 113, 694-701.

[32] Ruan, D., Fessler, J.A., Balter, J.M., Berbeco, R.I., Nishioka, S., Shirato, H. Inference of hysteretic respiratory tumor motion from external surrogates: a state augmentation approach. *Phys. Med. Biol.,* 53, 2923-2936 (2008).

[33] Seppenwoolde, Y; Berbeco, RI; Nishioka, S; Shirato, H; Heijmen, B. Accuracy of tumor motion compensation algorithm from a robotic respiratory tracking system: a simulation study. *Med. Phys.,* 2007 34, 2774-2784.

[34] Sharp, GC; Jiang, SB; Shimizu, S; Shirato, H. Prediction of respiratory tumour motion for real-time image-guided radiotherapy. *Phys. Med. Biol.,* 2004 49, 425-440.

[35] Sharpe, MB; Craig, T; Moseley, DJ. Image guidance: treatment target localization systems. *Front. Radiat. Ther. Oncol.*, 2007 40, 72-93.

[36] Shi, Y; Eberhart, R; Chen, Y. Implementation of evolutionary fuzzy systems. *IEEE Trans. Fuzzy Syst.,* 1999 7, 109-119.

[37] Shirato, H; Shimizu, S; Kunieda, T; Kitamura, K; Kagei, K; Nishioka, T; Hashimoto, S; Fujita, K; Aoyama, H. Physical aspects of a real-time tumor-tracking system for gated radiotherapy. *Int. J. Radiat. Oncol., Biol., Phys.,* 2000 48, 1187-1195.

[38] Shirato, H; Shimizu, S; Shimizu, T; Nishioka, T; Miyasaka, K. Real-time tumour-tracking radiotherapy. *Lancet,* 1999 353, 1331-1332.

[39] Su, M; Miften, M; Whiddon, C; Sun, X; Light, K; Marks, L. An artificial neural network for predicting the incidence of radiation pneumonitis. *Med. Phys.,* 2005 32, 318-325.

[40] Takagi, T; Sugeno, M. Fuzzy identification of systems and its application to modeling and control. *IEEE Trans. Syst. Man, Cybern.*, 1985 15, 116-132.

[41] Torshabi, A; Pella, A; Riboldi, M; Baroni, G. Targeting accuracy in real-time tmor tracking via external surrogates: a comparative study. *Technol. in Cancer Res. Treat.*, 2010 in press.

[42] Trofimov, A; Rietzel, E; Lu, HM; Martin, B; Jiang, S; Chen, GTY; Bortfeld, T. Temporo-spatial IMRT optimization: concepts, implementation and initial results. *Phys. Med. Biol.,* 2005 50, 2779-98

[43] van de Water, S; Kreuger, R; Zenklusen, S; Hug, E; Lomax, AJ. Tumour tracking with scanned proton beams: assessing the accuracy and practicalities. *Phys. Med. Biol.*, 2009 54, 6549-6563.

[44] Vedam, SS; Keall, PJ; Docef, A; Todor, DA; Kini, VR; Mohan, R. Predicting respiratory motion for four-dimensional radiotherapy. *Med. Phys.,* 2004 31, 2274–2283.

[45] Wei, Z; Wan, G; Gardi, L; Mills, G; Downey, D; Fenster, A. Robot-assisted 3D-TRUS guided prostate brachytherapy: system integration and validation. *Med. Phys.,* 2004 31, 539-548.

[46] West, JB; Fitzpatrick, JM; Toms, SA; Maurer, CR Jr; Maciunas, RJ. Fiducial point placement and the accuracy of point-based, rigid body registration. *Neurosurgery,* 2001 48, 810-817.

[47] Zadeh, LA. Fuzzy algorithms. *Info. & Ctl.*, 1968 12, 94-102.

[48] Zadeh, LA. Fuzzy Logic. *Computer,* 1998 1, 83-93.

[49] Zadeh, LA. Fuzzy Sets. *Information and Control*, 1965 8, 338-353.

[50] Zadeh, LA. Knowledge representation in fuzzy logic. *IEEE Transactions on Knowledge and Data Engineering,* 1989 1, 89-100.

[51] Zadeh, LA. Outline of A New Approach to the Analysis of of Complex Systems and Decision Processes. *IEEE Trans. Systems, Man and Cybernetics*, 1973 3, 28-44.

In: Brain Cancer, Tumor Targeting and Cervical Cancer
Editor: Elena K. Salvatti
ISBN: 978-1-61122-738-3

Chapter XI

Review of Tumours of the Paranasal Sinuses Based on 14 Cases*

***Alaitz Santamaría Carro*[1,†] *Sergio Pinar-Sueiro*[1], *Adolfo Vergez Sánchez*[2] *and Roberto Víctor Fernández Hermida*[1]**

[1]Department of Ophthalmology (Orbit and Oculoplasty Unit). Hospital of Cruces, Barakaldo, Spain

[2]Department of ENT, Hospital of Cruces, Barakaldo, Spain

Opthalmologists and ENT specialists are able to obtain good surgical access by the mediofacial approach to tumours of the paranasal sinuses with optimal aesthetic and functional outcomes

Abstract

Purpose

To review tumours of the paranasal sinuses and to analyse the characteristics of the surgical approach using lateral orbitotomy.

---

* Submitting institution Ophthalmology Department (Orbit and Oculoplasty Unit). Hospital of Cruces, Barakaldo (Spain).
This study was partially presented in the XII National Meeting of the Spanish Medical Residents in Ophthalmology (La Toja, 2007), being awarded a prize for the best presentation.

† Author for correspondence: Dr Alaitz Santamaría Carro. C/Ganguren Nº6, 5ºB, Ortuella 48530 (Vizcaya), Spain. Telephone: 692133038; E-mail: alaitz_sc@yahoo.es

Material and Methods

We present a series of 14 patients with various types of tumours of the paranasal sinuses, treated in our centre. We present a descriptive, retrospective study including the clinical picture of the patients, diagnostic methods used in consultations and imaging scans, medical-surgical management and long-term prognosis.

Results

All the subjects included in the study underwent paranasal tumour resection through a lateral approach. In two cases, this was complemented by superomedial and transpalpebral orbitotomy, and one of the patients required a bilateral approach. Three patients had tumour relapses; two of them being successfully operated on, with subsequent remission of the tumours, while the other patient did not undergo surgery due to the degree of systemic involvement and prognosis.

Conclusion

The lateral approach allows good surgical access to tumours of the paranasal sinuses. Acceptable aesthetic and functional outcomes are obtained with the current techniques of reconstruction of the facial skeleton after radical resection.

Keywords: Paranasal sinuses, epidermoid carcinoma, midfacial approach, lateral orbitotomy, embolisation, Medpor®

Introduction

The tumours of the paranasal sinuses require multidisciplinary management in which Ear, Nose & Throat (ENT), Ophthalmology, Maxillofacial Surgery and Plastic Surgery specialists are involved. The psychosocial impact of the neoplastic process for the patient can be anticipated; loss of visual function being among its notable repercussions. Current treatments aim not only to save the patient´s life, but also to preserve vision and achieve acceptable aesthetic outcomes to improve the patient´s quality of life.

The aim of this study is to describe and classify this pathology of the paranasal sinuses through a descriptive and retrospective study of 14 subjects suffering from neoplasia of the paranasal sinuses and to illustrate the characteristics of the midfacial surgical approach, examining differences with respect to other techniques.

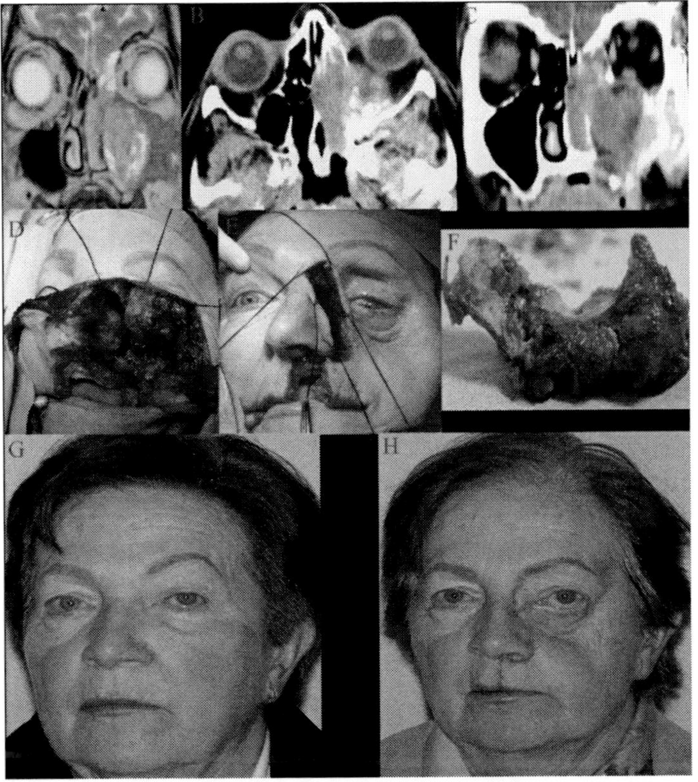

Figure 1. Epidermoid carcinoma of the left maxillary sinus with ethmoid sinus involvement, left nasal cavity and medial wall and floor of the orbit. A: coronal MRI (T2-weighted sequence). B: axial CAT scan (bone setting). C: coronal CAT scan (bone setting). D, E: Midfacial approach with en bloc resection of the lesion. F: Surgical sample including a section of the zygomatic arch. G: Preoperative photograph of the patient. H: Postoperative photograph showing surgical asymmetry, pending reconstruction.

Methods

We present a descriptive and retrospective study that includes 14 patients assessed and treated in our centre. All the cases corresponded to people with tumours of the paranasal sinuses, in various locations and with different aetiologies and pathogenesis, who required monitoring and multidisciplinary treatment by specialists from the ENT Department and Orbit and Oculoplasty Unit.

By reviewing the clinical histories of the patients, data was collected on their epidemiological and clinical characteristics (age, sex, general clinical history and specifically any eye problems), on the clinical episodes that were the focus of the study (advance of the illness and clinical signs and symptoms), and on the physical examinations of the eyes and orbits (visual acuity, Ishihara test, pupillary response, anterior and posterior chamber biomicroscopy, tonometry, Schirmer test, tear duct irrigation test, Hess Lancaster screen test, profile, position and movements of the eyelids, extraocular muscle function test, ocular and periorbitary sensitivity, palpation of orbital masses and inspection of regional lymph node basins). In addition, key digital images were assembled, including both the MRI and CT scans, and intraoperative and pre-post operatory photographs of patients.

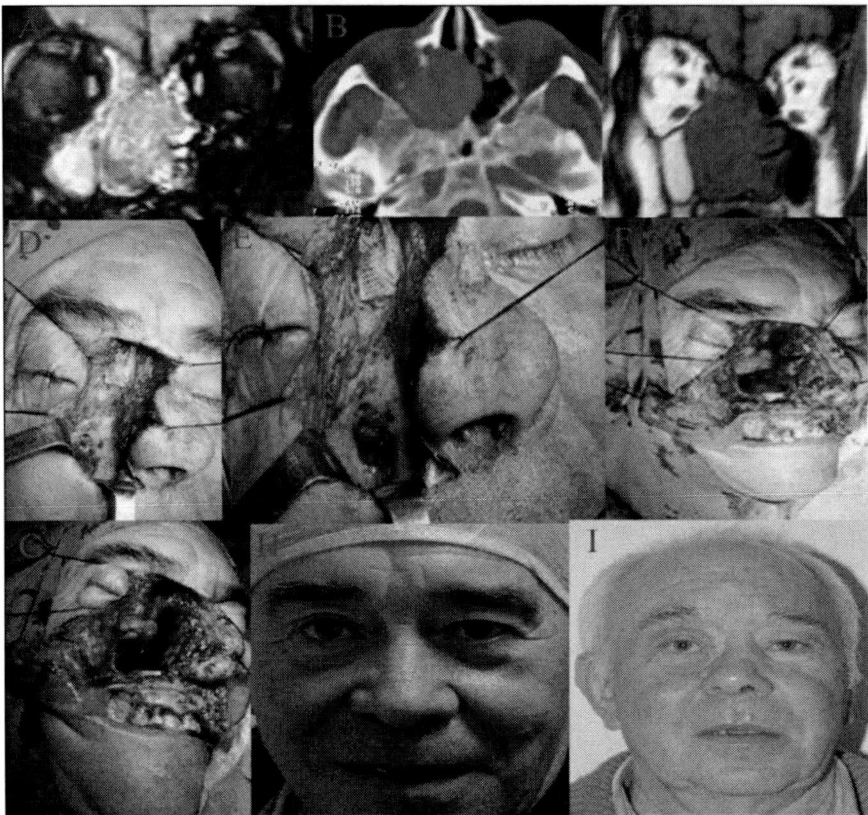

Figure 2. Adenocarcinoma of the right ethmoid and maxillary sinuses with involvement of the right nasal cavity, medial wall and floor of the right orbit. A: coronal MRI (T1-weighted sequence with gadolinium). B: axial CAT scan (bone setting). C: coronal MRI (T2-weighted sequence). D, E, F, G: Midfacial approach with en bloc resection of the lesion. H: Preoperative photograph. I: Postoperative photograph.

Patients freely agreed to the use of the clinical images, and signed the corresponding informed consent form which has enabled us to edit the images for their inclusion in this manuscript.

Results

We present a series of 14 patients with tumours of the paranasal sinuses, aged between 30 and 79 years. The following tumours were reported: one case of epidermoid carcinoma of the maxillary sinus (Figure 1), one case of ethmoid sinus carcinoma, one of adenocarcinoma of the ethmoid sinus, two adenocarcinomas of the ethmoid and maxillary sinuses (Figures 2, 3 & 4), one of the maxillary sinus, poorly differentiated (Figure 5), one meningioma of the ethmoid sinus (Figure 6), one ossifying fibroma of the ethmoid and sphenoid sinuses (figure 7), one osteoma of the ethmoid sinus (Figure 8), one osteoma of the frontal sinus, one frontal sinus mucocele, one maxillary sinus mucocele (Figure 9), one rhabdomyosarcoma of the nasal cavity (Figure 10) and one undifferentiated nasosinusal carcinoma with neuroendocrine features.

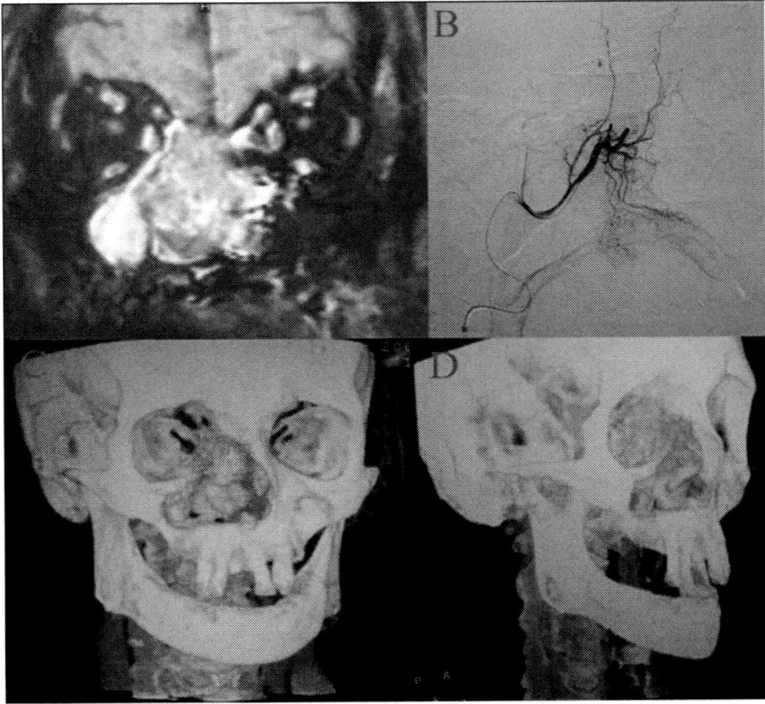

Figure 3. Adenocarcinoma of the right ethmoid and maxillary sinuses with involvement of the right nasal cavity, medial wall and floor of the right orbit. A: axial MRI (T1-weighted sequence with gadolinium). B: Arteriography through the external carotid artery, prior to embolisation of the tumour blood vessels. C, D: 3-dimensional reconstruction from the cranial CAT scan and of the facial skeleton, revealing the condition of the bone structure after combined treatment with surgery and radiotherapy.

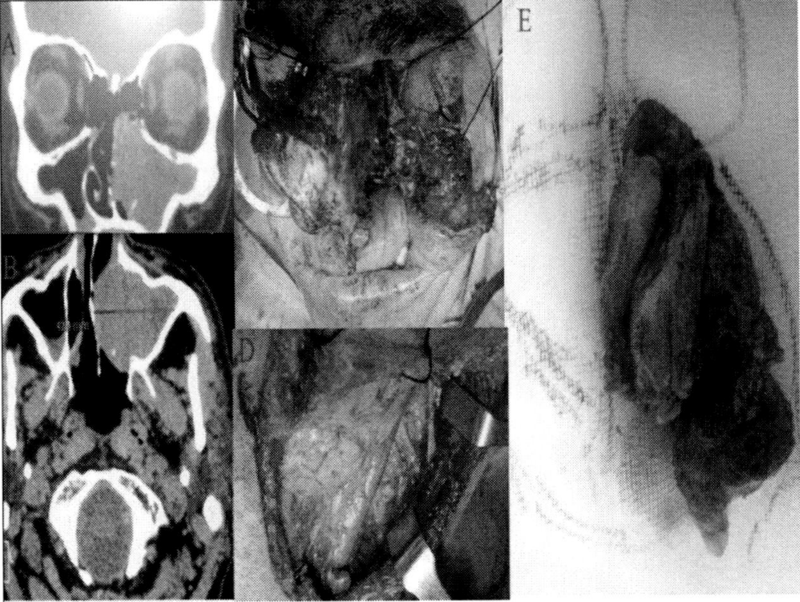

Figure 4. Adenocarcinoma of the left ethmoid and maxillary sinuses with involvement of the orbital floor. A: coronal CAT scan of the orbit. B: Axial CAT scan of the orbit. C: Surgical approach. D: Lymphadenectomy of the cervical lymph node basins. E: Mass removed.

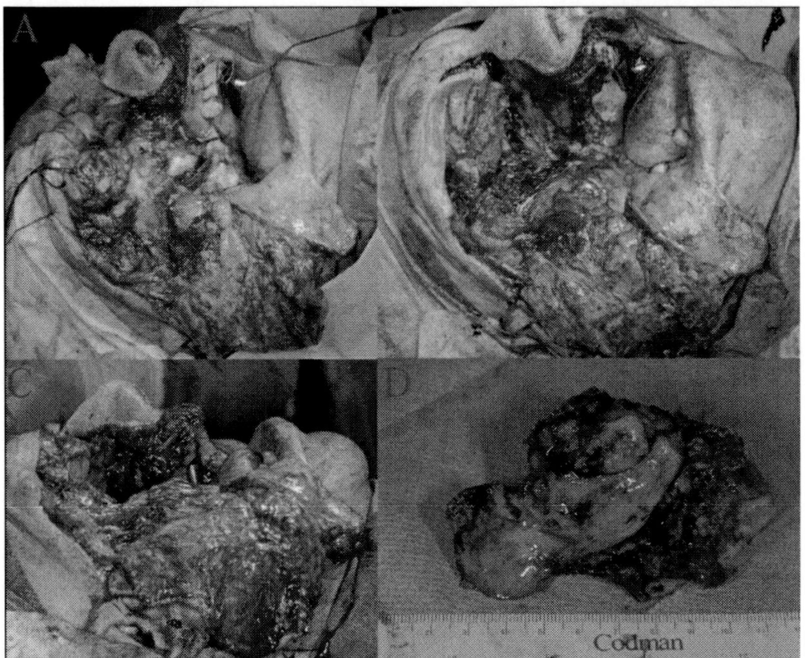

Figure 5. Adenocarcinoma of the maxillary sinus poorly differentiated. A: Midfacial approach with en bloc resection of the tumour and application of a titanium mesh for reconstructing the floor of the orbit. . B: Removed lesion (surgical piece).

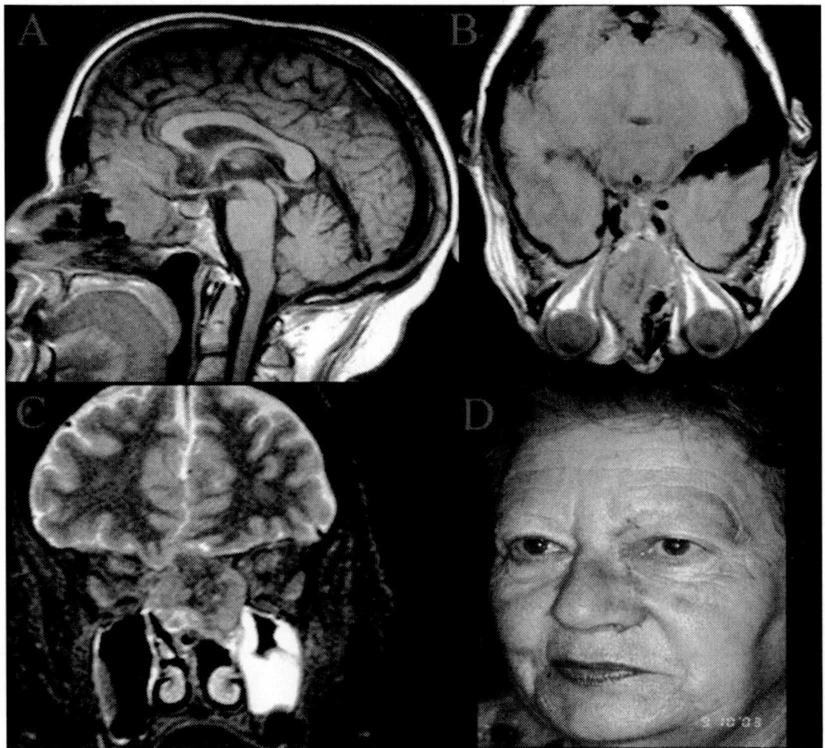

Figure 6. Meningioma of the ethmoid sinus. A: sagittal MRI (T1-weighted sequence). B: axial MRI (T1-weighted sequence). C: coronal MRI (T1-weighted sequence). D: Postoperative photograph of the patient.

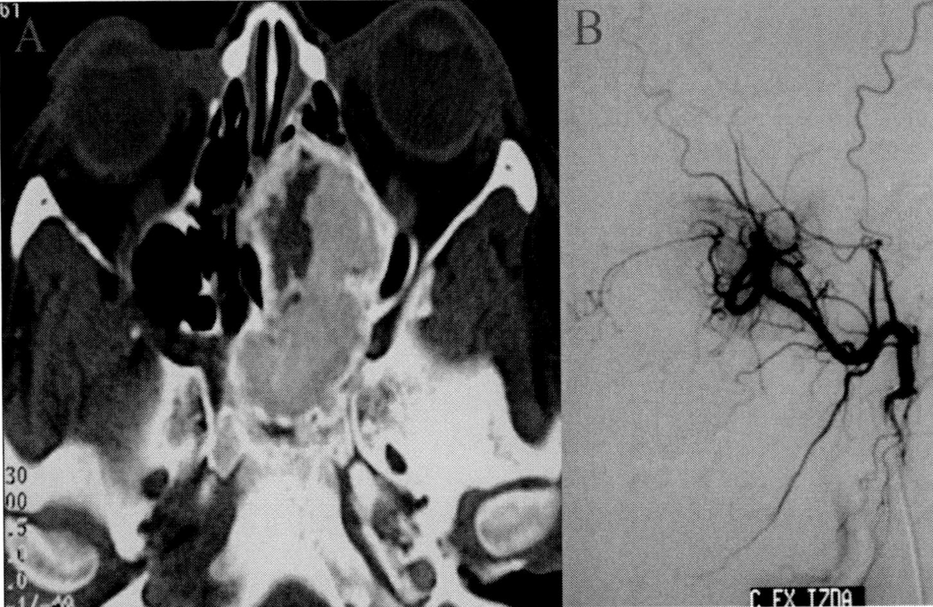

Figure 7. Ossifying fibroma of the ethmoid and sphenoid sinuses. A: axial CAT scan (bone setting). B: Arteriography through the external left carotid artery, prior embolisation of the tumour blood vessels.

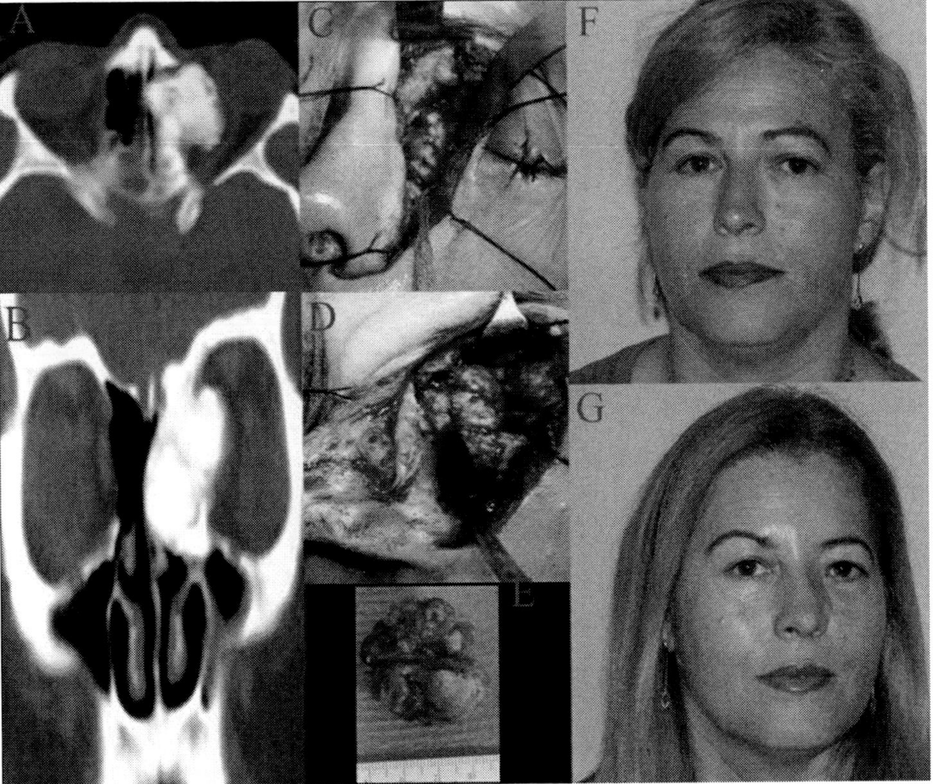

Figure 8. Osteoma of the left ethmoid sinus. A: Axial CAT scan of the orbit (bone setting). B: coronal CAT scan of the orbit (bone setting). C, D: Midfacial surgical approach for en bloc removal of the lesion. E: Surgical sample. F: Preoperative photograph. G: Postoperative photograph.

Clinical Patterns

All the patients showed a slow progression and had masked and vague signs and symptoms, which made an early diagnosis difficult. The common symptoms of patients were cephalea, retro-ocular pain, respiratory insufficiency and recurrent sinusitis. In terms of ophthalmologic manifestations, two cases of mucocele caused episodes of intermittent diplopia. Five cases presented clear displacement of the eye ball: with frontal sinus mucocele and epidermoid carcinoma of the ethmoid sinus causing downward and inferotemporal displacement respectively. The slow course of the lesions appeared to allow time for neuro-ophthalmological compensatory mechanisms thus avoiding the onset of permanent diplopia. Subjects with epidermoid carcinoma of the maxillary sinus, poorly differentiated maxillary sinus adenocarcinoma and osteoma of the ethmoid sinus were found to have more than 2 mm of asymmetry using the Hertel exophthalmometer. No deterioration in visual acuity was observed nor were there changes in the anterior segment or the fundus of the eye in any of these cases. However, on examination, five patients were found to have developed complete obstruction of the ipsilateral tear duct (Table 1).

In five of the cases it was possible to palpate the neoplasia or detect the lesion by direct visualization, be it through the superior medial angle of the eyelid (osteoma of the ethmoid sinus and meningioma of the ethmoid sinus, rhabdomyosarcoma of the nasal cavity) or the nasal cavity (epidermoid carcinoma of the ethmoid and maxillary sinuses, ossifying fibroma of the ethmoid and sphenoid sinuses, and rhabdomyosarcoma of the nasal cavity).

Table 1. Clinical signs and symptoms of patients with tumours of the paranasal sinuses

	FM	MM	ECM	ECE	ACE	ACEM1	ACEM2	PDAM	FO	EO	ME	OFE	RSFN	UCN
Headache	+					+	+	+	+	+	+	+	+	+
Recurrent sinusitis		+		+	+	+	+	+	+				+	+
Respiratory insufficiency			+	+	+	+	+			+		+	+	+
Intermittent diplopia	+	+				+	+	+					+	+
Eyeball displacement	+			+			+	+					+	
Exophthalmus			+					+		+				
Mass in the medial angle										+	+		+	
Mass visible in the nasal cavity				+								+	+	
Tear duct obstruction			+				+	+					+	+

FM: frontal sinus mucocele; MM: maxillary sinus mucocele; ECM: epidermoid carcinoma of the maxillary sinus ; ECE: epidermoid carcinoma of the ethmoid sinus; ACE: adenocarcinoma of the ethmoid sinus; ACEM 1 & 2: adenocarcinoma of the ethmoid and maxillary sinuses; PDAM: poorly differentiated adenocarcinoma of the maxillary sinuses; FO: frontal sinus osteoma; OE: osteoma of the ethmoid sinus ; ME: meningioma of the ethmoid sinus; OFE: ossifying fibroma of the ethmoid sinus; RSNC: rhabdomyosarcoma of the nasal cavity; UCN: undifferentiated carcinoma of the sinuses with neuroendocrine features

Preoperative Embolisation

Eleven subjects (Table 2) underwent embolisation of the tumour-feeding vessels prior to surgery in order to reduce the tumour mass by deprivation of blood supply. The vessels targeted were various branches of the external carotid, depending on the tumour location and vascularisation.

Table 2. Treatment and postoperative evolution of patients with paranasal sinus tumours. Abbreviations as defined for Table 1

	Approach	Embol.	Aden.	RTP	Recon-struction	Recur.	Secuelae
FM	Superomedial orbitotomy	-	-	-	-	-	-
MM	Lateral paranasal	-	-	-	-	-	-
ECM	Lateral paranasal	+	+	-	Orbital floor Titanium mesh	+	Reconstruction of the zygomatic arch Involvement of the tear duct
ECE	Lateral/superior medial paranasal	+	+	-	Medial wall /orbital floor Medpor® sheet	-	Macroscopic remains in the cribriform plate
ACE	Lateral paranasal	+	+	+	Medial wall /orbital floor Medpor® sheet	-	Involvement of the tear duct
ACEM1	Lateral paranasal	+	+	+	Medial wall /orbital floor Medpor® sheet	-	Involvement of the tear duct
ACEM2	Lateral paranasal	+	+	+	Medial wall /orbital floor Medpor® sheet	-	Involvement of the tear duct
PDAM	Lateral paranasal	+	+	+	Medial wall /orbital floor Titanium mesh	-	Involvement of the tear duct
FO	Through the upper eyelid	-	-	-	Upper/medial wall Medpor® sheet	-	-
EO	Lateral paranasal	+	-	-	Medial wall of the orbit Medpor® sheet	-	-
ME	Lateral paranasal	+	-	+	Medial wall of the orbit Medpor® sheet	+	Involvement of the tear duct
OFE	Lateral paranasal	+	-	-	-	-	-
RSFN	Lateral paranasal, bilateral	+	+	+	Medial wall of the orbit Medpor® sheet	+	Involvement of the tear duct
UCN	Lateral paranasal	+	+	+	Medial wall /orbital floor Medpor® sheet	-	Involvement of the tear duct

Surgical Technique

In most cases access to the tumours was obtained through a lateral orbitotomy, extended to the superomedial orbital region and/or the nasal cavity in the cases which required a wider surgical field. In the cases of frontal sinus mucocele and osteoma of the frontal sinus the approaches taken were superomedial orbitotomy and through the upper eyelid, respectively, since there were masses located superomedially to the orbit. In the case of rhabdomiosarcoma of the nasal cavity, the lesion was removed en bloc through a bilateral midfacial approach (Figure 10). Neoplasia that infiltrated the neighbouring bone structures (maxillary bone, zygomatic bone, nasal septum, nasal bones and etmoid bone, among others) required removal of the affected sections of bone using a chisel and/or curettage as well as of the tumour mass en bloc. As far as possible, supra and infraorbital neurovascular bundles, the tear duct, the eyeball and other intraorbital tissues were left intact.

In individuals in which invasion had reached the orbital walls, reconstruction was necessary, using *Medpor®* sheets, microplates or titanium meshes (Table 2) (Figure 5), during the same surgical procedure. We have not recorded any cases of iatrogenic diplopia in two years of monitoring. Five patients presented with tear duct obstruction before surgery, while eight developed this problem after intervention and three of these were treated surgically. In one case it was not possible to avoid leaving macroscopic tumour remnants (Table 2), while in another case there was clear postoperative facial asymmetry, which required surgical reconstruction at a later stage (Figure 1).

Lymphadenectomy (Table 2, Figure 4)

Eight patients underwent lymphadenectomy at an early stage with midfacial approach.

Radiotherapy (Table 2)

Six patients were given postoperative adjuvant treatment using radiotherapy.

Relapses

Tumour relapse occurred in three patients. Of these, two appeared posterosuperior to the original tumour: the epidermoid carcinoma of the ethmoid sinus with involvement of the lamina papyracea, cribriform plate and frontal sinus (recurrence one year after the first surgical intervention), and meningioma located in the nasal cavity which had spread to the ethmoids (relapse two years after the initial surgery) (Table 2, Figure 3). These individuals underwent surgery again, in combination with preoperative embolisation and radiotherapy, and are currently stable. The third patient did not undergo further surgery or other adjuvant treatments given her short life expectancy (Figure 10).

Discussion

Tumours of the nasal cavity and paranasal sinuses represent between 0.2% and 1% of all tumours detected and between 3 to 6% of head and neck neoplasias [1,2,3]. Specifically, epidermoid carcinoma of the maxillary sinus is the most common, while in general tumours of the maxillary sinus are the most common group (70-80%) followed by those of the ethmoid sinus (10-20%), [1,2].

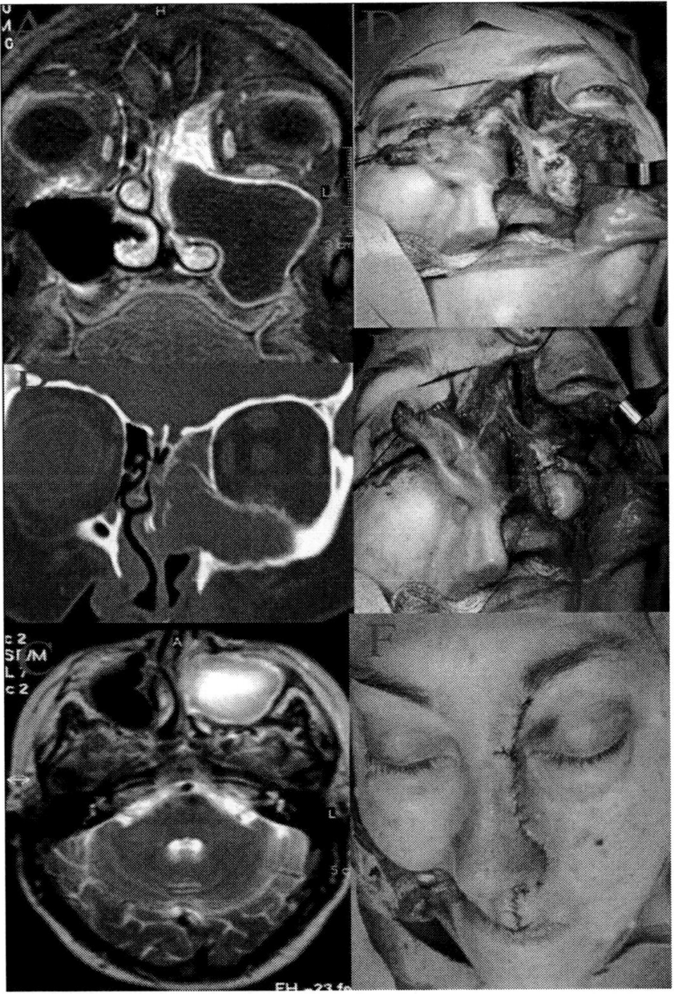

Figure 9. Left maxillary sinus mucocele with involvement of the left ethmoid sinuses and nasal cavity. A: coronal MRI (T1-weighted sequence). B: coronal CAT scan (bone setting). C: axial MRI (T1-weighted sequence). D, E: Midfacial approach for removal and curettage of the lesion. F: Postoperative outcome.

A higher incidence has been observed in Asian countries (specifically Japan and Indonesia), and there seems to be a relationship between exposure to saw dust and to formaldehyde in the tanning process and adenocarcinoma of the paranasal sinuses. In addition, women with long-term exposure to dust from textiles also have a greater risk of suffering form adenocarcinomas; while, on the other hand, squamous cell carcinoma in men has been linked to exposure to asbestos and work in textile and leather industries, as well as in construction and handling of organic waste [4]. Other studies have concluded that for the relation of these cancers with tobacco and alcohol there is not as epidemiological evidence as there is in the case other head and neck tumours [2]..

Since Denoix established the TNM staging system for cancer, several other classifications have been created [2]. In 1933, Ohngren classified the tumours of the maxillary sinus taking as a reference the imaginary line that runs from the medial canthus of the eyelid to the angle of the mandible.

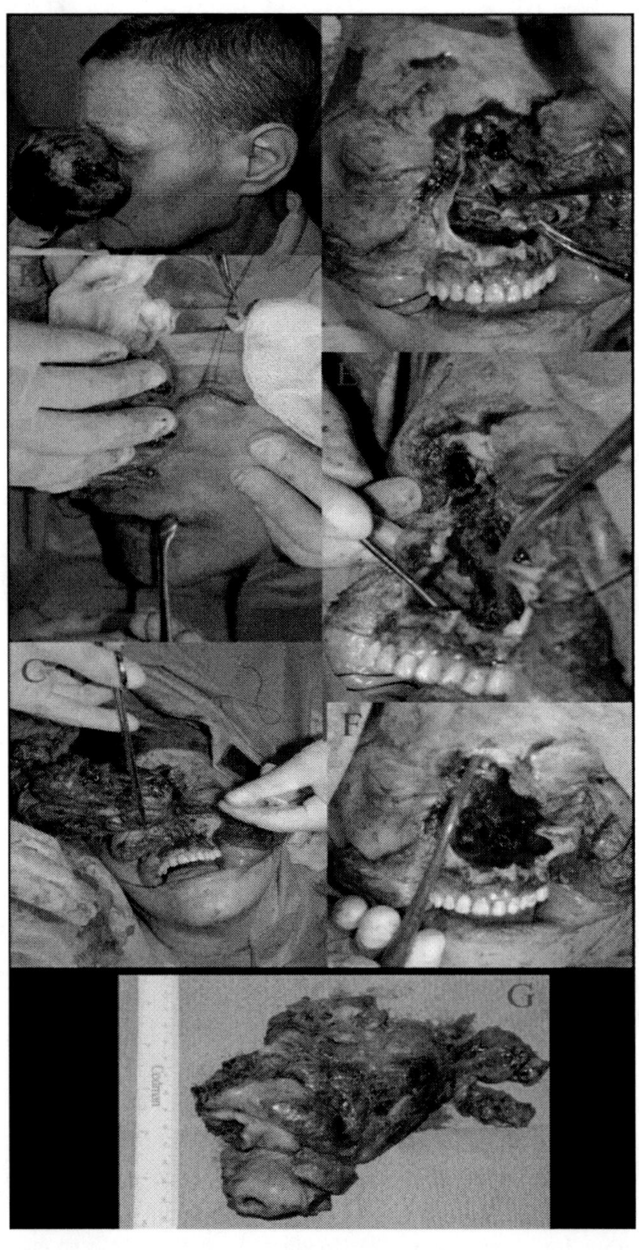

Figure 10. Rhabdomyosarcoma of the nasal cavity. A: Preoperative picture. B: Palpebral fixation with 2/0 Prolene® prior to bilateral midfacial approach (C, D, E, F). G: Surgical sample.

The concept of tumour of the paranasal sinuses (specifically of the maxillary sinus) was originally published in the fourth edition of the *TNM Classification of Malignant Tumors of the Union Internationale Contre le Cancer (UICC)* and the AJCC Cancer Staging Manual of the *American Joint Committee on Cancer (AJCC)*. In 1977, the UICC established "the ethmoid sinus" as a cancer site, and in 2002, it included cancer of the "nasal cavity" (Table 3 and 4) [2,5].

Table 3. TNM classification for tumours of paranasal sinuses

TNM CLASSIFICATION
T
Tx Primary tumour cannot be assessed
To No evidence of primary tumour
T1 Tumour restricted to the sinus with no bone damage
T2 Tumour with erosion or destruction of the infrastructure including the hard palate or the middle nasal meatus
T3 Tumour invades the skin of the cheek, posterior wall of the sinus and floor of the orbit and anterior ethmoid sinus
T4 Tumour invades the orbit and cribriform plate, posterior part of the ethmoid bone, sphenoid sinus, nasopharynx, soft palate, temporal fossa or the base of the skull
N
Nx Lymph node involvement cannot be assessed
No Tumour cells absent from regional lymph nodes
N1 Metastasis in only one ipsilateral regional lymph node, smaller than or equal to 3cm
N2a Metastasis in only one ipsilateral regional lymph node, 3-6 cm N2b Several ipsilateral lymph nodes < 6 cm N2c Bilateral or contralateral lymph nodes < 8 cm
M
Mx The existence of metastasis cannot be assessed
M0 No metastasis
M1 Metastasis to distant organs

According to the scientific literature, there is great variability in the degree of orbital involvement in paranasal tumours (35-74%). This can be attributed to the different definitions given to "orbital invasion". Some authors include the tumours that are close to the orbit but which do not involve its walls. On the other hand, Ianetti et al., establish some degree of erosion of the wall of the bone as a key requisite [5,6]. In our series, 78.57% of patients (11 out of 14) had orbital involvement, according to Ianetti´s criteria.

With regards to tumour staging, several studies have confirmed a lower rate of survival in stage T4, referring in particular to carcinoma of the maxillary sinuses, especially if there is orbital and/or ethmoid invasion, lymph node involvement or distant metastasis. Guntinas-Lichius et al. recommend radical surgery in stages I and II while proposing multimodal treatment for stages III and IV cases despite their grave prognosis [7,8,9].

Table 4. TNM staging system for tumours of paranasal sinuses

	N0	N1	N2	N3	M1
T1	Stage I	Stage III	Stage IVb	Stage IVb	Stage IVc
T2	Stage II	Stage III	Stage IVb	Stage IVb	Stage IVc
T3	Stage III	Stage III	Stage IVb	Stage IVb	Stage IVc
T4	Stage IVa	Stage IVa	Stage IVb	Stage IVb	Stage IVc
M1	Stage IVc	Stage IVc	Stage IVc	Stage IVc	Stage IVc

The clinical manifestations of the tumours depend on their location, the involvement of nearby structures and the evolution of the disease. Generally, early diagnosis is difficult because initial signs and symptoms are mild and vague. In our series, cephalea and pain behind the eyes, soft tissue oedema and sinusitis and/or episodes of recurrent orbital cellulitis were common. Tumours of the frontal sinus may cause downward displacement of the eyeball, unilateral facial numbness and visible and/or palpable masses. Ethmoid sinus tumours vary according to their location within the sinus. Those located in the anterior part of the sinus cause swelling of the medial angle of the eyelids, excessive tearing due to tear duct involvement and regional hypoesthesia. On the other hand, those occurring in the posterior part of the sinus can cause inferotemporal displacement of the eyeball and may invade the extraocular muscles and optic nerve, causing ptosis, diplopia and decreased visual acuity. Tumours of the nasal cavity tend to start with respiratory insufficiency and epistaxis and by local invasion may spread to affect the tear duct and orbital structures. With regards to the maxillary sinus tumours, the clinical signs and symptoms are determined by their growth and proximity to local structures. They can affect the eyeball (upward displacement, diplopia, orbital apex syndrome, etc.), the tear duct, and sensitivity of the cheek, nasal cavity and palate. Spread towards intracranial structures may cause various neurological manifestations, such as superior orbital fissure syndrome, cavernous sinus syndrome, Foster-Kennedy syndrome, and intracranial hypertension, among others [11].

The slow evolution and vague signs and symptoms mean that the patients suffering from paranasal sinus cancer seek medical help late, making early diagnosis difficult. The taking of detailed medical histories and careful examinations, as well as the carrying out the relevant medical tests, are key when there is any suspicion of cancer. Antero-inferior tumours have a better prognosis, since they cause signs and symptoms in the nasal cavity and palate that can be detected relatively early on and they seldom affect intracranial vital structures. On the other hand, tumours that are located in the postero-superior area have a grave prognosis, since they often invade the orbital cavity, the sphenoid sinuses and the base of the skull. In our series, the symptoms were hidden until diagnosis, the extent of lesions only coming to light at the time of diagnosis.

CAT and MRI are most useful imaging techniques. They reveal the size and density of the tumour, involvement of neighbouring structures and destruction of bone walls and/or osteoblastic reactions in cases of slow progression. It is also possible to study the behaviour of the tumour after injection with radiopaque contrast media. However, the main objective of these scans is to identify the surgical boundaries of the tumour, in cases where it is resectable.

It is common to perform arteriography to analyse the vascularisation of the tumour, making it possible to plan embolisation of the blood vessels of the tumour prior to surgery, in order to decrease its size and maintain haemostasis during surgery, and ultimately improve prognosis.

We should not forget the need for further cancer screening tests to look for lymph node and distant metastases, since the therapeutic strategy will depend on the tumour staging.

Prior to taking a surgical approach to paranasal neoplasia, it is essential to assess the extent of the lesion and the existence of potential concomitant conditions which might influence the general state of the patient. The objective of curative surgery is to remove the whole tumour, as well as neighbouring structures that are affected. Radical resections that were classically carried out on individuals with periorbital involvement can be now avoided thanks to the use of radiotherapy, chemotherapy, and other techniques such as the frozen section-guided tumour resection procedure introduced by McCary et al [11]. The current tendency is to opt for conservative curative surgery to achieve acceptable results in terms of functionality and aesthetics. However, tumours with intracranial involvement beyond the dura mater, metastasis in regional lymph nodes or distant spread are considered to be unresectable, hence the importance of a detailed study of the extent of the tumour prior to undertaking surgery.

In general terms, anterior and inferior paranasal lesions are easier to approach, as opposed to those in posterior and superior areas, due to the difficulty of preparing the surgical site and the involvement of vital structures.

In our series, we almost exclusively used the lateronasal approach. This technique consists of extending downwards, towards the ala of nose, the classic Lynch orbitotomy, resulting in a vertical incision approximately 5-6 mm from the medial canthus of the eyelid (towards the nose), and about 2-3 cm long. Lateral orbitotomy allows access to tumours in the maxillary and ethmoid sinuses and, in the case of tumours of ethmoid and frontal sinuses, can be combined with orbitotomy from the superior medial angle (following the superomedial orbital ridge) or transpalpebral orbitotomy (along the superior palpebral fold), as found in the cases of the frontal sinus mucocele and osteoma of the frontal sinus in our series. If the periostium of the paranasal sinus and/or orbit were not affected, we carried out tumour dissection leaving the periostium intact and attempting to preserve the neighbouring neurovascular bundle, the tear duct and the infraorbital structures intact. Some of these interventions require reconstruction of the tear drainage system and/or the facial skeleton during the same operation or subsequent surgery. It is essential to replace the wall of the orbit in order to achieve good aesthetic and functional results and to avoid diplopia. In our study no cases of iatrogenic diplopia were reported.

Lymph node basin metastases of these tumours are rare. Cantú et al., in a review of 704 patients with paranasal tumours, and considering the associated poor prognosis, proposed a prophylactic lymphadenectomy in squamous cell carcinoma of the maxillary sinus and undifferentiated carcinoma of the ethmoid sinus [12].

It should be noted that, in addition to local surgery and regional lymph node approaches, radiotherapy may be used and helps to improve local control of this type of lesions [13,14].

All the patients in our series are currently in remission, although it is vital to continue regular clinical follow ups and tomography scans for early detection of a possible recurrence.

To conclude, tumours of the paranasal sinuses require a multidisciplinary approach bringing together the knowledge of ENT, ophthalmology, radiology and oncology specialists. There is a great variation the histological pattern of tumours of the paranasal sinuses, and while there are no comparative clinical trials of the effectiveness of the various combined therapies, it seems that combination of surgery and radiotherapy provide the best local control of this disease and offer the longest life expectancy.

A joint approach involving ENT specialists and ophthalmologists using a Lynch incision extended towards the ala of the nose permits optimum access for the midfacial dissection of tumours of the paranasal sinuses, with very acceptable functional and aesthetic outcomes.

References

[1] Rodríguez Céspedes W. Cáncer de la fosa nasal y senos paranasales. In: Salaverry García O, ed. *Neoplasias malignas de cabeza y cuello*. Lima: Fondo Editorial de la Universidad Nacional, 2000:60-74.

[2] Gras Cabrerizo JR, Orús Dotú C, Montserrat Gili JR, Fabra Llopis JM, De Juan Beltrán J. Análisis epidemiológico de 72 carcinomas de senos paranasales. *Acta Otorrinolaringol Esp* 2006;57:359-63.

[3] Myers LL, Nussenbaum B, Bradford CR, Teknos TN, Esclamado RM, Wolf GT. Paranasal sinus malignancies: A 18-year single institution experience. *The Laryngoscope* 2002;112(11):1964-69.

[4] Comba P, Battista G, Belli S, de Capua B, Merler E, Orsi D, Rodella S, Vindigni C, Axelson C. A case control study of cancer of the nose and paranasal sinuses and occupational exposures. *Am J Ind Med* 1992;22(4):511-20.

[5] Cantù G, Lazzaro Solero C, Mariani L, Mattavelli F, Pizzi N, Licitra L. A new classification of malignant tumors involving the anterior skull base. *Arch Otolaryngol Head Neck Surg* 1999;125:1252-57.

[6] Ianetti G, Valentini V, Rinna C, Ventucci E, Marianetti TM. Ethmoido-orbital tumors: our experience. *Craniofacial Surg* 2005;16(6):1085-91.

[7] Carrillo JF, Guemes A, Ramirez-Ortega MC, Onate-Ocana LF Prognostic factors in maxillary sinus and nasal cavity carcinoma. *Eur J Surg Oncol* 2005;31(10):1206-12.

[8] Liétin B, Mom T, Avan P, Llompart X, Kemeny JL, Chazal J, Russier M, Gilain L. Adenocarcinomas of the ethmoid sinus: retrospective analysis of prognostic factors. *Ann Otolaryngol Chir Cervicofac* 2006;123(5):211-20.

[9] Guntinas-Lichius O, Kreppel MP, Stuetzer H, Semrau R, Eckel HE, Mueller RP. Single modality and multimodality treatment of nasal and paranasal sinuses cancer: a single institution experience of 229 patients. *Eur J Surg Oncol* 2007;33(2):222-8.

[10] Pérez Moreiras JV. *Patología orbitaria*. 2nd ed. Vol 2. Barcelona: Edika Med; 2002.

[11] Scott McCary W, Levine PA, Cantrell RW. Preservation of the eye in the treatment of sinonasal malignant neoplasms with orbital involvement: a confirmation of the original treatise. *Arch Otolaryngol Head Neck Surg* 1996;122:657-9.

[12] Cantù G, Bimbi G, Miceli R, Mariani L, Colombo S, Riccio S, Squadrelli M, Battisti A, Pompilio M, Rossi M. Lymph node metastases in malignant tumors of the paranasal sinuses. *Arch Otolaryngol Head Neck Surg* 2008;134(2):170-177.

[13] Dulguerov P, Jacobsen MS, Allal AS, Lehman W, Calcaterra T. Nasal and paransal sinus carcinoma: are we making progress? *Cancer* 2001; 92(12):3012-3029.

[14] Katz TS, Mendenhall WM, Morris CG, Amdur RJ, Hinerman RW, Villaret DB. Malignant tumors of the nasal cavity and paransal sinuses. *Head Neck* 2002; 24(9):821-829.

Index

A

abstraction, 18
accelerator, 309
accessibility, 25, 190
accounting, x, 48, 269, 273, 308
acid, 3, 18, 22, 25, 26, 35, 36, 41, 115, 138, 214, 220, 224
acquisitions, 313
active site, 249
acute myeloid leukemia, 142
adaptability, 77, 212
ADC, 203
adenine, 24
adenocarcinoma, 175, 177, 178, 233, 245, 324, 328, 329, 333
adenovirus, 13, 24, 94, 124, 152, 153, 154, 155, 210, 214, 217, 220, 221, 253, 255, 260, 262
adhesion, 15, 56, 65, 86, 124, 152
adhesive interaction, 56
adiponectin, 59
adipose tissue, 88
adjustment, 98
adolescents, 269
adrenal gland, 58
adulthood, 101
adverse event, 175, 260
aesthetics, 337
aetiology, 96
African Americans, 129
aggregation, 6, 10, 17, 19, 21, 24, 73
aggressive behavior, 133
aggressiveness, 200, 275
AIDS, 48, 62
albumin, 5, 37
alcohol consumption, 159
algorithm, 314, 316, 319
allele, 249
alopecia, 50, 64
alters, 35, 75, 256, 261
amino acid, 249, 256
analgesic, 235
anaphylaxis, 241
anatomy, 160, 161, 165, 191, 259, 306
anchoring, 30, 73
angiogenic process, 55, 59, 62, 63
angiography, 202
antibody, 7, 18, 33, 39, 44, 58, 64, 68, 69, 70, 71, 76, 85, 114, 116
anticancer activity, 27, 68
anticancer chemotherapeutics, 50
anticancer drug, 42
antigen, 13, 20, 22, 23, 30, 37, 57, 60, 76, 87, 95, 119, 120, 135, 144, 145
antigen-presenting cell, 57
anti-inflammatory drugs, 59
antioxidant, 86
antisense, 53, 79, 252
antisense oligonucleotides, 79
antitumor, 35, 36, 37, 76, 88, 91, 212, 251
aorta, 161
APC, 94, 96, 106, 107, 136
apex, 336
aphasia, 250
apoptosis, 4, 12, 22, 33, 49, 50, 53, 54, 59, 66, 69, 74, 80, 83, 103, 110, 120, 148, 210, 213, 217, 221, 248, 249, 256, 261, 262, 265
aqueous solutions, 6
AR, 17, 36, 37, 40, 41, 42, 88, 132, 133, 135, 141, 154
arginine, 249
argon, 191
arrest, xiii, 57, 118, 247, 248, 249, 256, 261, 265
arrhythmia, 225
arterial blood gas, 241
arteries, 35, 63, 185
arteriography, 337
artery, 161, 169, 230, 234, 325, 327

articulation, 184, 237
artificial intelligence, 185
asbestos, 333
ascorbic acid, 18
aseptic, 204
Asian countries, 239, 333
assessment, 119, 127, 141, 162, 201, 280, 318
asthma, 62, 83
astrocytes, 102, 118, 123, 126, 143, 146, 149, 213, 248, 273
astrocytoma, xiv, 218, 262, 267, 268, 271, 273, 279, 280
asymmetry, 323, 328
asymptomatic, 248, 277
atherosclerosis, 63
ATP, 69
attachment, 23, 75, 91
Au nanoparticles, 36
Austria, 160
autopsy, 175, 264
autosomal dominant, 268, 274
avoidance, 50
awareness, 105

B

bacterial infection, 62
barriers, 67, 168
basal cell carcinoma, 96, 277, 280
basement membrane, 7, 61, 62, 72
basophils, 57
BBB, 207
beams, 317, 319
beneficial effect, 85, 251
benefits, x, 48, 69, 165, 168, 182, 186, 190, 205, 211, 234, 235
benign tumors, 277, 278
bioactive agents, 24
bioavailability, 6
biocompatibility, 73, 74
biodegradability, 74
biological activity, 21
biological behavior, 132, 269, 272, 276
biological media, 17
biomarkers, 30, 55
biomolecules, 17, 19, 215
birth weight, 97, 234, 268
black hole, 16
bleeding, 60, 64, 165, 166, 173, 204, 239
blood flow, 9, 63
blood stream, 56, 63
blood supply, 37, 59, 203, 329
blood transfusion, 166, 169, 174, 177, 178, 231, 235
blood vessels, x, 4, 7, 11, 43, 47, 55, 56, 58, 59, 60, 63, 65, 67, 85, 87, 211, 276, 325, 327, 337
blood-brain barrier, 86, 203, 207
bloodstream, 73
BMI, 174
BMS-275291, 69
body mass index, 177, 178, 180, 182, 236
bonds, 15
bone marrow, 50, 52, 56, 58, 68, 81, 82, 154, 211, 261, 275, 277
bone marrow transplant, 58
bones, 331
bowel perforation, 64
brachytherapy, 164, 180, 223, 228, 233, 308, 318, 319
brain cancer, 95, 130, 149
brain tumor, 89, 97, 121, 129, 130, 131, 132, 133, 134, 137, 139, 141, 143, 144, 147, 148, 149, 152, 216, 218, 219, 220, 224, 250, 255, 258, 259, 263, 264, 269, 270, 271, 272, 274, 275, 278, 279, 280, 281, 282
brainstem, 139, 275
branching, 62, 63
Brazil, 227
breast cancer, 9, 12, 19, 21, 27, 31, 33, 34, 36, 38, 52, 56, 57, 58, 69, 70, 76, 80, 81, 82, 83, 84, 96, 114, 143, 231, 239
breast carcinoma, 18, 84
breathing, 306, 307, 309, 310, 311, 313, 314, 318
BTC, 17
bystander effect, 218

C

caesarean section, 270
calcification, 272
calcium, 12, 139
caliber, 228
cancer cells, x, 12, 17, 19, 20, 21, 24, 30, 31, 32, 35, 36, 40, 41, 48, 50, 51, 54, 57, 58, 59, 75, 77, 82, 112, 118, 120, 154, 255
cancer progression, 36, 50, 51, 56, 82
cancer screening, 337
cancer stem cells, 50, 143, 144, 146, 147, 213, 219
candidates, 105, 107, 125, 128
capillary, 60, 64
carbohydrate, 29, 40
carbohydrates, 75
carbon nanotubes, 73
carboxyl, 25, 76
carboxylic groups, 75
carcinogenesis, 59, 80
carcinogenic human papillomavirus, 159
cardiovascular disease, 233

cartilage, 272
casein, 23
CAT scan, 323, 324, 325, 326, 327, 328, 333
catalytic activity, 259
catheter, 169, 183
cavernous sinus syndrome, 336
cell culture, 66, 86, 215
cell cycle, 102, 110, 117, 125, 139, 210, 249, 256, 257
cell death, 17, 22, 36, 49, 51, 67, 75, 155, 210, 220, 252, 256, 257, 262
cell differentiation, 105
cell fate, 104, 148, 261
cell killing, 34
cell line, 12, 13, 18, 19, 23, 31, 36, 84, 104, 126, 127, 138, 147, 151, 154, 155, 221, 256, 262
cell membranes, 22
cell metabolism, 79
cell surface, 15, 19, 23, 24, 25, 28, 31, 65, 75, 82, 86, 106, 107, 124
cellular immunity, 13
cellulitis, 336
central nervous system, ix, xii, xiii, 97, 129, 130, 132, 134, 135, 143, 145, 146, 151, 152, 154, 200, 213, 219, 221, 222, 267, 268, 279, 281
central nervous system XE "central nervous system" primitive neuroectodermal tumours, 97
cerebellar development, 103, 107, 112, 135, 139
cerebellum, xi, 94, 96, 98, 101, 102, 105, 106, 107, 108, 110, 111, 112, 137, 138, 139, 140, 256, 261, 268, 275, 276
cerebral edema, 270
cerebral function, 202
cerebral hemisphere, 102, 204, 273, 275
cerebrospinal fluid, 117, 145, 207, 270, 274
cerebrum, 98
cervical cancer, iv, ix, xi, xiii, 158, 159, 160, 161, 162, 163, 164, 165, 166, 167, 168, 169, 171, 172, 173, 174, 175, 176, 177, 178, 179, 180, 181, 182, 183, 184, 190, 191, 192, 193, 194, 195, 196, 197, 198, 227, 231, 232, 234, 235, 236, 237, 238, 239, 240, 241, 242, 243, 244, 245, 246, 284, 286, 287, 294, 295, 296, 297, 298, 299, 300, 301, 302, 303, 304
cervical conization, 178
cervix, 159, 160, 164, 177, 178, 190, 191, 192, 197, 228, 229, 231, 232, 233, 235, 238, 240, 241, 242, 243, 244, 246
cesarean section, 270
chemical, 2, 29, 65, 67, 73, 96
chemical stability, 73
chemiluminescence, 38
chemokines, 58, 59, 62
chemoprevention, 80, 82
chemotaxis, 9
chemotherapeutic agent, xiii, 207, 247
chemotherapy, ix, xiii, 1, 4, 50, 52, 58, 64, 68, 85, 87, 100, 120, 121, 133, 134, 163, 169, 176, 206, 207, 208, 214, 215, 217, 219, 223, 225, 227, 229, 231, 239, 242, 243, 247, 249, 250, 252, 271, 274, 275, 276, 280, 284, 295, 297, 298, 299, 301, 337
chicken, 62
childhood cancer, 97, 100, 131, 134
chitosan, 43
cholesterol, 19, 20, 39, 91, 117, 144, 145
choroid, xiv, 267, 268, 269, 271, 274, 279, 280
chromosomal alterations, 136
chromosome, 98, 104, 107, 109, 112, 115, 136, 249, 256, 258, 276, 277
chronic lymphocytic leukemia, 79
cilium, 145
circulation, 3, 5, 31, 56, 74, 76, 90, 91
cirrhosis, 62
classes, 111, 160, 161, 191
classification, 98, 99, 103, 129, 130, 132, 138, 141, 142, 160, 161, 162, 168, 200, 201, 216, 218, 221, 222, 250, 269, 272, 276, 281, 335, 339
cleavage, 15, 117
cleft lip, 271
cleft palate, 271, 277
clinical application, 240
clinical assessment, 129
clinical presentation, 109, 129, 269
clinical trials, x, 48, 52, 68, 69, 100, 108, 118, 122, 128, 210, 211, 212, 251, 253, 255, 338
cloning, 116, 145
closure, 8, 175, 184, 185
cluster of differentiation, 54
clustering, 314
clusters, 81, 314
CNS, xii, xiii, 94, 95, 97, 98, 99, 104, 105, 108, 111, 114, 122, 124, 131, 134, 135, 151, 153, 199, 200, 202, 203, 206, 207, 211, 213, 215, 247, 248, 250, 255, 256, 258, 259, 268, 269, 270, 271, 272, 273, 274, 275, 276, 278, 281, 282
CNS-PNETs, 97, 98, 104, 111
cobalt, 18
coding, x, 22, 24, 48, 51, 55, 254, 256
coherence, 214, 216
colectomy, 235, 244
collagen, 56
colon cancer, 59, 64, 85, 143, 244
colonization, 54, 57
colorectal cancer, 38, 72
common signs, 270
common symptoms, 328

communication, 29, 160
community, 50
compatibility, 25
compensation, 307, 309, 310, 317, 319
competition, 9
compilation, ix
complement, 208
complexity, 79, 111, 163, 168, 183, 312
complications, xii, xiii, 73, 80, 158, 162, 163, 164, 166, 167, 168, 172, 173, 174, 175, 176, 177, 178, 179, 180, 181, 182, 183, 184, 185, 190, 196, 202, 227, 228, 236, 237, 238, 239, 250, 272, 274
composition, xii, 56, 57, 73, 138, 200
compounds, ix, 1, 7, 13, 14, 17, 24, 26, 27, 36, 68, 75, 76
compression, 270
computed tomography, 160, 178
computerised tomography, 202
computing, xiv, 306
configuration, 189
confinement, 306
congenital brain tumors, 270, 271, 273
conjugation, 9, 27, 29, 31, 41
conjunctivitis, 13
connective tissue, 7
consensus, 17
conservation, 243
construction, 333
contralateral hemisphere, 211
control group, 167
controlled trials, xi, 158, 223
controversies, 260
conversion rate, 174
coordination, 21, 187
copolymer, 29
coronary heart disease, 234
corpus callosum, 271, 278
correlation, xiv, 26, 33, 35, 84, 103, 137, 206, 233, 273, 279, 306, 307, 310, 312, 313, 314, 315, 316
correlation analysis, 312
cortex, 102, 111, 135
cortical pathway, 202, 206
cost-benefit analysis, 238
counseling, 270
Coup, 62
covering, 60
coxsackievirus, 152
cranial nerve, 270, 276
craniopharyngioma, 271, 277, 282
CSCs, 50, 52, 55, 112, 113, 114
CSF, 254, 264, 270, 274, 275, 276
CT, 149, 201, 202, 203, 250, 286, 301, 306, 309, 324
cues, 121, 184
culture conditions, 123
current limit, 125
cycles, 212, 232, 234, 250, 307
cycling, 120
cyclins, 106
cyclooxygenase, 83
cyst, 274
cysteine, 17
cystitis, 228
cystotomy, 177
cytogenetics, 111
cytokines, 58, 59, 65, 251
cytometry, 113
cytoplasmic tail, 116
cytosine, 210, 218, 219
cytoskeleton, 56
cytotoxic agents, 54, 209
cytotoxicity, 16, 36, 77, 79, 127, 209

D

danger, 19
data analysis, 307, 311
data set, 110
database, xiv, 174, 306, 309, 313, 315, 316
death rate, 159
debulking surgery, 179
defects, 122, 139, 155, 278
defence, 212
degradation, 14, 15, 52, 56, 70, 71, 73
dehiscence, 175, 181
demographic data, 174
dendritic cell, 57, 58, 75
deposition, 25, 306
deprivation, 27, 232, 329
depth perception, 163, 168, 184, 235
deregulation, xi, 94, 98, 103, 104, 106, 108, 113, 117, 118
derivatives, 13, 16, 19, 21, 30, 32, 75, 139
desensitization, 27
destruction, 7, 38, 60, 67, 335, 337
detection, xiv, 18, 20, 21, 25, 35, 74, 109, 116, 246, 269, 306, 338
developing brain, 127, 278
developing countries, 95, 228
diabetes, 62, 83
diarrhea, 64
diet, 96, 159
differential diagnosis, 270
diffusion, 22, 63, 186, 203, 206, 225, 255
dimethylsulfoxide, 26
diplopia, 328, 329, 331, 336, 338
direct adsorption, 76
direct measure, 309

disability, 250
disease progression, 141
dislocation, 272
displacement, 27, 328, 329, 336
dissociation, 107
diversity, 66
DNA damage, 50, 120, 146, 216, 249
DNA ligase, 122, 149
DNA repair, 210, 213, 249
DOI, 220, 223
dosage, 4, 206, 210
down-regulation, 27, 116
drainage, 3, 5, 63, 240, 337
Drosophila, 107, 114
drug delivery, 3, 8, 14, 25, 30, 32, 33, 36, 38, 39, 40, 43, 51, 72, 73, 74, 76, 87, 90, 91
drug discovery, 65
drug release, 74
drug resistance, 51, 120, 154, 252
drug targets, xi, 93
drug testing, 128
drug therapy, 12
dura mater, 337
dyes, 18, 21, 34, 37, 222
dysplasia, 114, 144, 272

E

ECM, 14, 15, 62, 94, 102, 329, 330
ECs, 50, 55, 56, 57, 60, 62, 63, 65, 66, 67, 68, 75
edema, 84, 183, 203, 250, 254, 270
editors, 131, 132, 133, 135
electric current, 204
electrocautery, 204
electrolyte, 277
electromagnetic fields, 96, 130
electron, 18, 90
electron microscopy, 90
electrons, xiv, 305, 307
electroporation, 73, 90
ELISA, 29
elucidation, 98
EM, 138, 195, 245, 263
emboli, 181
embolus, 56
embryogenesis, 96, 102, 256
embryonic stem cells, 101, 102, 126, 135, 146, 150, 152, 154
emission, 16, 24, 40, 215
encoding, 104, 108, 210
endocrine, 21, 27
endocrine system, 27
endoderm, 123
endometriosis, 62, 239
endoscope, 187, 217
endothelial cells, 7, 8, 9, 10, 11, 12, 35, 38, 43, 45, 50, 81, 83, 85, 86, 88, 89
endothelium, 10, 54, 57, 59, 62, 63, 65, 86, 88, 90
energy, 4, 15, 16, 18, 76
energy transfer, 15, 18
engineering, 74
enteritis, 228
environmental change, 51
environmental factors, 121
environmental influences, 96
enzymes, 15, 17, 209
eosinophils, 57
EPC, 62
ependymal, 97, 149, 248
ependymal cell, 97, 149, 248
ependymoma, 110, 138, 140, 143
epidemiologic studies, 249
epidemiology, 129
epithelial cells, 58, 115, 121, 144, 145, 274
epithelial ovarian cancer, 44
epithelium, 86, 140
EPR, 6, 17, 75, 91
equilibrium, 57
equipment, xii, 158, 168, 184, 228
ergonomics, 186
erosion, 335
ESCs, 101, 102, 123, 124
estrogen, 232
ethical issues, 211
ethylene, 49
ethylene glycol, 49
etiology, 258
eukaryotic cell, 19, 42
European Community, 316
European Union, 210, 218
examinations, 323, 336
excision, 198, 276
excitation, 214, 220
exons, 115
experimental condition, 211
exploitation, 306
exposure, ix, 1, 13, 19, 96, 130, 131, 249, 333
external growth, 49
extracellular matrix, 7, 12, 14, 56, 57, 59, 62, 63, 102
extraocular muscles, 336
extravasation, 5, 57, 63

F

facial asymmetry, 331
false negative, 176, 241
family members, 68, 141
fascia, 187

fat, 272
FDA approval, 68, 69, 70, 71, 72, 172
ferritin, 41
fertility, 164, 177, 178, 179, 192, 197, 198, 234, 244
fetal growth, 131
fetus, 131, 270, 272
fever, 181
fiber, 13
fibrinogen, 56
fibroblast growth factor, 62, 72, 123
fibroblasts, 34, 50, 57, 58, 59, 61, 83, 84, 256, 257, 264
fibrosarcoma, 19, 23
fibrosis, 228
FIGO, 160, 162, 164, 165, 167, 168, 176, 180, 190, 236, 298
filament, 119
filtration, 184
first generation, 124, 260
fish, 115
fistulas, 240
fixation, 233, 334
flexibility, 63, 171, 172
fluctuations, 308
fluid, 63, 64, 169
fluorescence, 10, 14, 15, 16, 21, 24, 39, 88, 136, 137, 220
fluorophores, 19
fMRI, 202, 206, 217, 250
folate, 7, 24, 26, 33, 36, 37, 38, 39, 40, 44
folic acid, 24, 26, 41
Food and Drug Administration, 2, 171
Ford, 38, 154, 196
forebrain, 102, 112, 149, 220
formaldehyde, 333
fragments, 7, 31, 33, 73, 75, 76, 77
frameshift mutation, 144
France, 1, 131
freedom, 168, 235
freezing, 90
functional hierarchy, 213
functional imaging, 222
Functional Magnetic Resonance Imaging, 250
functional MRI, 202, 203, 219
funding, 316
fusion, 23, 52, 66, 91, 171, 306

G

gadolinium, 324, 325
gastroenteritis, 13
gastrointestinal tract, 50
GDP, 20
gene combinations, 128
gene expression, 55, 107, 109, 111, 123, 128, 132, 142, 147, 201, 250, 258, 262, 264
gene silencing, 219
gene targeting, 24, 262
gene therapy, ix, xiii, 13, 125, 152, 153, 210, 219, 221, 222, 248, 251, 252, 253, 254, 255, 259, 260, 261, 263, 264
gene transfer, 42, 126, 139, 151, 152, 153, 154, 253, 260
genes, x, xiii, 48, 57, 77, 82, 96, 104, 106, 108, 109, 110, 111, 115, 122, 124, 127, 128, 129, 140, 142, 146, 149, 152, 209, 210, 213, 224, 247, 249, 252, 253, 255, 257, 261, 274
genetic alteration, 118, 132, 200, 212
genetic defect, 49, 59, 106
genetic disease, 130
genetic disorders, 190, 249
genetic factors, 96, 97
genetic information, 251
genetics, 131, 141, 147, 149, 218
genome, 100, 104, 108, 109, 124, 125, 150, 252, 255, 262
genomics, 100, 132
genotype, 155, 221
Germany, 157, 199, 267, 318
germline mutations, 96
gestation, 101, 102, 270
gestational age, 270
glial cells, 97, 101, 102, 106, 116, 135, 248
Glioblastoma, 94, 147, 203, 248, 249, 251, 252, 254
glioblastoma multiforme, xii, xiii, 107, 144, 148, 199, 216, 221, 223, 224, 225, 247, 258, 263, 264, 273, 279
glioma surgery, 204, 224
glucose, 44
glutamate, 25
glycerol, 257
glycogen, 107
glycol, 28, 76, 90
glycoproteins, 30
glycosylation, 116
gold nanoparticles, 73
GPC, 94, 108
grades, 200, 248
grading, 132, 200, 201
grouping, 314
growth arrest, 261, 262
growth factor, 7, 8, 15, 17, 34, 35, 36, 37, 38, 39, 40, 41, 42, 43, 44, 45, 54, 55, 56, 59, 62, 65, 72, 79, 81, 84, 85, 86, 89, 94, 95, 97, 121, 123, 131, 148, 212, 215, 249, 251, 259
guidance, xiv, 192, 305, 306, 308, 318, 319
guidelines, 87, 160, 191

gynecologist, 174

H

haematoporphyrin derivative, 2
haemostasis, 203, 337
hair, 50
hair follicle, 50
half-life, 3, 73, 207
harvesting, 211
head and neck cancer, 3, 4
health care, 186
health status, 228
hematoma, 181
hematopoietic stem cells, 115
hematopoietic system, 130
heme, 28, 41, 139
heme oxygenase, 139
hemiparesis, 250, 270
hemisphere, 102, 250, 273
hemoglobin, 181
hemorrhage, 270, 274, 275
hemostasis, 36, 40, 224
hepatocellular carcinoma, 68
hepatocytes, 29, 36
hepatoma, 29
hernia, 181
herpes simplex, 221, 253, 263, 264
herpes simplex virus type 1, 264
heterogeneity, xii, 52, 75, 77, 97, 106, 112, 113, 127, 200
hippocampus, 106, 218, 223
Hispanics, 129
histogenesis, 129
histogram, 308
histology, 56, 100, 109, 111, 175, 180, 183, 272
HIV, 67, 151
HIV-1, 67
homeostasis, 56, 57, 162
hospitalization, 235, 237, 240
host, 7, 15, 58, 59, 66, 124
HpD, 2
HPV, xi, 158, 159, 190, 297
hTERT, 53, 54
human brain, 130, 143, 148, 213, 216, 220, 221, 223, 255, 258
human development, 140
human genome, 190
human immunodeficiency virus, 151
Hunter, 60, 264
hybrid, 44
hybridization, 136, 137
hydrocephalus, 270, 273, 274, 275, 276
hydrogen peroxide, 18
hydronephrosis, 272
hydrophilicity, 14
hydrophobicity, 14
hydroxyl, 18
hyperplasia, 248
hypertension, 64, 84, 336
hyperthermia, 74
hypothalamus, 273
hypothesis, 26, 60, 97, 113, 154, 219
hypoxia, 8, 55, 75, 121, 147, 249, 259
hypoxia-inducible factor, 249
hysterectomy, xi, xiii, 158, 159, 160, 161, 162, 163, 164, 165, 166, 167, 168, 169, 172, 173, 174, 175, 176, 178, 179, 181, 182, 183, 184, 185, 190, 191, 192, 193, 194, 195, 196, 197, 198, 227, 228, 229, 231, 232, 233, 234, 235, 236, 237, 238, 239, 241, 242, 243, 244, 245, 246

I

iatrogenic, 331, 338
ICAM, 36
icon, 12
ideal, 3, 4, 32, 65, 74
identity, 115, 117, 137, 214
IFN, 54, 253, 254
IFN-β, 253, 254
IL-8, 70, 71
illumination, ix, 2, 4, 18, 205, 214
images, 104, 170, 171, 214, 306, 307, 318, 324
imaging modalities, 215
immortality, 49
immune response, 7, 151, 209, 251, 252, 255
immune system, 54, 55, 57, 59, 77
immunocompromised, 113
immunodeficiency, 124, 260
immunogenicity, 76
immunoglobulin, 56
immunomodulatory, 64
immunoreactivity, 100, 145
immunosurveillance, 50, 53, 78
immunotherapy, 12, 38
impairments, 96
imperforate anus, 277
implants, 236
in situ hybridization, 136, 137
in utero, 96, 268, 269, 270
in vivo, 9, 12, 13, 14, 16, 18, 19, 21, 22, 24, 25, 26, 29, 31, 32, 36, 37, 38, 40, 43, 52, 55, 66, 76, 83, 86, 88, 89, 90, 115, 120, 124, 125, 126, 127, 128, 129, 148, 150, 152, 153, 154, 222, 256, 262
incidence, 51, 55, 59, 64, 95, 96, 129, 130, 149, 164, 183, 228, 233, 236, 238, 239, 248, 249, 270, 272, 319, 333

independence, 64
indirect effect, 7
individuation, 307
Indonesia, 333
inducer, 9, 261
induction, 10, 22, 69, 123, 126, 175, 210, 256, 270
inefficiency, 81
infancy, xiv, 267, 279, 281
infant mortality, 278
infants, 134, 270, 273, 278, 280, 282
infection, 13, 124, 125, 126, 159, 163, 174, 181, 190, 203, 207, 209, 225, 235, 240, 289, 291, 293, 294
infectious agents, 96, 260
inflammation, 13, 50, 54, 57, 59, 74, 78
inflammatory responses, 254
informed consent, 324
infrastructure, 335
inguinal, 162, 240
inheritance, 145
inhibition, 19, 71, 72, 77, 80, 118, 123, 155, 211, 256
inhibitor, 53, 54, 60, 64, 68, 69, 70, 71, 72, 79, 108, 123, 140, 146, 149, 211, 249
initiation, 7, 104, 108, 117, 122, 128, 195
insertion, 79, 94, 118, 308
institutions, 184, 268
insulin, 23, 42, 131
integration, x, 48, 100, 101, 119, 124, 150, 152, 208, 217, 253, 319
integrin, 12, 13, 14, 35, 66, 69, 70, 71, 82, 88, 155
integrins, 13, 56, 67, 124
interference, x, 9, 40, 48, 65
interferon, 54, 70, 72, 90, 153, 254, 264
interferons, 251
intermolecular interactions, 24
internalised, 124
internalization, 21, 25, 27, 37
International Federation of Gynecology and Obstetrics, 160
intervention, xi, 51, 94, 127, 186, 280, 331
intestine, 64
intracranial pressure, 204, 225, 249, 270, 274
intravenously, 63
intravesical instillation, 28
invasive lesions, 159
ionizing radiation, 307
ions, xiv, 305, 307
IP-10, 70
ipsilateral, 329, 335
iron, 28
irradiation, 2, 5, 6, 28, 76, 206, 223, 239, 273, 276, 308, 310
ischemia, 67
isochromosome, 136
isolation, 113, 114, 119, 145, 146, 224
isomers, 29
Italy, 157, 185, 305

J

Japan, 2, 160, 333

K

karyotype, 249
karyotyping, 136
kidney, 57, 70, 71, 275
kill, 12, 36, 68, 209
kinetic studies, 34
kinetics, 40, 73
knots, 185, 186

L

labeling, 37, 261
lactose, 29
laparoscope, 184, 185
laparoscopic colectomy, 235
laparoscopic surgery, xi, 158, 162, 163, 164, 165, 167, 168, 180, 181, 190, 235, 236, 237
laparoscopy, xi, 158, 161, 163, 164, 165, 167, 169, 171, 172, 173, 175, 176, 177, 178, 179, 180, 181, 182, 183, 184, 185, 186, 187, 189, 195, 198, 231, 237, 238, 240, 245
laparotomy, 161, 166, 167, 172, 173, 174, 176, 177, 178, 179, 180, 181, 182, 183, 184, 185, 186, 190, 191, 194, 195, 198, 231, 235, 237, 238, 240, 245
lasers, 6
latency, xiv, 267, 307
LDL, 19, 20, 43
lead, 9, 10, 50, 103, 107, 108, 121, 122, 210, 269, 274
leakage, 60
learning, 162, 163, 168, 172, 184, 186, 187, 237, 313, 318
learning process, 187
leptin, 59
lesions, 3, 4, 6, 18, 61, 190, 201, 202, 207, 211, 232, 233, 250, 274, 317, 328, 337, 338
lethargy, 270
leucine, 95, 119
leukemia, 48, 54, 64, 70, 131
life expectancy, 235, 332, 338
lifetime, 13, 18, 21
ligament, 191
ligand, 11, 12, 17, 19, 22, 24, 26, 38, 42, 54, 62, 68, 69, 70, 71, 72, 75, 76, 116, 218
light conditions, 205, 214
limbic system, 256

linear model, 313
lipids, 76
lipoma, 268, 277
lipoproteins, 5, 19, 33
liposomes, 6, 28, 30, 35, 37, 39, 42, 66, 73, 74, 75, 76, 77, 79, 84, 90, 91
liver cancer, 70, 80
localization, 9, 13, 15, 22, 23, 29, 32, 44, 57, 140, 223, 306, 308, 310, 319
low-density lipoprotein, 3, 34, 38, 39, 44
lumen, 10, 11
lung cancer, 4, 15, 35, 37, 39, 69, 72, 234, 317
lung metastases, 57, 64
Luo, 78, 83, 140, 279
luteinizing hormone, 36, 44
lymph, xiii, 64, 74, 85, 160, 161, 162, 163, 166, 168, 171, 172, 173, 174, 175, 176, 177, 178, 179, 182, 183, 196, 227, 228, 229, 230, 231, 232, 234, 235, 236, 239, 240, 241, 242, 246, 275, 277, 324, 326, 335, 337, 338
lymph node, xiii, 64, 74, 85, 160, 162, 163, 166, 169, 171, 172, 173, 174, 175, 176, 177, 178, 179, 182, 196, 227, 228, 229, 230, 231, 232, 234, 235, 236, 239, 240, 241, 242, 246, 275, 277, 324, 326, 335, 337, 338
lymphangiogenesis, 64, 85
lymphatic system, 75, 175, 181
lymphedema, 180, 181, 240
lymphocytes, 41, 57, 58
lymphoma, 48, 54, 68, 71, 94, 118, 207
lysine, 26

M

mAb, 19, 30, 31, 76
machinery, x, 48
macromolecules, 5, 22, 23, 39
macrophages, 57, 58, 76, 153, 220
macular degeneration, 2, 4, 44, 72
magnetic resonance, 160, 234, 269
MR, 269
magnetic resonance imaging, 160, 234
magnitude, 28
majority, 3, 8, 112, 124, 128, 214, 257, 271
malignancy, ix, 1, 4, 54, 149, 163, 182, 274
malignant cells, 27, 28
malignant melanoma, 90
malignant tissues, 272
malignant tumors, 61, 248, 339
mammalian brain, 107, 123, 134, 138, 149
management, xiii, xv, 100, 129, 130, 160, 161, 163, 167, 169, 172, 190, 197, 220, 227, 228, 239, 318, 322
mandible, 333
manipulation, xi, 74, 89, 94, 123, 127, 128, 149, 184, 185, 235, 236
mantle, 68, 71
mapping, 74, 86, 87, 178, 202, 205, 246
marimastat, 89
marketing, 218
marrow, 64
mast cells, 57, 58
matrix metalloproteinase, 8, 14, 15, 36, 39, 45, 59, 62, 68, 72, 155
maturation process, 102
maxillary sinus, 323, 324, 325, 326, 328, 329, 332, 333, 334, 335, 336, 338, 339
MB, x, 9, 30, 82, 93, 95, 96, 97, 98, 99, 100, 101, 103, 104, 105, 106, 107, 108, 109, 110, 111, 112, 113, 114, 117, 118, 119, 120, 121, 126, 127, 128, 318, 319
media, 114, 119, 123, 337
median, xiii, 164, 166, 167, 172, 174, 175, 177, 178, 179, 232, 233, 236, 247, 276
medical care, x, 48, 51
medication, 96, 183
medulloblastoma, xiv, 95, 104, 113, 121, 128, 130, 131, 132, 133, 134, 135, 136, 137, 138, 140, 141, 142, 146, 147, 148, 149, 154, 155, 267, 268, 271, 277, 281, 282
melanoma, 18, 19, 27, 33, 38, 58, 80, 239, 262, 264
membership, 314
membranes, 19, 22, 30, 34, 39, 41, 75, 77, 234
meninges, 275
meningioma, 325, 329, 331
menopause, 232
menstrual cycles, 232, 234
mesencephalon, 273
mesenchymal stem cells, 121, 126, 260
mesenchyme, 98
mesoderm, 123
metabolism, 28, 51, 74
metal ion, 18
metalloproteinase, 36
metals, 19
metastasis, 14, 38, 40, 42, 81, 82, 83, 84, 85, 122, 127, 148, 178, 182, 230, 231, 232, 233, 239, 243, 245, 262, 275, 335, 337
metastatic disease, 51, 56, 99, 141
methodology, 65
methylation, 201
Mexico, 247
Miami, 191
mice, 9, 12, 18, 19, 21, 29, 38, 43, 58, 83, 84, 113, 114, 118, 128, 139, 140, 142, 143, 144, 145, 148, 149, 153, 155, 216, 219
microarray technology, 109

microinjection, 73
microscope, 205, 214
microscopy, 20, 33, 88, 220
midbrain, 139
Middle East, 222
migration, 7, 12, 35, 56, 60, 62, 69, 70, 71, 102, 139, 206, 218, 221, 223, 251
mimicry, 60
misunderstanding, 185
mitochondria, 22
mitosis, 110
MMP, 15, 35, 36, 42, 49, 59, 62, 69, 70, 71, 72, 83
mobile phone, 96
model system, 126, 128
modelling, 126, 127, 129
models, xi, 11, 15, 39, 52, 60, 85, 94, 96, 105, 106, 108, 111, 112, 118, 121, 124, 126, 128, 129, 148, 149, 151, 155, 189, 210, 215, 307, 310, 312, 313, 314, 316
modules, 24
molecular dynamics, 26
molecular oxygen, ix, 1, 2, 4, 18
molecular structure, 21
molecular targeting, 51
molecular weight, 25
molecules, ix, 2, 3, 4, 5, 10, 15, 18, 25, 28, 30, 31, 55, 56, 62, 65, 73, 77, 79, 84, 86, 107, 116, 117, 121, 215, 254, 255, 257
momentum, 307
monoclonal antibody, 53, 64, 79, 86, 211
monosomy, 276
Moon, 90, 139, 146, 193
morbidity, xi, xiii, 158, 162, 164, 167, 168, 169, 175, 176, 177, 194, 196, 227, 232, 233, 236, 239, 240, 244
morphogenesis, 56, 139
morphology, 74, 97, 98, 105, 107, 126, 131
mortality rate, 167, 237, 269, 273
mother cell, 52
motif, 8, 12, 13, 17, 36, 62, 66, 88
moulding, 103
MP, 33, 39, 42, 139, 140, 147, 148, 174, 195, 280, 339
MRI, 49, 56, 202, 203, 205, 225, 250, 280, 306, 323, 324, 325, 327, 333, 337
mRNA, 69, 83, 190
mucin, 30
mucous membrane, 241
multiple factors, 313
multiple myeloma, 41, 68, 71, 72, 85
multiple regression analysis, 233
multiple sclerosis, 62
multipotent, 52, 101, 102, 118
multivariate analysis, 221
mutagenesis, 260
mutant, 111, 264
mutation, 57, 64, 96, 98, 104, 108, 109, 111, 112, 122, 140, 149, 201, 249, 274, 276, 279
mutations, 55, 96, 107, 108, 113, 118, 122, 136, 140, 213, 225, 249, 255, 259, 274, 277, 279
myeloid cells, 82
myelomeningocele, 277
myocardial infarction, 175, 211
myofibroblasts, 59

N

nanoparticles, x, 6, 10, 21, 25, 30, 32, 34, 38, 39, 43, 44, 48, 73, 74, 77, 79, 80, 89, 90, 91, 215
nanosystems, 51
nanotechnology, ix, x, 32, 48, 78, 90
narcotic, 235
nasopharynx, 272, 335
National Institutes of Health, 153
natural killer cell, 36
nausea, 249
navigation system, 206
necrosis, 4, 12, 22, 67, 211, 218, 248, 250, 251, 259
negative effects, 270
neocortex, 135, 139
neonatal intracranial tumors, 271
neonates, 277
neoplasm, 2, 207, 279
neoplastic tissue, 3
neovascularization, 8, 41
nerve, 161, 162, 164, 175, 181, 191, 230, 232, 243, 273
nervous system, xiv, 94, 95, 132, 145, 146, 151, 152, 213, 267, 281
Netherlands, 2
neural development, 101, 118
neural network, 313, 318, 319
neurobiology, 134
neuroblastoma, 66, 87, 153
neurogenesis, 101, 102, 107, 117, 118, 135, 141, 145, 148, 155, 213, 216, 219, 222
neurogenic bladder, 272
neurologic symptom, 249
neurons, 97, 102, 106, 123, 126, 135, 143, 150, 153, 216, 222, 253, 255
neuropathy, 178
neuroscience, 114, 213
neurosurgery, 200, 202, 203, 205, 217, 219, 220, 221, 222, 223
neutrons, 76
neutrophils, 57
NH2, 26

NIR, 15, 18, 21
NMR, 26
nodes, xi, 158, 162, 164, 167, 169, 172, 174, 176, 179, 183, 230, 232, 236, 238, 239, 240
nodules, 103
non-Hodgkin's lymphoma, 27
nonlinear systems, 314
NSAIDs, 59
nuclei, 89, 103, 139, 274
nucleic acid, 75
nucleolus, 276
nucleoplasm, 20
nucleus, 13, 20, 22, 24, 42, 97, 107, 108, 276
null, 264
nurses, 243
nutrients, x, 7, 47, 49, 60

O

obesity, 62, 88, 171, 231
obstacles, 59
obstruction, 114, 239, 272, 329, 331
oedema, 203, 207, 336
oesophageal, 2
oligodendrocytes, 97, 102, 123, 126, 248
oncobiology, 49
oncogenes, 54, 77, 121, 137, 252
oncogenesis, 110, 140, 260
oophorectomy, 182
operations, 203
opportunities, x, 47, 62, 90
optic nerve, 273, 336
optical fiber, 26
optimization, 317, 319
oral cavity, 37
orbit, 323, 324, 325, 326, 328, 330, 331, 335, 337
orchestration, 62
organ, 63, 74, 86, 107, 153, 271
organelles, 22, 25
organic solvents, 24
organism, 251
osteoma, 325, 328, 329, 331, 337
osteoporosis, 233
ovarian cancer, 36, 37, 80, 89, 234
ovarian cysts, 233
ovarian failure, 233
ovaries, 198, 229, 232, 233, 239, 243
overlap, 112, 149
overproduction, 8
ovulation, 63
oxidation, 38
oxidative damage, ix, 2, 4, 5, 18
oxidative stress, 50
oxygen, x, 2, 4, 7, 14, 16, 41, 44, 47, 49, 60, 62, 64, 91
oxygen sensors, 62

P

pain, xii, 158, 162, 165, 181, 183, 185, 235, 328, 336
palate, 272, 335, 336
palladium, 33
palliative, 3, 204
palpation, 56, 324
pancreatic cancer, 71, 85, 89, 143
paradigm shift, 213
parallel, 5, 101, 103, 112, 117, 118, 150
parenchyma, xiii, 102, 247, 248, 255
patents, 79, 149
pathogenesis, xi, 93, 96, 98, 101, 103, 108, 113, 114, 118, 121, 128, 234, 259, 323
pathology, 97, 144, 153, 174, 322
pathways, 49, 50, 51, 55, 57, 60, 81, 82, 106, 109, 111, 112, 113, 114, 117, 119, 128, 139, 212, 214, 224, 261, 274, 279
PCA, 313
PDGFR, 49, 53, 54, 62, 64, 65, 71, 146, 212, 251, 297
PDT, 2, 4, 5, 6, 7, 8, 9, 11, 12, 14, 16, 17, 21, 23, 24, 25, 28, 29, 30, 31, 32
pelvis, 165, 175
penetrance, 105, 122
peptide chain, 26
peptides, ix, 2, 7, 14, 15, 22, 24, 27, 41, 57, 66, 67, 73, 75, 77, 87, 88
perfusion, 63, 64, 202, 203, 225, 271
pericytes, 57, 62, 63
perinatal, 271, 273, 274, 277
peripheral nervous system, 102, 136, 152, 153
permeability, 5, 34, 35, 63, 74, 75, 84, 90, 91, 152
permission, iv
permit, 184, 190
Perth, 93, 104
PET, 176, 202, 217, 306
phage, 65, 66, 77, 87, 91
pharmacokinetics, 43, 73
pharmacology, 51
phenotype, 35, 50, 54, 56, 97, 104, 108, 113, 118, 119, 122, 147, 150, 155, 221, 262
phenotypes, xii, 57, 112, 200
phosphatidylethanolamine, 75
phosphatidylserine, 65
phosphorylation, 23, 261
photodynamic therapy, ix, 2, 13, 32, 33, 34, 37, 38, 39, 40, 41, 42, 43, 44
photographs, 324
photons, xiv, 305, 307

photosensitivity, 3, 6, 19, 40
photosensitizers, ix, 1, 2, 4, 5, 6, 8, 9, 10, 13, 16, 17, 18, 19, 20, 22, 24, 25, 29, 30, 31, 33, 37, 38, 41, 42
phototoxicity, 5, 18, 23, 27, 30, 31
physiology, 50, 259
pilot study, 192, 193, 195, 198, 244, 245
pituitary gonadotropes, 27
Piver–Rutledge–Smith, 160, 161
placebo, 89
placenta, 131
plaque, 126
plasma membrane, 17, 65, 115, 144, 145
plasma proteins, 5, 41
plasminogen, 72
plasticity, 74
Platelet-Derivated Growth Factor Receptor, 251
platform, 90, 109, 110, 171
plexus, xiv, 267, 268, 269, 271, 274, 279, 280
pneumonia, 181
pneumonitis, 319
point mutation, 109, 133
poison, 207
polarity, 114, 117, 144
polarization, 58
polymer, 25, 29, 73, 75, 90
polymers, 75, 217
polymorphisms, 96
polypeptide, 23, 35
pons, 273
porphyrins, 2, 12, 16, 24, 29, 30, 31, 33, 34, 39, 41
Portugal, 47, 78
positive correlation, 312
positron, 160, 178, 196
positron emission tomography, 160, 178, 196
precancer, 18, 43, 190
precursor cells, 95, 101, 102, 105, 106, 118, 125, 150, 151, 154, 220
pregnancy, 63, 81, 96, 102, 106, 269, 270
preparation, iv, 21, 57, 91, 270
prevention, 80, 83
primary brain tumours, xii, 131, 199, 200
primary tumor, xiii, 49, 57, 155, 221, 240, 247, 248, 255
primate, 151, 152
probability, 63, 309, 314, 315
probability density function, 309
probability distribution, 315
probe, 15, 21, 40, 109, 110
prodrugs, 14
progenitor cells, 57, 67, 101, 102, 105, 106, 112, 114, 119, 128, 134, 138, 146, 150, 152, 153, 213, 220, 222
prognosis, xiii, xv, 8, 14, 33, 38, 39, 57, 82, 100, 103, 104, 109, 137, 141, 160, 200, 201, 208, 215, 216, 221, 232, 245, 246, 247, 267, 268, 269, 271, 272, 274, 275, 277, 278, 280, 322, 335, 336, 337, 338
project, 130
prolapse, 181
proliferation, 7, 28, 45, 49, 55, 56, 57, 60, 62, 63, 68, 69, 70, 71, 72, 103, 106, 107, 108, 109, 112, 114, 115, 121, 125, 126, 128, 134, 138, 139, 140, 141, 154, 212, 248, 249, 250, 251, 256, 257, 262
promoter, 62, 128, 201, 257
propagation, 7, 314
prophylactic, 338
prostate cancer, 58, 77, 80, 83, 87, 143
proteins, 8, 12, 13, 15, 17, 20, 22, 23, 29, 33, 57, 65, 66, 88, 106, 107, 117, 120, 121, 124, 148, 251, 254, 261
proteolysis, 14, 56
proteolytic enzyme, 7
proteome, 151
prothrombin, 45
protons, xiv, 26, 305, 307
proto-oncogene, 124, 139
prototype, 91
pruning, 62, 63, 64
psychosocial stress, 219
PTEN, 201, 258
ptosis, 336
pulmonary embolism, 175, 181, 211, 236
purity, 123
pyelonephritis, 181

Q

quality of life, xii, 2, 4, 100, 158, 185, 228, 231, 232, 238, 322
quantum dot, 33, 73, 216
quantum dots, 33, 73, 216
quantum yields, 14

R

rabies virus, 216
radiation therapy, xiv, 133, 134, 228, 229, 231, 233, 234, 241, 242, 276, 305, 317
radical hysterectomy, xi, xii, xiii, 158, 160, 161, 162, 163, 164, 165, 166, 167, 168, 169, 172, 173, 174, 175, 178, 179, 181, 182, 183, 185, 190, 191, 192, 193, 194, 195, 196, 197, 198, 227, 228, 229, 231, 232, 233, 234, 235, 236, 237, 238, 239, 241, 242, 243, 244, 245, 285,뱀296
Radical Hysterectomy, 159, 165, 179, 231, 232, 235, 237, 239

radical mastectomy, 231
radical vaginal trachelectomy, 234
radicals, 18
radio, 68, 169, 214, 239
radiopaque, 337
radiopaque contrast, 337
radioresistance, 146, 206, 216
radiotherapy, ix, xiii, xiv, 8, 52, 64, 99, 100, 120, 121, 133, 148, 174, 176, 206, 207, 208, 211, 216, 222, 223, 225, 227, 228, 238, 239, 242, 247, 250, 267, 269, 274, 276, 305, 306, 307, 309, 316, 317, 318, 319, 325, 331, 332, 337, 338
radium, 206, 215, 223
reactions, 33, 83, 337
reactive oxygen, 2, 4
reactivity, 5
reading, 161, 256, 258
reagents, xi, 24, 94, 113, 124
reality, 74, 79, 218
receptors, 7, 8, 9, 12, 13, 15, 17, 20, 21, 24, 27, 29, 31, 34, 41, 44, 55, 58, 63, 65, 68, 79, 106, 107, 108, 251, 256
recognition, 17, 38, 51, 66
recombination, 258
recommendations, iv
reconstruction, xv, 322, 323, 325, 331, 337
recurrence, xii, 54, 166, 167, 175, 179, 180, 200, 204, 206, 214, 233, 234, 235, 236, 239, 276, 331, 338
recycling, 25, 40
redundancy, 213
regenerate, 113, 114, 119, 121
regeneration, 5, 83, 211, 213
Registry, 258
regression, 19, 52, 64, 68, 72, 78, 81, 122, 210, 212, 218, 223, 273, 313
regrowth, 68, 127
rehabilitation, 162, 165
relapses, xv, 322
relevance, 80, 122, 136, 259, 272
remission, xv, 322, 338
remodelling, 60
renaissance, 185
repair, 201
replication, 58, 110, 125, 155, 210, 250, 255, 260, 263, 264
requirements, 168, 213
researchers, 105, 160
resection, xi, xv, 99, 158, 161, 164, 202, 204, 205, 206, 207, 214, 215, 216, 221, 223, 224, 228, 232, 274, 275, 276, 322, 323, 324, 326, 337
residual disease, 168
residual error, 308
residues, 17, 27, 29, 168
resistance, 36, 64, 72, 120, 122, 146, 147, 149, 208, 212, 214, 221, 251
resolution, 100, 202, 215, 269, 306
resources, 233
respiratory dysfunction, 171
reticulum, 22, 256
retinoblastoma, 39, 116, 220
retroviruses, 124, 210, 255, 260, 263
reverse transcriptase, 54
RH, 140, 141, 232, 282
ribonucleotide reductase, 54
right hemisphere, 204
risk assessment, 128
risk factors, 239, 243
risks, 168, 181, 185, 202, 211, 224, 234
RNA, 9, 49, 53, 54, 55, 88, 124
robotics, ix, xi, 158, 163, 171, 173, 174, 179, 181, 182, 184, 195, 237
rodents, 256
rubber, 206

S

salpingo-oophorectomy, xi, 158, 159
scale system, 90
scaling, 237
scattering, 214
sclerosis, xiv, 267, 268, 273, 279, 280
second generation, 124
secondary brain tumors, 258
secondary damage, 306
secrete, 43, 59, 121
secretion, 7, 29
seed, 55, 57
seizure, 278
selectivity, ix, xiv, 2, 3, 5, 6, 8, 13, 20, 25, 26, 29, 30, 32, 35, 37, 50, 51, 63, 74, 75, 107, 305, 307
self-sufficiency, 50
semiconductor, 215
sensation, 186, 235
sensing, 15, 40
sensitivity, 4, 10, 121, 252, 316, 324, 336
sensitization, 175
sensors, 37, 205
septum, 331
sequencing, 66, 150
serum, 19, 28, 114, 119, 123, 127, 155, 221, 256
sex, 95, 119, 323
sexual activity, 228
shape, 63, 74
shear, 56
shock, 125

side effects, xi, xiv, 6, 9, 30, 51, 64, 93, 202, 207, 276, 305
signal peptide, 256
signal transduction, 72, 82
signaling pathway, xiii, 32, 62, 63, 65, 67, 77, 78, 79, 139, 247, 252, 256, 257
signalling, 12, 49, 62, 87, 89, 103, 104, 106, 107, 108, 110, 112, 113, 117, 118, 121, 122, 128, 139, 141, 212
signals, 22, 32, 50, 53, 67, 80, 85, 138
signs, xiv, 98, 99, 103, 105, 232, 249, 267, 269, 270, 274, 276, 323, 328, 329, 336
silica, 30, 34
silicon, 21, 33
silver, 204
simulation, 317, 319
Singapore, 131
sinuses, ix, xiv, xv, 321, 322, 323, 324, 327, 329, 332, 333, 334, 335, 336, 337, 338, 339
sinusitis, 328, 329, 336
siRNA, 49, 57, 80, 88
skeleton, xv, 322, 325, 337
skin, 3, 4, 19, 57, 163, 174, 175, 241, 275, 335
skin cancer, 3, 4
smoking, 96, 159
smooth muscle, 7, 63, 139
smooth muscle cells, 63
sodium, 2, 40, 71, 79
solid tumors, 14, 18, 59, 64, 75, 84, 131, 133, 219
solubility, 6, 17, 24, 63
somatic cell, 52, 122, 251
somatic mutations, 260
Spain, 78, 222, 321
specialists, 321, 322, 323, 338
specialization, 60, 62, 63
species, 2, 4, 22, 125
spectroscopy, 39, 40, 225
speculation, 5, 105, 108, 117
spinal cord, 102, 224, 248, 268, 275
spinal cord tumor, 248
spindle, 117, 276
spine, 135
spleen, 26
sponge, 206
sprouting, 62
squamous cell carcinoma, 15, 40, 168, 177, 178, 333, 338
stabilization, 62
staging systems, 160
standard deviation, 312
states, 52, 58
stem cell differentiation, 150
stem cell lines, 150
stenosis, 164, 181, 228, 236, 272
stent, 169, 175
stillbirth, 268, 270
stochastic model, 52, 112, 113
storage, 28
stratification, xi, 93, 100, 134, 135, 136, 201, 259
stroke, 211
stroma, x, 48, 55, 56, 57, 70, 71, 79, 127, 278
stromal cells, 67, 261
structural changes, 104
structural protein, 13
subcutaneous emphysema, 236
subgroups, 98, 105, 109, 110, 111, 128, 132, 147
submucosa, 28
substitution, 19
success rate, 177
suicide, 52, 252, 253, 255
sulfate, 71
Sun, 6, 35, 38, 42, 43, 83, 148, 150, 194, 218, 319
superficial bladder cancer, 2
suppression, 58, 59, 62, 64, 67, 107, 155
surgical intervention, 186, 187, 331
surgical resection, xii, xiii, 99, 100, 199, 206, 247, 264, 265, 271, 273, 277
surgical technique, 161, 162, 165, 178, 182, 191, 204, 236
surrogates, xiv, 306, 307, 310, 312, 313, 316, 318, 319
survival, ix, x, xi, xii, xiii, 1, 2, 4, 18, 47, 49, 51, 54, 55, 58, 60, 63, 64, 65, 66, 67, 69, 70, 71, 72, 78, 80, 81, 85, 99, 100, 109, 121, 125, 127, 129, 130, 134, 141, 158, 164, 167, 175, 176, 179, 181, 182, 183, 199, 204, 206, 207, 208, 209, 211, 212, 213, 214, 215, 221, 227, 228, 229, 231, 234, 235, 236, 239, 245, 247, 250, 251, 253, 254, 256, 260, 265, 271, 273, 275, 276, 277, 280, 335
survival rate, xii, xiii, 100, 164, 179, 182, 199, 207, 214, 227, 231, 234, 236, 239, 271, 273, 275, 276, 277
survivors, 100, 129, 271, 274, 275, 276, 277, 280
susceptibility, 96, 97
sustainability, 59
suture, 185, 186
swelling, 336
symptoms, 232, 233, 249, 269, 270, 323, 328, 329, 336
syndrome, xiv, 96, 104, 107, 122, 130, 136, 267, 268, 274, 276, 277, 278, 279, 280, 336
synergistic effect, 256
synthesis, 12, 13, 16, 20, 21, 22, 24, 25, 26, 29, 34, 35, 37, 41
systemic mastocytosis, 70

T

T cell, 254
T lymphocytes, 88
T regulatory cells, 58
team members, 168
technetium, 40, 240, 241
technologies, xiv, 50, 60, 66, 186, 305, 306, 307
technology, xii, 32, 66, 111, 113, 158, 163, 168, 180, 184, 185, 190, 216, 237, 307
telephone, 258
telomere, 80
temperature, 22, 74, 91
tendon, 78
tension, 56, 84
testing, 126, 149, 213
textiles, 333
TGF, 49, 54, 62, 69, 70, 71, 79
thalamus, 273
theatre, 205
therapeutic agents, x, 6, 30, 39, 44, 48, 63, 78, 211
therapeutic approaches, x, xiii, 27, 48, 64, 200, 212, 247, 278
therapeutic interventions, 65
therapeutic targets, 44, 58, 141
therapeutics, 4, 7, 34, 52, 62, 64, 72, 88, 89, 90, 260
thrombin, 45, 56
thrombosis, 33, 36, 64, 85, 181
thymoma, 54
time periods, 316
TIMP-1, 40
tissue homeostasis, 152
titanium, 326, 331
TLR, 49
TLR2, 58, 82
TLR4, 58
TNF, 49, 54, 58, 71, 211
tobacco, 333
tonometry, 323
toxicity, 5, 23, 30, 41, 50, 63, 124, 125, 133, 153, 207, 211, 219, 252, 259
tracks, 205
training, 162, 163, 168, 186, 187, 194, 310, 312, 313, 314, 316
traits, 213
trajectory, 205, 308, 309
transcription factors, 107, 108, 139
transcripts, 111
transducer, 122
transduction, 89, 153, 210, 253, 255, 263
transection, 166
transfection, 82, 88, 124
transferrin, 28, 34, 37, 39, 41, 86
transformation, 30, 52, 118, 119, 128, 129, 211, 212, 272
transforming growth factor, 17, 54, 62
transfusion, 181
transgene, 124, 151, 251, 252, 253, 254, 255, 257, 265
translation, 7
translocation, 20, 23, 107
transmembrane glycoprotein, 12
transmission, 142, 151
transplantation, 115, 128, 142, 224
transport, 5, 20, 22, 23, 28, 38, 60, 65, 151
trauma, 162, 235
tremor, 177, 237
trial, 41, 64, 89, 125, 133, 178, 190, 211, 217, 220, 221, 222, 223, 224, 225, 235, 242, 244, 262, 263, 264
triggers, 15
tropism, 125, 126, 153, 255
true/false, 314
tumor cells, x, 7, 9, 11, 12, 13, 15, 19, 20, 21, 24, 27, 28, 29, 31, 36, 38, 39, 42, 48, 49, 50, 51, 52, 55, 56, 57, 58, 60, 63, 64, 68, 74, 75, 77, 87, 89, 250, 251, 252, 255, 256, 257, 275
tumor development, 58
tumor growth, 7, 9, 11, 50, 55, 56, 57, 60, 66, 68, 83, 88, 143, 153, 155, 262, 265
tumor necrosis factor, 218, 251
tumor progression, x, 48, 49, 51, 52, 55, 56, 58, 65, 77, 82, 220
tumorigenesis, 41, 81, 96, 101, 105, 112, 117, 121, 122, 127, 128, 135, 138, 155, 251
tumour growth, 15, 112, 113, 121, 143, 211, 212
tunneling, 185
turnover, 28
tyrosine, 8, 17, 62, 72, 79, 212, 256, 261

U

ultrasound, 56, 74, 91, 269, 271, 277
umbilical cord, 260
underlying mechanisms, 97
uniform, 11, 160
unique features, x, 48, 60, 314
United Kingdom, 130, 134
updating, xiv, 306, 312, 313, 315, 316
ureter, 161, 175, 181, 230
urinary tract, 168, 174, 232, 243
urine, 18
urothelium, 28
uterine cancer, xi, 158, 159
uterus, 229, 234, 238, 270

V

vaccine, 58, 264
vagina, 161, 181, 185, 228, 229
validation, 21, 128, 189, 315, 319
variables, 159, 231, 313
variations, 74, 109, 190, 306, 308, 309, 313, 314
vascular cell adhesion molecule, 65
Vascular Endothelial Growth Factor Receptor, 8, 49, 251
vascular system, 32, 83, 181, 240
vasculature, 3, 6, 7, 11, 38, 64, 66, 67, 84, 85, 87, 88, 89
vasomotor, 232
vector, 17, 125, 151, 152, 153, 255, 259, 262, 263, 308, 309
VEGF expression, 35
VEGF protein, 12
VEGFR, 8, 9, 33, 42, 49, 54, 62, 64, 68, 70, 71, 72, 85, 145, 251
vehicles, 13, 22, 42, 74, 153, 209, 255, 257, 260
vein, 9, 85, 166, 181
velocity, 56
ventricle, 97, 272, 273, 274, 276
versatility, 55, 77, 186, 189
vessels, 9, 10, 56, 60, 63, 64, 68, 75, 89, 162, 171, 204, 230, 329
viral infection, 124, 209
viral vectors, 73, 124, 152, 153, 252, 255
viruses, 66, 124, 159, 209, 210, 221, 225, 255
vision, 170, 171, 184, 185, 187, 235, 237, 322
visual acuity, 323, 328, 336
visualization, 163, 177, 329
vitamins, 75, 96
voiding, 169, 175, 178
vomiting, 64, 202, 249, 270
vulnerability, 96, 103
vulva, 236

W

Washington, 309
waste, 333
water, 6, 63
weakness, 64
wear, 187
white matter, 101
wild type, 58, 124, 210
windows, 96, 131
withdrawal, 62
Wnt signaling, 139, 140
workers, 20, 31, 66
World Health Organisation, 98
worldwide, xi, 158, 159, 228
wound healing, 63, 205
wound infection, 163

X

xenografts, 19, 55, 57, 66

Y

yeast, 67
yield, 123, 166, 169, 173, 208
yolk, 272
young adults, 130, 134
young women, 164

Z

zinc, 14
zygomatic arch, 323, 330